Ecology of
Tropical Oceans

ICLARM Contribution No. 389

Ecology of Tropical Oceans

Alan R. Longhurst

Department of Fisheries and Oceans
Bedford Institute of Oceanography
Dartmouth, Nova Scotia, Canada

Daniel Pauly

International Center for Living
Aquatic Resources Management
Manila, Philippines

ACADEMIC PRESS, INC.

Harcourt Brace Jovanovich, Publishers

San Diego   New York   Berkeley   Boston
London   Sydney   Tokyo   Toronto

ACADEMIC PRESS, INC.
1250 Sixth Avenue, San Diego, California 92101

United Kingdom Edition published by
ACADEMIC PRESS INC. (LONDON) LTD.
24–28 Oval Road, London NW1 7DX

Library of Congress Cataloging in Publication Data

Longhurst, Alan R.
  Ecology of tropical oceans.

  Bibliography: p.
  Includes index.
  1. Marine ecology—Tropics.   I. Pauly, D. (Daniel)
II. Title.
QH95.59.L66   1987   574.5'2636   87-1453
ISBN 0–12–455562–4 (alk. paper)

PRINTED IN THE UNITED STATES OF AMERICA

87 88 89 90   9 8 7 6 5 4 3 2 1

Contents

Chapter 10 Population Biology of Large Marine Invertebrates 307

Preface

Our reason for writing this book is simple. By far the largest part of the oceans lies in the tropics, but there is still a lack of general texts devoted to tropical oceanography, biology, and fisheries. Given the wealth of books devoted to the polar oceans, this is really a paradox. What we have tried to write is the book each of us wanted to read when we first began to work in tropical seas.

It seemed to us useful at this time to write a comprehensive book on the ecology of tropical oceans because of the bias in popular perceptions concerning the nature of tropical marine environments. From what is obtainable on the bookshelves, it is understandable that the tropical oceans should generally be perceived as bordered entirely by coasts bearing coral reefs or mangrove forests and the open tropical ocean as featureless if not peopled by coral atolls. As we hope to make more widely known, the truth is different: tropical marine environments are far more diverse than most people—even marine scientists—believe them to be.

Though our principal motivation in writing is, of course, to improve the possibilities of managing tropical fisheries, we do not directly address the question of how best to influence the interactions between fish and fishermen in warm seas. It is now clear that North Sea solutions will not suffice for the tropical fisheries, and during the last decade there has been a search for new ways of understanding the complexities of fisheries based on many more as well as more diverse species of fish than the great high-latitude fisheries, where management procedures are now relatively well established. One of the most consistent results of research on the management of tropical fisheries is that the manner in which fish are supported by tropical marine habitats—even when fishing pressure is absent—is not well understood.

It seemed to us, therefore, that a useful contribution might be a simple but comprehensive examination of what is currently known about the structure of all tropical marine habitats and how fish make a living in them under natural conditions. That is what this book is intended to be. We have covered most of the important aspects of this subject, though we have deliberately ignored some. For instance, we have not discussed the role of marine mammals or of seabirds in tropical ecosystems, though both are important there. Nor have we discussed the biology of the deeper parts of the tropical oceans, because this has already been well and comprehensively reviewed in several recent books. Finally, we have not discussed any issues related to pollution of the tropical marine environment: we both believe that this is ultimately not a scientific, but an economic and political issue.

The taxonomy of tropical marine organisms is by no means fixed, and we have thought it better to follow the original authors' usage rather than attempting to harmonize all the specific and generic names we quote into line with current usage; in doing so we would risk adding yet more confusion. Only where it seems to us to have been essential to do so have we updated our quotations to what seems to be the presently accepted taxonomy.

The reader will find that each of us has contributed to the text differently, though we have worked on all chapters together. The first eight chapters are a general review of the geography, oceanography, and biology of tropical oceans with some emphasis on fish, written from the point of view of a marine ecologist. The last two chapters are an integration of individual studies of the population biology of fish and marine invertebrates into a simple model describing the reproduction, growth, and mortality of these populations, written from the point of view of a fisheries biologist.

We hope that these two approaches will complement each other, and together will provide a useful introduction to the biological basis of tropical fisheries. Because we want to make the large, but often neglected literature on tropical marine environments more accessible to those who need it, we have used references in our text especially generously, perhaps more so than is customary in a book of this kind. Nor have we avoided reference to the "gray literature," which requires more than usual energy to acquire, because so much of the good work fostered by the international agencies exists only in this form.

We are often surprised at the general lack of awareness of how much research in the marine sciences is being done now in tropical countries and in tropical oceans by scientists based in colder seas. In fact, the level of expertise in tropical laboratories has improved greatly in the last 10 years for a variety of reasons, and tropical marine biology has now come of age. We think that this book, which originated in our experiences in tropical marine ecology and fisheries research, is timely. We hope it may be useful to marine ecologists everywhere, and will at least widen their perceptions of the tropical oceans.

Alan R. Longhurst
Daniel Pauly

Ecology of
Tropical Oceans

Chapter 1

Introduction

This book is about life in tropical oceans on our cool, interglacial planet. Cool, that is, compared with distant geological eras when the polar oceans were warm and there were no ice caps, but not as cold as during glacial epochs. Tropical oceans during the persent interglacial period cover a large part of our planet's surface, yet their warmth is only a thin layer above a great cold water mass. As we shall see, the fact that most of the water of the present-day ocean, even in the tropics, is at polar temperatures has profound significance for tropical biology.

The general increase in diversity of all life forms from high to low latitudes is well known, and is an important but poorly understood factor in tropical fisheries biology. We also accept easily that vertical gradients for diversity occur in the oceans just as they do on mountains, but it is much less well understood that the overall diversity of marine faunas varies also longitudinally, both between and within oceans, depending on the age and recent glacial history of each. That the tropical Atlantic marine fauna is much less diverse than the Pacific is still not generally realized, and the consequences of this difference for tropical fisheries biology are largely unexplored.

The actual fishery production from the tropical ocean is a difficult statistic to calculate, and its exact value is unimportant. However, the tropical ocean includes almost 50% of the total area of all open water and 30% of the total area of continental shelf, but produces only about 16% of global fish production; this is an enigma we need to understand. In the warm tropical oceans, life apparently marches to a faster beat and its high diversity enables life forms to occupy every niche imaginable, so that, on the face of it, we might be forgiven for expecting the tropical oceans to be the fish-baskets of the world: The reality, we know, is otherwise.

TABLE 1.1. Landed Weights of Marine Fish from Tropical Seas, Rearranged from FAO Landing Statistics for 1983[a]

	Anadrom.	Clupeoids	Level ground demersal	Reef demersal	Small pelagics	Tuna	Rays and sharks	Total	% Totals
Indian Ocean, Western	15,251	452,650	383,374	31,979	370,093	230,991	121,699	1,606,037	16.91
Indian Ocean, Eastern	18,908	172,747	162,249	6,927	157,751	103,693	53,427	675,702	7.12
Pacific, Western Central	18,376	713,172	344,338	159,828	1,009,508	717,336	72,555	3,035,113	31.96
Pacific, Eastern Central	41	278,255	22,671	12,693	181,255	327,722	25,899	848,536	8.94
Pacific, South East	0	226,708	54,000	6,700	93,705	8,987	14,960	405,060	4.27
Atlantic, Western Central	996	981,705	106,483	237,020	45,221	306,029	21,628	1,699,082	17.89
Atlantic, Eastern Central	16,961	337,908	493,752	27,063	11,303	286,082	52,719	1,225,788	12.91
Total for all tropics	70,533	3,163,145	1,666,050	482,210	1,868,836	1,980,840	362,887	9,495,318	
% Totals	0.74	33.31	17.55	5.08	19.68	20.86	3.82		

Anadromous — *Hilsa, Chanos, Lates, Alosa,* etc.
Clupeoids — *Sardinella, Sardinops, Chirocentrus, Dussumeria, Engraulis, Stolephorus, Brevoortia, Ethmalosa,* etc.
Level ground demersal — *Arius, Saurida, Nemipterus, Polynemus, Sciaena, Pseudotolithus, Leiognathus, Drepane, Dentex,* etc.
Psetodes, Cynoglossus, Rhombosolea, etc.

Reefs — *Lutjanus, Serranus, Epinephelus, Labrus, Acanthurus,* etc.
Small pelagics — *Caranx, Vomer, Selar, Sphyraena, Exocoetus, Trichiurus, Rastrelliger,* etc.
Tuna — *Acanthocybium, Scomberomorus, Auxis, Euthynnus, Katsuwonus, Thunnus, Xiphias, Makaira,* etc.
Sharks, rays — *Galeorhinus, Raja, Mustelus, Carcharinus,* etc.

Indian Ocean, Western — Mozambique to Sri Lanka
Indian Ocean, Eastern — Sri Lanka to Western Australia
Pacific, Western Central — Hainan to Gulf of Carpentaria to mid-Pacific at 180°W
Pacific, Eastern Central — Baja California to Ecuador to mid-Pacific at 180°W
Pacific, South East — Peru
Atlantic, Western Central — Florida to Northern Brazil to mid-Atlantic at 40°W
Atlantic, Eastern Central — Mauretania to Angola to mid-Atlantic at 40°W

[a] Data in tonnes.

Tropical fisheries, with certain exceptions to be discussed below, are small in global terms, difficult to manage, and very sensitive to the effects of a mechanised, modern fishery. Table 1.1 shows how tropical fish catches are distributed among ecological groups of fish; it is to be noted that the fisheries of coral reefs, so often taken to be a model for tropical fisheries, actually contribute only about 5% of all fish production in the tropics.

In approaching the problems of tropical fisheries, and understanding their biological basis, comparison with the better-documented high-latitude fisheries is almost inevitable, whether this involves comparison of fishery management techniques, or the analysis of ecological and physiological functions, or both. Such comparisons must, by their nature, be based upon ecological rules and generalizations, which are hard to reach and seldom sufficiently robust to be relevant to a whole climatic zone.

It is even harder to integrate the full diversity of each climatic zone into generalizations applicable to all of that zone. We want to show in this book that the tropical oceans, their coastlines and resources are richly diverse and much better explored than many believe. Indeed, it is largely work conducted in tropical oceans that is leading the global discipline of biological oceanography to a new level of understanding of the interaction between ocean physics and biological production. In fishery biology, similarly, methods developed for the analysis of tropical fish are beginning to find applications in high latitudes, once the sole source of methodology for the discipline.

Perhaps the greatest difficulty in understanding the dynamics of tropical resources is the unfortunate fact that most tropical fish stocks have already been much modified by just a few decades of industrialized fishing—in many cases with the support of international organizations. This modern pulse of mechanized fishing has in some cases destroyed the natural ecological balance, evolved over tens of millenia and preserved during many centuries of nonmechanized fishing; a decade of industrial fishing has even caused once-important species or stocks almost totally to disappear. Under such circumstances, the investigation of critical phenomena such as the variability of recruitment becomes a matter of great difficulty. It is our hope that this book will encourage well-documented accounts of the changes still occurring in exploited tropical resources, by providing an environmental background to such accounts.

Although our principal intention is to assist the development of management techniques appropriate to the tropical fisheries, we have deliberately avoided a direct approach to management problems. Instead, we have concentrated our attention, and the subjects we discuss, on the natural environment of tropical fish and their ways of making a living in that environment. We have chosen to do this because this aspect of the tropical fisheries problem seems to have been relatively neglected, yet a lack of understanding about the tropical marine ecosystems which produce their yield of fish may be at the root of many of the most difficult problems to be solved in tropical fishery management.

Chapter 2

Geography of the Tropical Oceans

PARTITIONING THE TROPICAL SEAS BY REGION AND LATITUDE ZONES

Too often in the ecological literature, when reference is made to tropical marine ecosystems, the only examples used are coral reefs or mangrove forests, implying that these ecosystems typify all life in tropical seas. This has perhaps occurred because in earlier times relatively few ecologists had the privilege of travelling widely in the tropics to study counterparts of the ecosystems of higher latitudes. In fact, ecologists from colder countries often sought out reefs and mangroves whenever they had the opportunity to undertake tropical studies: The attraction of novelty was too great! Unfortunately this bias has led to a number of important hypotheses concerning relative energy-flow in high- and low-latitude ecosystems to be formulated on insufficient evidence. Therefore, it is one of our purposes to present a more balanced description of the great range of marine habitats that actually occurs in the tropics.

We must start from the simple fact that the geometry of the globe, and the distribution of its land masses, means that 37% of the area of all seawater lies between 20°N and 20°S and that almost 75% lies between the 45° lines of latitude. The tropical seas are not, therefore, an insignificant equatorial band worthy only of anecdotal accounts: Rather, they are the major, central portion of the world ocean. To make it more graphic, consider that almost the whole of the Arctic Ocean would fit comfortably into just the Arabian Sea. Or that the whole of the great circumpolar Southern Ocean (looming so large in Gerardus Mercator's misleading projection) would actually fit quite nicely into the Indian Ocean (which

Mercator made so small) in the triangle between India, Madagascar, and Australia. That the greatest continental shelves are in northern high latitudes is well known, but these are especially large in very high, polar latitudes. It is far less well known that 30.3% (= 8.1 × 10⁶ km²) of all continental shelves are in tropical oceans, with the largest area in the western Pacific Ocean. Table 2.1, derived from Moiseeve (1969), locates the large areas of tropical continental shelf by region. This clearly shows how small the tropical shelf areas are along the eastern coasts of oceans, and how by far the largest shelf areas are in the western Pacific, lying in a great arc between northern Australia and Thailand. Once again for comparison, one could fit at least 11 Grand Banks of Newfoundland into this area, with another four or five up through the semitropical East China Sea.

TABLE 2.1. **Relative Areas of Continental Shelves and Open Ocean[a]**

Percentage areas	Open ocean area	Continental shelf area
Polar and boreal (45–90°)	26.6	40.9
Temperate (20–45°)	36.8	28.8
Tropical (0–20°)	36.6	30.3
Total areas	360.3 million km²	26.7 million km²

Some tropical continental shelf areas (million km²)

East China Sea	0.53		East Africa	0.13	
South China Sea	2.64		Mozambique	0.12	
Philippines	0.05		Madagascar	0.21	
Indonesia	0.49				0.46
Solomon Sea	0.23		Eastern India	0.19	
		3.94	Burma	0.25	
Coral Sea	0.41		Peninsula Malaysia	0.17	
Fiji Basin	0.14				0.61
Tasman Sea	0.24		Gulf of Mexico	0.64	
		0.79	Caribbean	0.44	
Mexico-Colombia	0.40		Antilles	0.31	
		0.40	Guiana	0.19	
Red Sea	0.18		Brazil	0.81	
Arabia	0.08				2.39
Persian Gulf	0.24		Mauretania–Liberia	0.34	
West India	0.32		Gulf of Guinea	0.39	
		0.82	Angola	0.10	0.83

Total = 10.24 million km²

[a] Data rearranged from Moiseev (1969).

Finally, in placing the dimensions of tropical seas in perspective, many people might be surprised to find that the Great Barrier Reef of northeast Australia spans 14° of latitude, and is as much as 250 km wide. If it lay on the east coast of North America, it would extend from the Bahamas clear up to Cape Cod and Maine, almost to the Canadian border. Off western Europe, it would extend from northern Denmark to the coast of Portugal.

REGIONAL CHARACTERISTICS OF CONTINENTAL SHELVES

Geomorphology of Tropical Shelves

It is not only the relative areas of continental shelves and the open tropical oceans we need to quantify if we are to understand the basis of tropical fish production. Rather, we must also know the geographical characteristics of continental shelves: Are they narrow or wide, flat or steep-to, what are their surficial deposits, what is the geology of the coastline that lies behind them, and do they lie on the east or the west of a continental mass? All these characteristics—quite apart from the circulation of water masses over them—determine kind and quantity of fish production.

Except in the areas in high latitudes where continental shelf morphology was directly modified by the action of ice during glacial epochs, the width and depth of continental shelves is determined by the same factors everywhere, including in the tropics (see, for example, Shephard, 1973): Tectonic activity causes shelves to be narrow on active, subducting margins as along the eastern coast of the Phillipines, and allows them to be wider along passive margins, as on the eastern coasts of the American continent. Island arcs protect shelves from wave action and allow the development of wide shelves, such as the Sunda and Arafura Shelves. The rift represented by the Red Sea may be too young for the full development of a continental terrace to have occurred.

Major rivers can modify the morphology of continental shelves: Indus, Ganges, Irrawaddy, Amazon, Orinoco, Niger, Zambezi, and Tigris-Euphrates rivers all have widened the adjacent continental shelves and have built fans of sediment down the continental slope. The Ganges fan extends thousands of kilometers into the Bay of Bengal. Figure 2.1 shows the astonishing manner in which a few major tropical river systems dominate the global transport of sediments from the continents to the ocean. The tropical oceans receive more than 11×10^6 t yr^{-1} of sediment from the few major rivers dominating the run-off from the continents: This total

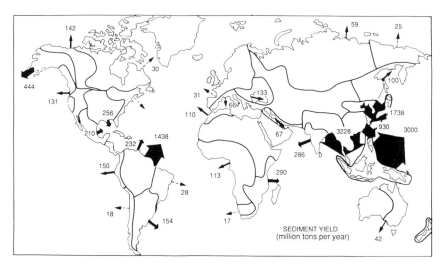

Fig. 2.1. River transport of silt to the continental shelf from the major drainage basins. (Modified from Milliman and Meade, 1983, *The Journal of Geology,* Copyright 1983 The University of Chicago Press.)

is more than 83% of all sediment discharged to the ocean globally. It is not surprising to find that this statistic is reflected in the distribution of the sediments of tropical continental shelves and seas, and in the high turbidity of coastal seas in the monsoon regions of the tropics.

The superficial geology of continental shelves is determined by processes that also determine coastal morphology, working in concert with several important biogenic processes of sediment production. Shelf sediments in the tropics comprise inorganic components such as sand derived from weathering of rock, and transported by wind, longshore currents, or rivers and also silts and clays that are mostly river-borne. The organic components of shelf sediments include terrestrial plant debris in all stages of decomposition which is either river-borne as particulates, or flocculated from dissolved organic material at the fresh/salt water interface in the estuaries. The remaining shelf deposits are oolitic sand formed by precipitation from dissolved calcium; shell–sand and larger particles of calcareous material derived from molluscan benthos; reef corals and coralline algae, produced largely *in situ* and not transported great distances; and finally, small particulate organic material from planktonic communities of plants and animals. The C/N ratios of sediments may be used to indicate their origin (Walsh, 1983); vascular plants produce detritus with high ratios ($C/N = >15$), while algae and marine invertebrates produce detritus with a low ratio ($C/N = <6$). Contrary to popular conceptions, tropical continental shelves dominated by the effects of mangroves or reef corals are more the exception than the rule. Mangroves are important only where a deltaic or low-lying coastal plain occurs, and corals are important only where negligible amounts of terrigenous material reach their habitat.

Atlantic Continental Shelves

Almost the whole western coast of Africa from Mauretania to the Cape of Good Hope is dominated by terrigenous deposits, with the exception of a small region off Ghana and the Ivory Coast. Coral occurs only on three islands in the Gulf of Guinea, at Annobon, Príncipe, and Sao Tomé. The very deep (to 400 m) shelf between Gabon and South Africa is dominated inshore by pelagic organic muds, and is sandy and rocky offshore. The discharges from the Congo, Niger, and Cameroons rivers dominate the shelf in the eastern Gulf of Guinea to the extent that even off the rocky coasts of the Cameroons, the sediment consists of black or blue muds of estuarine origin. Farther west, biogenic and mineral sands dominate as far northwest as Mauretania except off river mouths and close inshore where "green mud" occurs. The shelf widening at the Bissagos delta, off Guinea-Bissau, is a complicated area shaped by submerged channels, relict sand bars, remnant deltas,and many modern deposits of riverine silt and coastal green mud. Mangrove coasts occur principally at the Niger and Bissagos deltas. Three submarine canyons cut across the shelf near Abidjan, east of Lagos, and at the Congo mouth. Along most of the east–west coast of the Gulf of Guinea, linear fossil Holocene coral banks occur at the break of slope in 150–180 m (Allen and Wells, 1962; Longhurst, 1957); Buchanan, 1957; these are dominated by dead skeletons of *Madracis, Lophelia,* and *Dendrophyllia*. These assemblages resemble living coral banks occurring at similar depths from Cape Verde to Norway.

The western Atlantic shelf is much more complex, and dominated by terrigenous deposits principally near the mouths of the Magdalena, Orinoco, and Amazon rivers. The continental coastline of Central America is much folded and fractured by tectonic activity, and has basins formed by calcareous deposits. The wide, very shallow Yucatan shelf has offshore banks (notably Campeche Bank), at least one being of atoll form, and the shelf sediments are predominantly calcareous. Along the northern South American coast, which itself is much altered by large tectonic faults, a shelf is lacking in the west because of the outgrowth of the Magdalena estuary. Off Venezuela the shelf deposits are terrigenous and the shelf itself is complex, with the deep Cariaco Trench intruding. Offshore, the Antilles and Caribbean islands are rocky, or bear coral reefs, and the geological environment recalls that of the western Pacific, an impression strengthened by the occurrence off Belize of a barrier reef complex similar in structure if not in extent to the Great Barrier Reef of Queensland. Fossil coral banks, dominated like the Gulf of Guinea examples by *Dendrophyllia* and *Lophelia,* occur off the Orinoco coast on the upper continental slope (Allen and Wells, 1962).

Farther south and east, neither the Orinoco nor the Amazon have built deltas across the continental shelf, though the amount of terrigenous material transported to the shelf is enormous in each case; the Orinoco delta of 20,000 km^2 occupies a re-entrant in the coastline. About 1500 km of the northeast coast of South America are dominated inshore by terrigenous fluid mud. Regularly, every 30–60 km along this coast, thixotropic mud banks and mud shoals occur offshore. Each of these mud banks may be of order 20 km^2 in extent and be as much as 2 m thick. Because of their fluidity, these mud flats migrate slowly westward under the influence of the longshore current. The coastal mangrove forests colonize the intertidal flats, adding wedge-shaped pieces of land to the coastline (Wells, 1983). Beyond this spectacular mud regime, the outer shelf is covered with relict sand (van Andel, 1967).

From the eastern tip of South America southward to Argentina, the long straight coast backed by the Brazilian highlands has a continental shelf that is mostly narrow (about 20 km) and unusually steep-to, with slopes of up to 20° over a distance of 2500 km (Shephard, 1973). Along this coast, riverine muds gradually give way to calcareous deposits and biogenic shoals. Also some coral reefs occur, very disjuct with the main Caribbean center of western Atlantic coral reefs; at Queen Charlotte Bank off Brazil, a series of such reefs increases the width of the shelf considerably.

Pacific Continental Shelves

The western coast of the Americas resembles that of western Africa, though its morphology is dominated by an active subduction zone almost the full length of the continent. Coral reefs and banks are largely absent, and mangroves dominate only along parts of the central American and Columbian coasts. Despite the absence of major rivers, terrigenous material is very important. The shelf of western South America is very narrow, except off Ecuador, where it reaches 75 km near the Gulf of Guayaquil. Pelagic organic material is important in the upwelling regions of the Peru Current, where it forms anoxic muds deep on the shelf. Farther north, off Guayaquil, and in the Panama Bight, silts and terrestrial organic material contribute to muddy deposits, especially closer to the coast; here, the outer shelf tends to have sandier deposits. The shelf is very narrow or absent off Central America, where depths fall away down the continental slope to 6000 m in the Middle America Trench. Only off Panama and the Gulf of Tehuantepec is there a significant width of shelf; elsewhere, only in isolated bays and gulfs are there shallow areas resembling continental shelf habitat.

The western Pacific is a geologically most complex region and contains by far the largest and most important continental shelves anywhere in the tropics. The complexity of the region is due not only to previous movement of the Australian plate, and the presence of small isolated continental blocks, but also to the occurrence of abundant coral reefs and great river discharges. South of Taiwan, the Chinese coastline is at the northern end of the great continental shelf region that extends southeast to northern Australia. From Taiwan, the shelf widens southward until it reaches a width of 480 km off the Gulf of Tonkin and accommodates the island of Hainan. Throughout, deposits are a mosaic of relict sands, carbonate debris from coral reefs, and coralline algae, though over most of the southern part of this section muds and muddy sands tend to predominate. On the eastern coast of Hainan, there are reef corals, and coral banks (the Hsisha Ch'unto reefs) are based in deep water. Silts and muddy sands predominate along the relatively narrow east Vietnam shelf, but round into the very large Gulf of Thailand to the south deposits are more muddy, and sands and relict coral banks lie farther offshore (Emery, 1967; Emery *et al.,* 1972).

The Sunda Shelf of the South China Sea continues to the southeast through the Java and Timor Seas, and after an interruption by the deep basin of the Banda Sea, extends to the Sahul Shelf of Northern Australia. This shelf complex is of great width but is unusually shallow; the central part of the Sunda Shelf is less than 100 m, while the Gulf of Thailand is about 80 m. In the southeast, between Sumatra and Kalimantan (Borneo) the Sunda Shelf is only 10–40 m deep. Some of the passages between the main islands opened only during the early historical period, and the Gulf of Carpentaria is less than 10,000 years old. The Sunda Shelf is dissected by submerged, branching river channels trending northward toward the deep South China Sea between Kalimantan and the mainland, named (after their discoverer) the Molengraaf river system. Significant sediment flows still occur along the axis of the valleys of this system. Such are some of the tropical legacies of the low stand of the sea surface during the last, recent glaciation.

Sediments are especially sandy (and carry small coral reefs) in some straits between major islands, such as the Malacca Strait, and consist of soft mud in the deeper regions to the north, as the Sunda Shelf slopes into the South China Basin. The Pacific coasts of Thailand, Peninsular Malay-asia, Sumatra, Java, and the lesser Sunda Islands front onto this shelf as do the south and west coasts of Kalimantan. The Philippines, Sulawesi, the Moluccas, and the western peninsula of New Guinea stand clear of the Sunda Shelf, with their own, much narrower, shelves, especially steep-to around the Philippines. Isolated and aggregated coral formulations of all kinds are broadly distributed throughout this complex of narrow shelves, and muddy deposits are largely restricted to isolated bays and gulfs, the remainder of the shelf being of carbonate sands and coralline algae.

The deep basin of the South China Sea itself and the morphology of its surrounding shelves, including the Sunda Shelf, are the result of highly complex geomorphology, and include the drowned margins of the Asian continental block and the submerged island margins of Taiwan, Luzon, Palawan, and Kalimantan (Ben-Avraham and Emery, 1973). The Asian margin is associated with wide continental shelves, and the island margins to the east with narrow shelves. Running southwest to northeast across the southern part of the South China Sea are the "dangerous grounds," a shallow plateau, having an overall depth of 1500–2000 m, upon which many coral reefs have built during its progressive submergence, and into which many dendritic valleys have been cut extending from the Molen-graaf river system of the Sunda shelf. The central basin slopes gently southwards from 3400 to 4200 m depth, and has a number of volcanic seamounts in its central region. Coral reefs occur not only on the danger-ous grounds, but also on the continental and island margins, indicating submergence of these blocks also; the margins are deeply cut with valleys which can be traced upwards onto the shelf, and submerged wave-cut terraces are common.

The northern coasts of Australia all have broad continental shelves, although that along the Queensland coast is dominated by the Great Bar-rier Reef. Since Australia is a low-lying, low-rainfall continental mass, its rivers have slight channel gradients and intermittent flow, so they carry only small quantities of terrigenous materials to the sea. Shelf sediments, therefore, are dominated by biogenic material. On the northwest coast, from the Sahul Shelf to the western Arafura Sea, the deposits are largely sand-grain-sized biogenic calcareous material and coastal coral reefs are an important feature. However, from the eastern Arafura Sea eastward, deposits are dominated by terrigenous muds and silts, for here the terrain behind the coast is relatively steep, rainfall is high, and terrigenous mate-rial more abundant in river discharges. Coastal green mud therefore cov-ers essentially the whole of the Gulf of Carpentaria, and much of the shelf of southeastern New Guinea and the Torres Straits. Mangroves occur in large patches throughout this region: The south coasts of the Gulf of

Carpentaria and of New Guinea, and much of the coastline of southeast Asia and northern Sumatra are (or, at least, were until recently) mangrove coasts.

Continental Shelves of the Indian Ocean

In this ocean, a very wide range of conditions exists. The west coast of peninsular Malaysia, Thailand, and Burma, including the Andaman Sea, has a shelf that widens northwards. Deltaic muds from the Irrawaddy River almost entirely cover the shelf in this regon, though sands and shell-gravel cover large areas off western peninsular Malaysia. In the northern Bay of Bengal, Gangetic muds cover the whole shelf, though fine sands predominate inshore in shallow water. Needless to say, corals are absent in this region except at the Andaman Islands, and much of the coast is dominated by mangroves. The eastern coast of India, fronting the Bay of Bengal, has a narrow shelf dominated by terrigenous muds, giving place to calcareous oolitic sand in deeper water. The rather shallow shelf of Sri Lanka is dominated by calcareous biogenic material, and is much dissected by the heads of submarine canyons. The Palk Strait between Sri Lanka and India is extremely shallow, around 10 m in depth on the average, and much encumbered with coral heads on a sandy substrate.

The shelf off western India increases in width progressively north to the Arabian Sea, where it reaches more than 300 km wide. There is an offshore sequence from coastal mineral sands, through a midshelf muddy zone, to calcareous biogenic sands at the shelf edge. The midshelf muddy zone dominates much of the coast of western India, to the extent that during the southwest monsoon it appears as an offshore zone of fluid mud resuspended in the water column by wave action. This mud-zone can then act as a breakwater or "chakara" which protects the beaches behind from wave action and it is also greatly influencing the local biota and fisheries of this coast.

In the northeast Arabian Sea, the continental shelf off the Indus River is complex. It owes some of its characteristics to a major earth movement that occurred in the early nineteenth century; the Gulf of Cambay was formed when the sudden tectonic subsidence of the coastal plain allowed the sea to cover it. The subsidence area is now a large, extremely shallow arm of the sea south of Saurashtra, comparable with the Gulf of Kutch to the north. The Indus valley is continued as a steep-sided canyon cutting across the shelf north of Kutch.

From the northwest Arabian Sea, right around to Madagascar, the shelf is very narrow and in places nonexistent when the coast runs along fault lines. Only in the Persian (or Arabian) Gulf does the shelf widen. Within this Gulf, mud from the Tigris and Euphrates dominates the north and east, the remainder being sandy, calcareous material mixed with aeolian sand from the Arabian deserts. The narrow shelf of the Gulf of Oman along the coast of Baluchistan and Iran appears to be dominated by mud deposits, which are replaced farther west, along the south coast of the Arabian peninsula, by a mixed rocky and sandy substrate. On the other side of the peninsula, the Red Sea coastline has no river inputs and is dominated by aeolian sands, coralline algae, and coral debris from the abundant coastal reefs that extend from the Yemeni and Ethiopian coasts up to the Gulfs of Suez and Aqaba.

Along the East African coastline, the narrow shelf, where it exists at all, is mostly dominated by calcareous deposits and coastal reef formations, with mud being reported only from the immediate vicinity of estuaries—such as those at the mouths of the Tana, Ruaka, Zambezi, and Limpopo rivers. The only important mangrove habitats in the western Indian Ocean occur on the coast of Africa between Mombasa in Kenya and Dar-es-Salaam in Tanzania, around Beira in Mozambique, as well as on much of the coast of Madagascar. There are offshore banks on the continental shelf at 3°S (North Kenya Banks) and at 5–8°S associated with the islands of Pemba, Zanzibar, Lantham, and Mafia. These banks mostly have a hard, coralline substrate. Coral occurs as fringing reefs along much of the coastline, especially between 5° and 15°S. Only in Delagoa Bay, Mozambique, does the shelf widen significantly and soft sandy-mud deposits dominate the benthic environment.

FORMATION AND CHARACTERISTICS
OF COASTAL LAGOONS

In the tropics, as in higher latitudes, many stretches of coastline are backed not by solid land, but by coastal lagoons behind barrier islands, open to the sea to various degrees, with or without rivers, and of great significance in some coastal fisheries as nursery grounds (Yáñez-Arancibia, 1987). This family of coastal lagoons differs from those that are formed in association with coral reefs and will be discussed in the next section.

Lagoons not associated with coral reefs are most strongly developed on coasts with a history of submergence during the Holocene postglacial rise in sea level during the last 10,000 years or so (Emery, 1967). Lagoons may also form behind the cuspate spits that accrete at the mouths of open estuaries and delta mouths. These trend southward in the northern hemisphere and northward in the southern hemisphere. As we shall see, special evaporative mechanisms may also build lagoons in the arid tropics.

The formation and maintenance of a lagoon's coastal barrier depends on a balance between the supply of sedimentary material and its removal to deeper water by wave action. Sediment may be supplied to the outer side of the barrier by longshore currents and normal wave action, or may enter the lagoon itself by beach "washover" during storms. It may also accumulate by the settlement of river-borne silt, by flocculation of organic material at the salt water–fresh water interface, or by local estuarine plant production, especially by a mangrove ecosystem.

Coastal lagoon systems occur in all tropical oceans although they are significantly less important in the western Pacific than elsewhere; they are an especially prominent topographic feature on both the Atlantic and Pacific coasts of Mexico, Central America, and in the Gulf of Guinea. Both coasts of India, but especially the eastern, are backed substantially by lagoon systems, here often called backwaters. Along the shores of the South China Sea, of the Philippines, and of the remainder of the archipelagos lying to the north of Australia, major lagoon systems are rare.

The primary determinant for the existence of lagoon–barrier coasts is a relatively small tidal range (Nichols and Allen, 1981); in microtidal (<2 m) environments, wave action is especially important in maintaining linear barrier islands. As Macnae (1968) points out, very long stretches of the eastern coasts of Africa, Australia, and Madagascar carry linear dunes, reaching as much as 250 m in height, driven by the wind and wave action of the Trades. Lagoons form behind these dunes wherever they are breached by a river mouth.

There is a relationship between tidal range and the width and number of the mouths of tidal channels by which lagoons open to the sea. In microtidal regions there are few, narrow entrances, and barrier islands are long and narrow (Nixon, 1982). Also of great significance for lagoon morphology is the pattern and amount of river discharge that they receive and must carry to the sea; in many places, not only in the tropics, remains of terrestrial plant cover destroyed by deforestation are having important effects on the sedimentary regime within lagoons. Evaporation is a further important factor in the evolution of lagoons once formed, especially in the arid tropics.

Some of the lagoon systems in the humid tropics have developed into features that dominate great lengths of coastline. In the eastern Gulf of Guinea, an interconnecting waterway formed by several interacting processes exists behind about 600 km of coastline. In the west, there are typical sand-barrier lagoons supplied by longshore drift of sand and maintained by wave action. To the east, these sand barrier lagoons merge into the great waterway system of the Niger delta, many of whose entrances are protected by cuspate sand bars that have been built out by longshore drift of material in the classical manner. It is to be noted that such water-ways in the tropics tend to be dominated by a mangal ecosystem, as is the Niger delta, and that where this occurs the branching of the drainage channels is strictly dendritic, as in a high latitude salt marsh. However, unlike a salt marsh, a mangal contains an abundance of anastomosing channels forming a reticulate pattern; this is the consequence of the greater stabilization of the banks of meandering drainage channels by mangroves than by salt-marsh grasses (Webb, 1958; Allen, 1965).

The lagoon systems that dominate both Atlantic and Pacific coasts of Mexico are mostly formed behind coastal barrier islands (Mandelli, 1981). As in the Gulf of Guinea, the lagoons are in a state of continuing evolution, and many are already filled in by sand transport and the accumulation of biogenic material. Nevertheless, the remaining complex dominates the coastal ecology, and is an important factor in the coastal fisheries of Mexico (Yáñez-Arancibia, 1978). As in the case of some of the African lagoons, there is a natural alternation between dry-season conditions with little river discharge, when many of the tidal entrances become closed, and rainy season conditions when river discharge reopens the tidal entrances.

In highly arid tropical areas, with little rainfall, much evaporation, and a desert hinterland, lagoons have different origins and dynamics. Purser and Evans (1973) have described the formation and evolution of such systems in the Persian Gulf. A series of dunes, submerged by the Holocene rise in sea level, became cemented by the oolitic precipitation of calcium carbonate to build a coastal soft-rock formation (or "sabkha"). Possibly the high temperatures and evaporation also mediated the precipitation of calcium sulfate as gypsum and aided in the cementation of crusts over the submerged dunes. Rising sea levels left some sabkha isolated as offshore barrier islands, behind which the strong onshore winds and the longshore current combined to form hammer-shaped sand accumulations; soon, a lagoon system of the form shown in Fig. 2.2 began to develop. Within such lagoons, the production of biogenic material is intense and fills them relatively rapidly.

Even in the extreme environments of the early stages of formation of these lagoons, with very high temperatures and hypersaline conditions, an impoverished Indo-Pacific fauna develops: It is dominated by imperforate forams *(Quinqueloculina, Triloculina)*, corals, molluscs *(Cerithium)*, crustacea (cyprideid ostracods) and echinoderms (Evans *et al.,* 1973). These benthic organisms produce calcareous debris, over which grows a dense mat of calcareous red algae that consolidates it. The margins of the channels often bear strong growths of other macroalgae and the shallow flats carry sea-grass meadows. Throughout this biological system, the

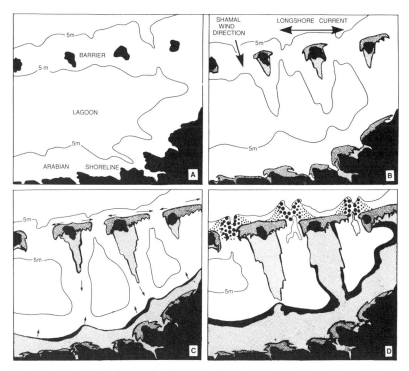

Fig. 2.2. Geological time series (A–D) to illustrate the evolution of coastal lagoons (sabkha) on the arid coast of Oman. (Modified from Purser and Evans, 1973.)

precipitation of inorganic calcium carbonate continues to accrete material. Eventually, the chemistry and shallowness of the lagoon is such (Krumbein, 1981) that most animals and higher plants are excluded. The development of microbial stromatolites commences, and mats of cyanobacteria begin to cover the flats as nitrogen and phosphorus become trapped and rapidly recycled within the ecosystem, which proceeds toward the final formation of a coastal gypsum lake (Krumbein and Cohen, 1977).

The southern Persian Gulf shows how such processes on desert coasts can be modified by their orientation in relation to prevailing weather. The western parts of the Trucial Coast are protected by the Qatar peninsula, and here there are wide intertidal mud flats having a high carbonate content. The central parts of this section of the coast are protected by the more distant Great Pearl Bank barrier, on and behind which longshore spits, offshore islands, and lagoons of varying dimension have been formed. It is in the eastern, more exposed part of this region that the most complex, hammer-shaped sabkha spits are formed. The northeastern parts of the coast are quite unprotected and receive the full effects of the long fetch of swell from the northern Persian Gulf, and of the northwest "shamal" wind; spits and lagoons on this coast are all longshore, are based on capes, and trend with the dominant wave direction.

GEOGRAPHY OF CORAL REEFS

Coral reefs are important throughout the tropics in all three oceans wherever water clarity is high and there is little turbidity from river discharges. They are a paradigm for tropical ecology and come in many forms: As well as atolls, which occur widely in the open oceans, masses of biologically produced calcium carbonate also occur as fringing reefs, barrier reefs, and patch or platform reefs, and are important features of continental shelf geography in low latitudes. Deep-water coral banks and thickets also occur in cold water below the lighted zone of the ocean, at about the break of slope of the continental shelf throughout tropical and temperate regions of the Atlantic, and probably also the Pacific.

Each reef contains, as Darwin observed during the voyage of *Beagle* almost 150 years ago, a record of the geological evolution of its site. Vertical growth of the whole reef community of animals and plants, and the consequent accretion of carbonate rock, can be rapid enough to keep up with interglacial rising sea levels, or with the subsidence of continental margin blocks. Though he was not able to know the true geological causes for changing sea levels, Darwin brilliantly deduced how a reef surrounding an isolated volcanic island might evolve—as the sea level rises to drown the original peak—into a typical atoll, and a fringing into a barrier reef; his "Coral Reefs" (Darwin, 1842) is available in modern reprints and should be the starting point for anyone wishing to understand coral reefs.

Darwin's account was greatly expanded in an exhaustive treatise by Davis (1928), and his observations were confirmed by greatly expanded data which supported his general hypothesis; rival but incorrect theories had caused the term "the coral reef problem" to be coined during the intervening period, and this is the title of Davis' book. Darwin's basic division of reefs into fringing reefs, barrier reefs, and atolls (or reef islands) has stood the test of time and is still the basis of modern classification of reef forms. In introducing coral reefs, we shall not stress their geological evolution but rather their ecology, and how their present-day morphology reflects present-day forcing by wind, waves, and tidal range.

Carbonate reefs, built by the activities of plants and animals, have existed since the Precambrian period ($>570 \times 10^6$ y BP), though the responsible organisms have gradually and significantly changed during geological time. Precambrian limestone reefs were formed as aggregations of stromatolites, built layer on seasonal layer by the growth of mats of cyanophytes, and the accretion of silt. Such formations are still built today, as in Shark Bay, Australia, in a hypersaline lagoon—the classic site for modern stromatolites—and also on the sandy ground of the off-shore Bahamas Bank, under conditions where sand grains are raised in suspension by strong tidal streams, so that the stromatolites grow by the accretion of grains at each tide. Palaeozoic reefs began to exhibit some of the diversity of modern reefs as the animal phyla progressively acquired the capability of forming hard skeletal parts. The first coral appears in the Ordovician, but it was not until the Mesozoic ($60–225 \times 10^6$ y BP) that modern scleractinians appeared and reefs began to have a modern appearance, though cornet-shaped rudist molluscs, now extinct, then formed a dominant component. Since the beginning of the Tertiary (60×10^6 y BP), and even as early as the Jurassic of eastern France, some reefs must have quite closely resembled those that occur today in the western Pacific, the center of modern reef diversity.

Coral reefs need rather special conditions for their growth. Sea-surface temperatures in excess of about 18°C appear to be required by hermatypic corals (those having symbiotic algae, or zooxanthellae, in their tissues and being the principal architects of modern reefs), though lower temperatures may be tolerated for short periods each winter at the northern and southern limits of their distribution. A small (usually <2 m) tidal range appears also to be correlated with the occurrence of hermatypic reef corals, which are also sensitive to salinities less than about 27%; unusually heavy and sustained rainfall may destroy coral on the more exposed parts of reefs, as may fluctuating or unusually high temperatures over about 29°C. Corals are killed by settlement of organic detritus or the clay fractions in turbid river discharges. For this reason, they do not occur where the larger tropical rivers discharge onto the continental shelf, and they respond quickly to local pollution and to new discharges from coastal engineering works or upland reforestation.

Leaving aside for the moment the deep-water oceanic atolls, fringing reefs may occur on continental shelves along or extending outward from the actual shore. If sea level is rising, fringing reefs may come to lie some distance offshore as a linear barrier reef with a lagoon between the reef and the shore. Such barriers differ widely in size, the largest being the great Barrier Reef that covers the wide continental shelf off Queensland. Here it forms a huge complex of reefs and low and high islands from Torres Straits south to the Northumberland Islands, progressively deepening and becoming more detached from the coastline to the south as it follows the line of the top of the continental slope. Unlike oceanic atolls, barrier reefs, as is the case with the Great Barrier Reef, may be borne on a substratum of sandy deposits, part of the normal series of sands and silts eroded from the adjacent continent and forming the continental slope. The Great Barrier Reef comprises about 2500 individual reefs and islands, up to 50 km in length. The barrier itself varies from 25 to 250 km wide, stretches across 10° of latitude, or about 1750 km, in length, and is separated from the coast by a lagoon 20–70 km in width. The present morphology of the Great Barrier Reef is the result of repeated changes in sea level during the Pleistocene glaciations, when the original reef was subject to the spread of riverborne silts, and to subaerial weathering (Bird, 1979). This reduced it to a chain of individual topographic features on which renewal of coral growth has occurred during each renewed postglacial rise of sea level.

Though the Great Barrier Reef is by far the largest, it is not unique. A barrier reef of about 650 km in length lies off the north coast of New Caledonia, extending northwest to southeast beyond each end of the island; north of the Fijian islands the Great Sea Reef runs east–west for about 260 km. A smaller barrier reef system occurs also in the Caribbean, principally off Belize on the eastern coast of the Yucatan peninsula. This extends from the Banco Chinchorro off Quintana Roo, south through Lighthouse Reef and Turneffe Island and into the Gulf of Honduras. Reefs continue eastward along the Honduran coast as the Islas de la Bahia. However, nowhere is this barrier so impervious as that off Queensland, which for much of its length forms an almost complete and continuous wall of coral reefs and islands.

Fringing reefs occur along much of the coast of eastern Africa, where both the reef and the coastal lagoon (where it occurs) have dimensions of only some hundreds of meters in width, yet occupy much of the extremely narrow continental terrace. Fringing reefs also occur on all suitable coasts in the rest of the western Indo-Pacific, and in the Caribbean–Antilles region; on relatively unsuitable coasts they occur only on rocky headlands. Similar structures occur also around some large oceanic islands, such as New Caledonia, Fiji, and the Hawaiian group. The longest continuous fringing reef, however, lies along the Red Sea coastline, over a total length of about 4500 km. Seawards, fringing reefs may have a very steep front with only minor terracing, and they plunge down to great depths in a very short distance; this is the case along much of the East African coastline.

Continental shelf atolls may arise by more rapid growth of corals around the edge of an emerging bank or old degraded platform reef. Where tectonic activity causes changes in elevation of the seafloor, raised or submerged atolls have been identified. Tinian, in the Marianas, with its uplifted rim, is an example of a raised atoll, while many examples of submerged atoll exist: The best known occur on the great series of flat limestone banks that comprise the Bahama Banks in the western Atlantic. The Campeche Bank in the western Gulf of Mexico is a neighboring and similar structure whose topography does not break the sea surface. On such offshore banks and on continental shelves, isolated tabular coral patches, platforms, and knolls also occur. Their form, like that of atolls, is determined by wave action, tidal streams,and the geology of their substratum; they may become linear or horseshoe shaped or, in quiet situations, circular. Patch and platform reefs are also an important feature of the Torres Straits, much of the Sunda Shelf, the coast of Sri Lanka, as well as in the Carribean region. Another series occurs from the Seychelles and Chagos banks southwards, where submerged reefs and atolls form a conspicuous physiographic feature in the eastern Indian Ocean.

Oceanic atolls are most numerous in the western Indo-Pacific, and have a variety of forms, though by definition all enclose a central lagoon; as Darwin rightly conjectured and geological drilling has confirmed, a typical atoll has its origin in a fringing reef around a volcanic peak or other midocean island, now submerged by rising sea levels, leaving a ring-shaped reef to mark its disappearance. In the great Pacific atoll province from the Marquesas to the Philippines, most atolls represent the summits of truncated volcanic cones submerged by rising sea levels. In the western Indo-Pacific, atolls rise from sinking fragments of continental plates; this is the case in the Maldives, Laccadives, and in the Coral Sea. In the Atlantic, the atolls east of Nicaragua are the largest group sited atop ancient volcanic peaks.

The growth of an atoll on its submarine platform is very plastic in relation to prevailing winds and wave action. It is only in very quiet regions, protected by other islands, that atolls may approach the diagrammatic circular form; instead, there is often an elongated spur at each leeward corner, with a relatively straight front facing the prevailing winds. Depending on the prevailing action of wind and waves, white carbonate sand may accumulate as a "cay" or small island on the leeward spurs or on the windward front. The reef flat is usually broken by sufficient channels to allow the internal lagoon to fill and drain at each tidal cycle, and the lagoon is frequently dominated by one principal entrance.

As we shall discuss in Chapter 4, the diversity and kinds of organisms that comprise coral reefs is very different in the Atlantic and Pacific Oceans; the greatest number of organisms occurs in the western Pacific, where much of the diversity of Tertiary reefs is preserved. Apart from having ecological consequences, these between-ocean differences in biotic diversity lead to differences in the general morphology of Atlantic and Pacific reefs: The seaward algal ridge and terrace (see below) is much less well developed on Atlantic reefs, and Pacific reefs have a much greater profusion of coral growth exposed between tidemarks on the reef-flat.

Unfortunately, coral reefs are now threatened throughout much of their range, especially where there are large coastal populations, such as East Africa, or where tourism (and the sale of decorative specimens of corals and molluscs) is economically important and unregulated. In the Philippines, for example, which is near the center of coral diversity, over 60% of the original coral reef cover has been destroyed in the last 20 years by dynamite fishing, by fish poisoning with domestic bleach, and by "muro-ami" or pounding-gears (Gomez et al., 1983).

COASTAL VEGETATION

In this section, we are concerned with three communities of plants that are important in coastal geography in the tropics: intertidal and sublittoral macroalgae on rocky shores, sea-grass meadows of lagoons, and mangrove forests on intertidal mud flats. Their ecological relations with the biota and fisheries of the continental shelf are discussed later.

Though the large brown seaweeds that dominate inter- and subtidal rocky reefs in higher latitudes are absent throughout the tropics, suitable substrates, ranging from exposed rocky shores to the reef-front of coral reefs, almost always carry a turf of smaller macrophytic algae—chlorophytes, rhodophytes, and phaeophytes. The most productive macroalgal communities in warm seas appear to be those dominated by the small brown algae *Sargassum* spp. For example, the infralittoral fringe at Bermuda is dominated by a dense turf of *S. bermudense* and *S. polyceratium* (Stephenson and Stephenson, 1972), and in the Antilles Wanders (1976) describes a turf of *S. platycarpum* below low-tide level around Curaçao. The genus *Sargassum* is ubiquitous in tropical seas and forms similar turfs in many places, from the Gulf of Guinea to Australia. Algal turfs in the tropical infralittoral fringe are frequently dense and include many genera and species; they are a source of commercial alginate and are important in the ecology of shore fishes on rocky coasts and other hard substrates. On bottoms of hard, clean coral sand algal turfs are often encountered together with sea-grass meadows; on the west coast of Florida, for instance, tufts of *Udotea* and *Halmidea* compete for space with sea-grasses (Dawes, 1981). Seaweeds of the genus *Eucheuma (E. cotti, E. spinosum)* are cultivated commercially in the Philippines (Smith and Pestaño-Smith, 1980). Elsewhere in the western Pacific, notably in Indonesia and Malaysia, culture of *Eucheuma* is gradually replacing the harvest of natural stands of algae, especially of *Gracilaria* and *Gelidiella* (Trono *et al.*, 1980).

Sea-grasses of two families of flowering plants, Potamogetonaceae and Hydrocharitaceae, occur widely throughout the tropics: Two groups of species can be recognized (McRoy and Lloyd, 1981)—shallow-rooted forms able to colonize oxidized and unstable sediments *(Halodule, Cymodocea, and Syringodium)*, and species that build a strong mat of rhizomes and form mature sea-grass meadows *(Thallassia* and *Posidonia)*. Sea-grass meadows occur as dense turfs in shallow water, where wave and tidal action is not excessive, and act as an accretion mechanism for suspended sedimentary material. Specific and generic specialization allows sea-grasses to occur in a wide range of ecological situations, even including the reef front of coral formations; they are familiar components of the ecosystems of coastal lagoons and of the lagoons within atolls or behind barrier reefs. The supple, strap-shaped leaves are able to withstand water movements and also regenerate rapidly in response to grazing by turtles, dugongs, manatees, and many species of fish in the same way as the sea-grass *Zostera* responds to grazing by geese in higher northern latitudes. Where the sediments are incapable of stabilization, the colonizing species persist; where they can be stabilized, permanent meadows of turtle-grass *Thalassia* frequently result. As with corals, and many other marine biota, the Atlantic sea-grass flora is less rich than that of the western Indo-Pacific, where about three-quarters of all species occur. The sea-grass ecosystem occurs most commonly throughout the Caribbean region, and also in the Indian Ocean and western Pacific.

It is often stated that mangroves replace salt-marsh vegetation *(Spartina, Salicornia)* in the tropics, and in general so they do; but in some more extreme environments a low turf of halophytic vegetation can occur on tropical salt-flats, especially as initial colonizers. However, it is mangroves of several families and many species that dominate the intertidal mud-flat zones in tropical seas and estuaries. Even in areas not already noted (Fig. 2.3) as being characterized by fringing mangrove forests, isolated plants and stands of mangroves will be found in suitable sites, from the sides of atoll lagoons to the mouths of small streams on sandy coastlines. It must be emphasized, however, that mangals are the dominant plant community occupying mud flats wherever estuarine conditions occur throughout the tropics.

Mangroves have a characteristic form common to most species, yet they are a numerous and taxonomically diverse flora. Dawes (1981) lists 80 species of 16 families of dicotyledons and 2 families of monocotyledons. These include the specialized Avicenniaceae and Rhizophoraceae, and also more generalized families such as Myrtacaceae, Tiliaceae, Ru-

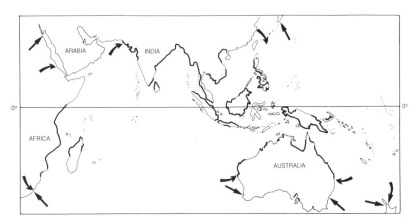

Fig. 2.3. Distribution of mangals, a thickened coastline indicating the presence of important mangrove areas. The northern and southern limits of *Avicennia* (straight arrows) and *Rhizophora* (curved arrows) are indicated. (Redrawn from Macnae, 1968.)

biaceae, and Plumbaginaceae. The most commonly known plants are of the genera *Rhizophora* (the red mangroves) and *Avicennia* (the black mangroves): Both are circumtropical. This diverse assemblage of plants, derived from mesophytic forest trees, has many physiological and structural traits in common. The diversity of mangrove plants gives a mangal (or mangrove ecosystem) the possibility of dynamic stabilization and growth over mud flats by a complex plant succession.

Mangroves develop most strongly in brackish or saltwater areas of high sedimentation, to which process they themselves contribute. They are able to conserve water and excrete salt, they can maintain themselves upright by a prop-root system in soft, anoxic mud, which precludes tap roots, and they can aerate their tissues by means of aerenchyma and specialized aerial roots (or pneumatophores) that stand erect from the mud surface. To establish themselves rapidly in an unstable soft muddy substrate, mangroves have evolved large viviparous seeds that implant themselves rapidly upon reaching the mud.

In the Atlantic (Dawes, 1981), the outer, colonizing fringe of a mangal is often dominated by the red mangove *(Rhizophora),* which is characterized by the possession of abundant prop-roots which enable a single plant not only to cover an increasingly large area, but actually to colonize forward over the sedimenting mud flat by vegetative means. In this zonation, the outer fringe of a mangrove system is a crowd of vertical, downcurving prop-roots, often bearing a mass of oysters. Inside the red mangrove fringe, somewhat higher in relation to tidal datum, is a zone of larger black mangrove trees *(Avicennia)*, mostly without prop-roots and growing on flat ground firm enough for walking. The spaces between the trees are covered by a forest of small erect pneumatophores sticking 30 cm or so above the surface. Further up yet, a zone of white mangrove, even larger trees, largely lacking either prop-roots or pneumatophores, gives way to the coastal dry-ground forest plant communities, in which *Nypa* palms are often conspicuous.

For the Indo-Pacific mangals, Macnae (1968) describes a more complex basic succession, which is fully expressed only in very heavy rainfall areas of the warmer parts of the tropics. Beyond the outer mangroves there is frequently a zone dominated by sea-grasses. The outermost mangal zone, not always present, is a park-like region of widely-spaced *Sonneratia* trees, the spaces between them occupied by their erect pneumatophores. This zone is progressively absent toward the cooler climates. Inside the *Sonneratia* zone is a seaward fringe of black mangroves *(Avicennia)* which may be especially wide (up to 750 m) in accreting situations, but reduced to a strip the width of only a few trees in stable conditions. The main forest lies within this and is dominated by the prop-roots of various *Rhizophora* species, of which three are important: *R. mucronata,* a salt-water-loving species, which is the commonest and extends from East Africa to New Guinea; *R. stylosa,* which occurs from Malaysia to Australia, and *R. apiculata,* which occurs in the same region, but also ranges west to India, especially in more brackish situations. These species form extensive forests in their areas of distribution. Of less interest in a fisheries context is the zone of *Bruguiera* or white mangrove lying shoreward of the *Rhizophora* forest; it can be divided into a series of subzones depending on drainage, rainfall, and the nature of the coastal forests.

The width of each zone on any shore depends on the tidal range and slope of the shore; in Sri Lanka, Macnae (1968) points out, the tidal range is only 75 cm and the zones are extremely compressed, while in some locations in Sumatra where the tidal range reaches 5 m each of the zones is several kilometers wide. Because, in general, the current necessary to suspend particles of silt from the deposits is stronger than those at which similar particles sink out, accretion occurs wherever plant or other obstructions slow down silt-bearing water sufficiently. This settling lag effect causes fine silt to be deposited far into the mangal. Accretion, supporting the advance of the seaward *Avicennia* fringe, occurs very rapidly under some circumstances. Macnae (1968) quotes a number of examples around the South China Sea of coastal accretion of between 25 and 200 m/y. Macnae also points out that in the tectonically active areas of Southeast Asia, especially near the straits between Peninsular Malaysia,

Sumatra, and Java, many of the most extensive mangrove areas occupy coastal lands that were submerged only in early historical times.

Somewhat like a coral reef, a mangal supports a host of animals and other plants. The prop-roots are host to a great variety of epiphytic algae and to molluscs and barnacles. Crustacea, especially sesarmid crabs, climb in the mangrove bushes and several specialized crustacea and fish are abundant on the mud surface itself, serving somewhat to aerate it with their burrows. However, Macnae (1968) and Walsh (1974) both suggest that much of the associated fauna responds not to plant species-specific niches, but more simply to gradients of salinity, radiation, and oxygen.

A modern account of the mangrove forests cannot avoid mention of the rapidity with which mangals are being destroyed or converted to other uses. The Indus delta is now almost completely denuded of mangroves, except along the edges of very well-flushed tidal channels. The coastal forests of southern Vietnam were devastated by U.S. defoliants during the recent war. Clearance for firewood, rice farming, urbanization, warfare, fish, and especially shrimp culture are the most frequent reasons for the current destruction of mangrove forests in Southeast Asia; the chainsaw, and arboricide chemicals are the tools that have caused it to happen so fast. Even apparently remote regions are not exempt; as we write, Japanese lumber interests are negotiating for the right to log the mangrove forests of southern Papua New Guinea. We shall return to this problem in relation to shrimp recruitment in Chapter 10.

Coastal habitats are very vulnerable to pollution by floating oil slicks, and mangals appear to be exceptionally vulnerable (Vandermeulen and Gilfillan, 1985). In most coastal systems, the effects of oiling are restricted to the intertidal front, but in mangals, this feature is carried into the very interior of the forest by the finest terminal creeks of the dendritic system of tidal channels. High spring tides will then carry the effects over the flats on which the trees stand. Such consequences have been observed in Puerto Rico where the *Zoe Colocotroni* and *Peck Slip* tanker spills caused heavy mortality to mangal flora and fauna.

The consequences of the rapid destruction of mangals for estuarine stability and the biological systems of the continental shelf remain difficult to calculate, but the next few decades will probably demonstrate what we cannot yet measure. A very comprehensive bibliography on mangroves has recently been published by UNESCO, and the reader will find it a convenient entry into the scattered literature on mangrove ecology and utilization (Rollet, 1981); the UNESCO volume on mangrove research methods (Snedaker and Snedaker, 1984) will also prove very useful to newcomers to this field of research.

Chapter 3

Circulation of Tropical Seas and Oceans

CONTINENTAL SHELF PROCESSES

Continental shelves are the shallow margins of the ocean basins, and their circulation and fish production cannot be understood except in that context. Shelf processes everywhere are heavily influenced by the eddying and variability of the major ocean currents, by global-scale oceanic tides propagating onto the shelf, by coastally trapped very long Kelvin waves, by atmospheric forcing, and by other mechanisms.

There is no systematic difference in relative distribution of microtidal (<2 m), mesotidal (2–4 m), or macrotidal (>4 m) coastal environments between high and low latitudes. As elsewhere, tidal ranges at tropical coasts are greatest where the shape of the coastline and the morphology of the shelf constrict the oceanic tides and increase their amplitude. Macrotidal environments occur widely in the tropical Indo-Pacific; in the northwest Arabian Sea the range reaches almost 11 m in the Gulf of Khambali, but ranges of 4–6 m are more usual in the Bay of Bengal, the Gulf of Tonkin, from northwest Australia to southern New Guinea, on both coasts of the straits between Madagascar and Africa, and finally in the Gulfs of Panama and Tehuantepec.

In the Atlantic, only a small region of the South American coast on either side of the Amazon mouth exceeds 4 m; the whole of the western coast of Africa has a tidal range of less than 2 m, being highest in the Bight of Benin. The whole Caribbean–Antilles region is microtidal, with many regions having tidal ranges of only about 1 m; similarly small tidal ranges are usual in the Hawaian archipelago and at other oceanic islands.

It is a common observation, both in the tropics and elsewhere, that there is a narrow strip of ocean above or slightly seaward of the break of slope at the edge of the continental shelf where there may be evidence of upwelling or at least ridging of isotherms toward the surface. Here one frequently encounters concentrations of seabirds, pelagic fish, and other evidence of enhanced biological production. This feature is very striking in the Gulf of Guinea during the dry season (November to March) when very large numbers of Palaearctic terns (*Sterna* spp.) occur along the edge of the continental shelf, together with yellowfin tuna and porpoises, because of the abundance of small pelagic fish and micronekton. These can be seen as a local thickening and vertical disturbance of the deep acoustic scattering layer just beyond the shelf break. Figure 3.1 shows this phenomenon at the continental edge off Mozambique in the western Indian Ocean. It is probably global in its occurrence.

Although most of the studies of these dynamic events at the shelf edge have been done in high latitude (see, for instance, Petrie, 1983; Pingree and Mardell, 1981; Sandstrom and Elliott, 1984), their findings are also applicable on tropical shelf-breaks. Transient wind-forcing, by moderate winds of two or more days duration blowing parallel to the shelf edge, can cause a response of sea level at the coast and of alongshore currents on the shelf. Associated with such events there may be strong vertical shear and upwelling along the shelf break from as deep as 400 m.

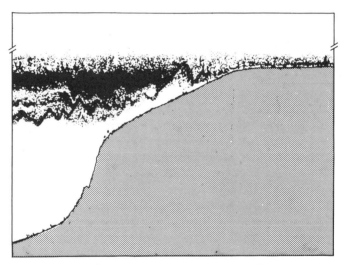

Fig. 3.1. Echogram taken passing over the edge of the continental shelf off Mozambique, showing layers of plankton "riding" internal wave trains. (Redrawn from Saetre and de Paula e Silva, 1979.)

Other studies of the interaction of the oceanic tide with the shelf edge have demonstrated the propagation of "packets" of internal waves normal to the line of the shelf-break which, under some circumstances, may rapidly travel coastward and dissipate tidal energy as solitary internal waves (solitons) with amplitudes of up to 50 m. These mechanisms have the ability to transport nutrients vertically across the thermocline by increasing vertical eddying, so as to enrich the mixed layer over the continental shelf with nutrients from below. It is likely that the propagation of trains of large-amplitude internal waves, which break as they pass onto the shelf from deep water, represents an important mechanism for maintaining the nutrient flux and supporting the high biological production in the photic zone on the continental shelf.

Often, dynamic processes over the shelf-break result in a linear frontal system separating the water of the continental shelf from oceanic or slope water. Such fronts have local consequences for the distribution of plankton and pelagic fish. The internal waves propagating along the horizontal density surfaces are frequently accompanied by linear, oily slick-lines at the sea surface parallel with the shelf-break. These alone are a good surface indication of the dynamic processes occurring below and have been widely observed in the tropical ocean, from the Timor Sea to the Gulf of Guinea.

Over the continental shelf itself, tidal dissipation is reflected in semidiurnal or diurnal tidal currents that are stronger in shallow water, around headlands, and in funnel-shaped estuaries and bays. Though the tidally induced movement of a water mass is oscillatory, it includes a net movement, or residual current, by which water is transported through the area. Where the turbulence imparted by bottom friction is sufficient, it breaks down the stability and stratification of the water mass over the continental shelf. This depends on the speed of the tidal current (u), the depth of the water (h), and the buoyancy imparted by solar heating, evaporation, and freshwater run-off. Mixing by tidal currents will overcome the effects of buoyancy when values of $\log_{10} (h/u^3)$ are less than about 2.0 (Simpson and Hunter, 1974). Tidally mixed and stratified areas on continental shelves are separated by a frontal region, important for biological production.

At such tidally induced fronts, the thermocline breaks the surface, nutrients are supplied to the euphotic zone, and biological production is enhanced. The location of fronts can be predicted from numerical simulations of tidal regimes, their location varying with the state of the tide: Displacement of a front may simply follow the semidiurnal tidal excursion where this is normal to the line of the front. Eddies and ring-structures of the scale of tens of kilometers may be associated with tidally induced fronts and are readily observed in satellite imagery of sea-surface chlorophyll or radiometer imaging of sea-surface temperature (e.g., Pingree *et al.*, 1982).

The existence of such fronts has been little investigated on tropical continental shelves and it is not yet clear if they have the same significance there as in high latitudes. However, Hunter and Sharp (1983), using regional tidal dissipation parameters have indicated where in the tropics we can expect tidal mixing to overcome buoyancy, and hence where tidal fronts occur: Such areas include the Gulf of Panama, Guyana to northern Brazil, Malacca Strait, Mindanao to New Guinea, northern Bay of Bengal, southern India, and the Persian Gulf. In the Persian Gulf, Hunter (1984) found that wind stress could also be an important factor in vertical stirring of continental shelf water. Forbes (1984) has identified regions within the Gulf of Carpentaria where tidal stirring can be expected to occur. Such areas occur not only around the coast of the Gulf, especially near headlands, but also occupy almost the whole of the Torres Straits and the south coast of New Guinea. Because, in winter, the whole of the Gulf of Carpentaria is mixed by wind-stress and convective overturn, tidally generated fronts may be formed only in the warm season.

Where the permanent thermocline is shallow, it is possible that the input of buoyancy at the surface by strong incident radiant energy may be enough to prevent tidal mixing. In extremely evaporative regimes, like the Red Sea and Persian Gulf, buoyancy may even be added at the surface by evaporation; this apparently anomalous situation is described for the Persian Gulf by Hunter (1984), who suggests that it is a common process in the dry tropics. The unstable situation, which results from near-coastal evaporation so that dense water overlies lighter, will overturn and cause some mixing as it does so. But the resultant water column may be more stable than in adjacent, less evaporative regions further offshore. An inverted estuarine circulation may then result with some vertical shear as the dense, saline water passes seawards to be replaced by fresher surface water. As we shall see, the whole Red Sea circulation is an example of this process on a grand scale.

While these shelf processes await urgent study by tropical oceanographers, it can be predicted that the identification of tidal fronts on continental shelves in low latitudes will prove to be of great importance in understanding local areas where anomalously high production appears to occur regularly.

Another shelf process that is probably of great importance to the reproduction of fish and the development of plankton communities is the formation of permanent or seasonally recurring coastal gyres in which the planktonic eggs and larval fish and other organisms may be retained in favorable habitats. Despite the theoretical requirement for such phenomena (Johannes, 1978; Bakun *et al.*, 1982), the empirical evidence for their utilization is rather sketchy. However, Saetre and da Sylva (1982) suggest the existence of eddy-like circulation over the wider part of the Mozambique off Beira, and this process could perhaps act as a larval retention mechanism. It is, of course, mostly in bays or behind headlands that small permanent closed gyres, running counternatant to the main stream offshore, may be anticipated.

WATER MASSES OF TROPICAL OCEANS

The water masses of tropical oceans are the basic environmental matrix within which fisheries ecosystems must be studied. To understand their characteristics, a good starting point is to emphasize that it is in tropical regions that the sharpest gradients in temperature and some other properties occur. Only 15–20 m below a canoe fisherman working under a tropical sun in the Gulf of Guinea, or off the west coast of Central America, the water begins to be cooler and by 40 m (or the height of a large tree in the tropical forest) there is cold water of only about 16°C. Thus warm, tropical surface water masses can be a very thin skin over the cold interior of the ocean. The two temperature regimes are separated throughout the tropics and subtropics by a sharp thermal discontinuity that resembles the summer thermocline of higher latitudes. Wind-induced turbulence mixes heat from the surface downwards through the mixed layer until a critical point is reached, within the thermocline, at which the density gradient (maintained by the equatorward transport of cold, high-salinity water) is sufficiently strong to overcome the downward mixing of surface heat. Normally, halocline and thermocline coincide in most tropical areas so that salinity increases downward across the discontinuity layer. Surface salinity is determined by evapo-precipitation which has a very strong but well-documented regional variability in the tropics (e.g., Baumgartner and Reidel, 1975).

Wyrtki (1964) recognizes three kinds of surface water in the tropics: (1) tropical surface water (25–28°C, 33-34‰), (2) equatorial surface water (20–28°C, 34–35‰), and (3) subtropical surface water (19–28°C, 35-36.5‰). This classification implies that the water above the tropical thermocline is warm and salty except where upwelling along the equator induces cooler temperatures, or where surface water from the major central oceanic subtropical gyres flows into tropical regions, becoming warmer and diluted with tropical rainfall, eventually reaching a state where it is indistinguishable from adjacent tropical surface water. The somewhat lower salinities characteristic of tropical surface water reflect the excess of precipitation over evaporation in the humid tropics. Any good global map of sea-surface salinities (such as Chart VI of Sverdrup, Johnson, and Fleming, 1942) will show those areas of the arid tropics where evaporation exceeds precipitation and where the tropical surface water becomes very salty.

The Red Sea is a special case of an evaporative basin with unique water masses. It is a 2000-km-long trench with a sill depth of only 110 m at Bab-el-Mandeb at its southern end; evaporation from the sea surface of about 2 m y^{-1} brings a constant inflow of surface water from the Arabian Sea. Winter cooling to 18°C in the northern parts of the Red Sea of water already at a salinity of 42‰ causes convection cells which form the deep water mass of the basin. These have a temperature of about 21.5°C and a salinity of about 40.6‰. Thus, at Bab-el-Mandeb, deep water having these characteristics flows out below the inflowing Arabian Sea surface water. This outflow is not negligible, being about half that of the deep outflow at the Straits of Gibraltar, and its origin is a somewhat similar process to what occurs within the Mediterranean Sea. The existence of an almost isothermal deep ocean basin in the tropics having bottom water warmer than 20°C could provide a model for the warm Mesozoic Ocean, and it is therefore important that its plankton ecology should be fully worked out; a start has been made by Weikert (1980) and Beckman (1984).

On some continental shelves, the tropical surface water mass becomes much influenced by river discharges and diluted by monsoonal rains. This may occur as a general "estuarization" of the continental shelf, or through the existence of a discrete plume of river discharge water. Surface salinity maps show the general areas where continental shelf estuarization is important: the Bay of Bengal, South China Sea, Gulf of Panama, and Gulf of Guinea. Plumes from major rivers such as the Congo and Amazon, and also more locally from small rivers and creeks during the rainy season, remain identifiable even within these estuarized regions, and the regions of low-salinity tropical surface water tend also to have a relatively high turbidity and muddy deposits.

The Bay of Bengal is a classical site of continental shelf estuarization, and salinities below 20‰ occur over much of the region at the end of the southwest monsoon in September–October induced by river discharges and regional excess of precipitation over evaporation. After the onset of the northeasterly (dry) monsoon, salinity begins to rise as evaporation exceeds the supply of fresh water from all sources. This reversal of the monsoon also reverses the gyral circulation of the Bay of Bengal so that as its salinity rises during the northern winter, the brackish mass water is progressively removed from the Indian coast and driven back down the eastern coast of the Bay. The same process seems to occur in the Bight of

Biafra, the extreme corner of the Gulf of Guinea, where the gross geography and wind patterns are somewhat similar to the Bay of Bengal. Satellite-tracked drifting buoys indicate surface drift during the southerly monsoon into the Bight of Biafra from south of the equator, crossing the conjunction of the South Equatorial Current and the Guinea Current (Piton and Fusey, 1982).

The Java Sea, at the southern end of the South China Sea, offers another good example of estuarization. Although no major river has its mouth in this shallow sea, its surface waters have a salinity that usually does not exceed 30–32‰. The region of lowest salinity is driven back and forth between the islands of Kalimantan and Java, depending on the monsoon regime (Wyrtki, 1961; Soegiarto and Birowo, 1975).

Ryther, Menzel, and Corwin (1967) discuss the discharge from the Amazon of 2×10^5 m^3 sec^{-1} of fresh water into the ocean. A single plume does not seem to be formed, as it normally is from lesser rivers, but lenses of freshened water 300–500 km in diameter occur northwest along the coast from the Amazon mouth. Within each lens, nutrients, except silicate, are low, as is biological productivity. The existence of a lens may cause upslope of isotherms inshore of it, with some local upwelling.

Below the thermocline, the subsurface water masses present in the tropics are all formed outside the region, principally by sinking near the subtropical convergence. Primary water masses formed in this way frequently extend from below the thermocline to about 500–600 m, and they may become secondarily modified by mixing. However, the subsurface water masses are, on the whole, more uniform than the surface water masses. They frequently contain a salinity maximum at their core, and also—more importantly for fisheries—an oxygen minimum layer. The latter is present almost everywhere in the tropical oceans at depths corresponding to the main oceanic thermocline. In the eastern basins of the Atlantic and Pacific, and especially in the northwest Indian Ocean, this feature is very strongly developed. Extremely low oxygen values may extend over depths of several hundred meters, and H_2S may be present although in low concentrations. The O_2 minimum usually occurs near a depth of very limited horizontal motion (Wyrtki, 1962a). The coastal upwelling that occurs seasonally along the west coast of India brings waters of very low oxygen content onto the continental shelf, and this has the important biological consequence of displacing the demersal fish fauna and benthic crustacea (Banse, 1968). Entrapment of biota in oxygen-poor upwelled water on this coast has been implicated in mass mortality of fish and crustacea. A similar phenomenon apparently occurs off Burma, with surfaceward crowding of the isolines and the ascent of O_2-poor water onto the shelf (Strømme et al., 1981).

Below the tropical subsurface water masses lie the intermediate waters that originate in high or even in very high latitudes, and that have little relevance to tropical fisheries, or to the biology of the upper ocean in the tropics. We shall discuss neither these nor the deep water masses in detail.

THERMOHALINE AND WIND-DRIVEN CIRCULATION

The wind-driven Ekman circulation and the density-driven thermohaline circulation combine to produce the surface and deep currents of the oceans whose basic description is now more or less complete. For reasons that will become clear below, there is currently a great concentration of oceanographic research effort in the tropical oceans. Perhaps the single most important attribute of the tropical ocean is its ability to adjust rapidly to a change in wind stress (Philander, 1985); the time required for such adjustment is correlated with latitude so that while the monsoon alternation can reverse the direction of the Somali Current in a few weeks, it would take about a decade to generate the Gulf Stream in midlatitudes from rest. This ability of the tropical ocean rapidly to change state and redistribute heat is what makes possible the El Niño–Southern Oscillation (ENSO) phenomenon that we shall discuss later in this chapter, and also the highly seasonal nature of much of the equatorial circulation.

Though offset to the north of the equator and modified by land masses, the current systems of the tropical oceans form two mirror images in the northern and southern hemispheres. This pattern is principally determined by the Hadley circulation through which heat is transferred to high latitudes by rising air masses over the equatorial doldrum regions of the ocean, so driving the permanent trade-wind systems to bring surface air equatorward, and by the change in sign of Coriolis' force at the equator, and its increasing influence poleward from the equator. Figure 3.2 shows schematically how these two forces account for the basic pattern of westward-flowing North and South Equatorial Currents, and eastward-flowing North and South Counter Currents and the Equatorial Undercurrent. Standard current atlases, such as will be found in any general text on oceanography, demonstrate how this pattern is translated into actual current systems, and especially the influence within the tropics of the eastern boundary currents that complete the gyral circulation, which is initiated in the tropics by the trades and supported at higher latitudes by the planetary westerlies.

Thus, the principal circulation of surface water in tropical oceans is by

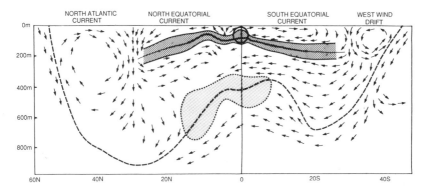

Fig. 3.2. Atlantic section to show the vertical and meridional transport within the main zonal current systems. The main thermocline appears as a bold dashed line and the tropical, subtropical thermocline is shaded, as is the zone of very low O_2 below. The eastward-flowing equatorial undercurrent is circled. (Simplified from Dietrich *et al.*, 1980. From "General Oceanography, An Introduction," Copyright © 1980 John Wiley & Sons, Inc.)

zonal (east–west) currents. The North and South Equatorial Currents are driven by the prevailing trade winds and, as Philander (1985) reminds us, it was an early triumph of dynamical oceanography that showed that these currents are driven by the curl, rather than the direction, of wind stress (Sverdrup, 1942). The eastward-flowing Equatorial Counter Currents, on the other hand, flow contrary to the prevailing winds, serving to maintain the oceans in geostrophic balance.

As Fig. 3.3 shows, water transport into the tropics by the eastern boundary currents is not large in relation to the recirculation of water within the tropical region by the westward North and South Equatorial Currents, balanced by the eastward Counter- and Under-Currents. It is remarkable that the Equatorial Under-Current, whose eastward flow balances the whole westward transport of the South Equatorial Current (about 35×10^{12} cm^3 sec^{-1}), was not discovered until the 1950s (Cromwell *et al.*, 1954). Yet this current, which occurs in all three oceans, is one of the most remarkable currents anywhere in the oceans, being driven by a pressure gradient itself caused by the elevated sea surface and deep mixed layer of the western parts of each ocean basin. That a narrow, remarkably straight river of thermocline water should flow rapidly eastward, directly under the Equator, has many implications not only for the biology of tropical oceans, but also for fishing operations there. Like the shelf-break front, the Equatorial Undercurrent is accompanied by the propagation of internal waves along the density discontinuity layer, which can be seen as "oily" linear slicks or as alternate calm and ruffled patches of water on

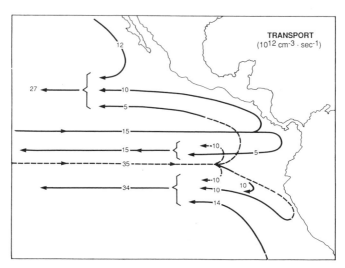

Fig. 3.3.   Mass transport of water by the equatorial current system; subsurface current flow is shown as a dashed line. (From Wyrtki, 1967.)

the surface of the ocean, often extending east–west from horizon to horizon.

THERMOCLINE TOPOGRAPHY AND MIDOCEAN DIVERGENCE

Some secondary manifestations of the general ocean circulation, such as the associated topography of the thermocline, and the existence of convergence and frontal zones are important for fisheries and oceanic biota. The global current systems are reflected in varying depths of the mixed layer above the thermocline, where shear occurs between surface currents and the subsurface water masses. The broad features of thermocline topography, hence of mixed layer depth in the tropics, are primarily a result of the dynamics of geostrophic water motion and only secondarily of wind stress and downward mixing. The ridges, domes, and troughs in this topography reflect the geostrophic balance of the principal current systems: Currents flow along the flanks of ridges, the ridge being on the left of the current in the northern hemisphere and on its right in the southern. Domes in thermocline topography represent circular motion anticlockwise in the northern hemisphere; conversely, large clockwise gyres in the northern hemisphere, such as the major oceanic subtropical gyre of the north Pacific deepen toward their centers. Such features of the thermocline topography as ridges and troughs extend many thousands of kilometers, usually trending zonally rather than meridionally.

In addition to topographic features at this scale, the even larger feature of the westward deepening of the mixed layer in all three oceans is also of importance to fisheries (Fig. 3.4.). Wind forcing of the sea surface by the tropical trade-wind systems normally results in an elevation of the sea surface of about 20 cm in the western Pacific (except in El Niño years) and a progressive deepening of the mixed layer westwards. Thus, in the eastern parts of the tropical Atlantic and Pacific, the mixed layer is generally only 20–40 m deep and lies above a very sharp thermocline, while in the Caribbean or east of the Philippines, the mixed layer is usually 100 m deep or more and thermocline gradients are much less steep so that the thermostad is reached only at 300–400 m.

As everywhere in the oceans, net production of organic matter in the euphotic zone depends on the rate of vertical transport of inorganic ions to replace the organic material lost from the euophotic zone by sedimentation, and no longer available for recycling there. Vertical transport occurs by a variety of mechanisms: at the centers of thermocline domes, where divergence brings the density discontinuity layer within the euphotic zone; by vertical turbulent mixing across the thermocline in the open ocean, depending on the intensity of wind stress; by midocean divergence

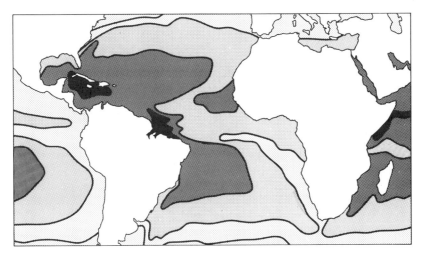

Fig. 3.4. Temperature at 100 m depth, to illustrate westward deepening of the thermocline in each ocean. Isotherms at 10°C, 15°C, 20°C, and (black shading) >25°C. (Redrawn from Gorshkov, 1978.)

due to reversal in sign of Coriolis' force along the line of the Equatorial Under-Current; by the island-wake effect; and by a variety of mechanisms causing divergence and upwelling at coastlines. The limits of regional organic production can only be defined once quantification of the vertical flux of nutrients by these mechanisms has been achieved.

In midocean, the principal centers of new production are the thermocline domes referred to above, and the wind-driven Ekman divergence along the equator that occurs in all three oceans. Major thermocline domes have been identified and explored in the eastern Pacific off Costa Rica (Wyrtki, 1964; Blackburn, 1981), and in the eastern Atlantic, off Guinea and off Angola (Voituriez and Herbland, 1982). These are caused by the interaction of the Pacific and Atlantic Equatorial Counter Currents and the coastline, and express the resultant anticyclonic gyres. The Costa Rica Dome is a quasi-permanent feature and a semipermanent upwelling center, though it constantly changes location and shape it remains in the same general region. The anticyclonic gyral action within the Costa Rica Dome may be so strong as to completely remove the mixed layer water laterally, causing the thermocline to break the surface in the center of the dome with important biological consequences.

At the southern end of the Mozambique Channel, Menaché (1961) found a dome and divergence with upwelling isohalines and much evidence of biological enrichment (birds, breezing schools of fish). This doming occurs in November at a boundary between the retrocurving African coastal current, returning north along the west coast of Madagascar and the southerly Aghulhas Current sweeping across the southern end of the Mozambique Channel. As Zahn (1984) was able to show by simulation modeling, these dynamic effects are linked not only to ocean circulation but also to bottom topography, and it now seems that the gyral patterns are caused both by the curl of wind stress and also by the existence of rises on the sea floor in the central Mozambique Channel.

However, by far the largest regions of upwelling occur quasi-permanently along the equator in all three oceans. Figure 3.5 shows how a wedge of cool surface water exists in non-El Niño years, symmetrically arranged on either side of the equator, and dominates the surface temperature chart of the eastern Pacific; similar patterns exist, at least seasonally, in the comparable regions of the Atlantic and Indian Oceans.

This figure also shows that this wedge of cool water is not simply the result of entrainment of cool water from the upwellings in the eastern boundary currents as the South Equatorial Current sweeps northwest across the equator. Rather, though the process by which cool water is upwelled at the equator is very complex, it has a simple basis. The change

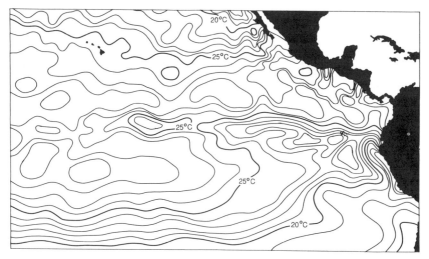

Fig. 3.5.   Eastern Pacific sea surface temperatures during a non-El Niño situation with upwelling in the eastern boundary currents and at the equator. (From Fishing Information, NOAA-NMFS, SW Fishery Center, La Jolla, California.)

of sign of the Coriolis force at the equator between the northern and southern hemispheres causes Ekman drift to diverge on either side of the equator itself—to the right in the northern hemisphere, to the left in the southern hemisphere. However, the structure of the water column at the equator is extremely complex because of the existence of the equatorial undercurrents; these have their velocity maximum in the upper parts of the thermocline, where much vertical turbulence is introduced that spreads the isopycnals by mixing both upwards and downwards, producing the very characteristic pattern (Fig. 3.6) that is seen for many properties in any transequatorial section of sufficient length. The eastward shoaling of the thermocline in each ocean contributes to the intensification of the effects of the equatorial divergence. The origin of the upwelled water is within the layer of maximum density discontinuity and thus the whole length of the equatorial divergence brings relatively nutrient-rich water to the surface. The equatorial undercurrents themselves actively and continually entrain water meriodinally from the westward flowing water masses to the north and south with the result that the actual residence time of water within the current is rather short—probably only a few weeks, and less long-distance eastward transport occurs than might be inferred from the velocity field.

SOUTH   NORTH

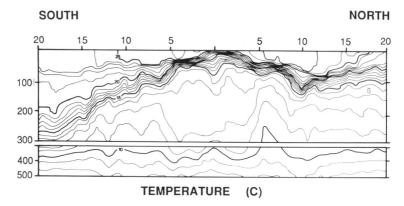

TEMPERATURE   (C)

Fig. 3.6. Temperature section across the equatorial current system, eastern tropical Pacific at 110°W. Original, based on EASTROPAC data (Love, 1971).

ECKMAN DIVERGENCE AND OTHER FORMS OF COASTAL UPWELLING

Because of its great importance for the support of fisheries in the tropics, perhaps the most written-about enrichment process is coastal upwelling. In fact, there has been a tendency to regard coastal upwelling as being important only in tropical seas because there are greater physical and chemical differences between upwelled and surface water masses in the tropics than in higher latitudes. In fact, of course, physical conditions apt to drive upwelling occur in all latitudes. The classical upwelling regions of the oceans are, however, in the subtropical eastern boundary coasts of Morocco–Mauretania, Angola–Namibia, Peru–Chile, California–Mexico, and western Australia; of these, the most biologically significant occurs in very low latitudes (5–15°S) off Peru.

In addition to these eastern boundary upwelling systems, coastal divergence occurs also in many other tropical locations, and driven by a variety of mechanisms: upwelling occurs off the Gulfs of Tehuantepec and Panama, off northeast Venezuela and Brazil south of Cabo Frio, from Ghana to Togo in the Gulf of Guinea, on the Somali coast and off southern Arabia, on the western coast of Sumatra and Java, and along much of the western coastline of India. It is important to understand that upwelling in the tropics, as a mechanism for nutrient enrichment of fisheries, is not confined to eastern boundary currents. The mechanisms driving these various upwelling locations are far from all worked out, but what we already know suggests that a variety of processes are implicated.

Some of these situations are relatively simple and driven by local conditions. The Gulf of Tehuantepec on the Pacific coast of southern Mexico lies west of a 40-km gap in the Sierra Madre through which extremely strong winds blow from October to March, driven by Atlantic anticylonic systems located in the Gulf of Mexico or over the southern United States. These winds induce strong mass transport divergence, and also downward mixing of the thermocline, to produce a cool water mass that extends westwards as a major plume. "Tehuantepecers," as these winds are termed, are violent events capable of moving great volumes of water westwards: temperatures as low as 16.5°C have been recorded at Salina Cruz at the head of the Gulf (Roden, 1961). This process is somewhat assisted by the existence of an anticyclonic gyre lying quasi-permanently off the Gulf (Fig. 3.7). A similar upwelling, though apparently not produc-

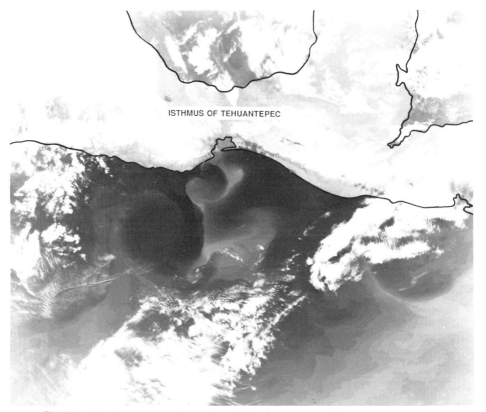

Fig. 3.7. Wind-driven upwelling off the Gulf of Tehuantepec, Mexico; grey scale in sea indicates surface temperature, lighter being warmer. (IRT satellite image from U.S. National Climatic Center, NOAA.)

ing such low surface temperatures, occurs in the Gulf of Panama, induced by Atlantic winds passing through a gap in the cordillera, and assisted by an offshore anticyclonic doming of the thermocline (Smayda, 1966).

A somewhat similar situation, though driven by much less strong winds, occurs on the northeast coast of Venezuela, where the Gulf of Cariaco, which has a 75-m sill at its mouth, is occasionally ventilated by transient effects of the northeast trades from January to April (Griffiths and Simpson, 1967). Upwelling is induced, mostly along the southern side of the Gulf, by the winds that run approximately parallel to the coast, though vertical transport is not from very great depths and brings water to the surface only 2°C cooler than surface temperatures. Elsewhere along the northern coast of South America other centers of coastal upwelling occur, though upwelling in the Carioca trench seems to be the most intense.

Much farther to the south, off the aptly named Cabo Frio in northeast Brazil, a coastal upwelling exists and brings water seasonally from about 150 m to the surface; this is a process that may occur in all poleward-flowing western boundary currents but it is not yet well understood. Further, such a phenomenon may be looked for on any coast where the continental slope is steep and the coastal current narrow and swift. In such situations, intrusions of cold bottom water onto the shelf occur, and—if the depth of origin is suitable—pump nutrient-rich water into the euphotic zone. It is possible, as Garrett and Munk (1979) have done, to find a theoretical basis for this mechanism both in the interaction of meanders or isolated mesoscale eddies with the continental slope, and also by wind-forcing. It is not yet clear which mechanism functions off northeast Brazil, but it is notable that this is one of the areas richest in mesoscale eddy formation in the tropical ocean.

Along the northern coast of the Gulf of Guinea, upwelling occurs weakly in January–Feburary and strongly during the rainy season in July–September. The causes of this upwelling are quite complex: Several mechanisms appear to be involved, including remote forcing. The principal upwelling season corresponds with greatest eastward flow of the Guinea Current along the coast, with most consistent upwelling winds running parallel with the coast, and with least solar heating due to the extremely heavy monsoon cloud cover (Wauthy, 1983). However, these processes are now thought not to account fully for the temperature cycle of the mixed layer over the shelf seen in Fig. 3.8. As Merle (1980) has shown, the mixed-layer temperature signal in the Atlantic is remotely driven and has a greater amplitude in the Gulf of Guinea than off South America because of the shallowness of the mixed layer in the eastern half of the ocean and its greater depth westwards. Seasonal intensification in

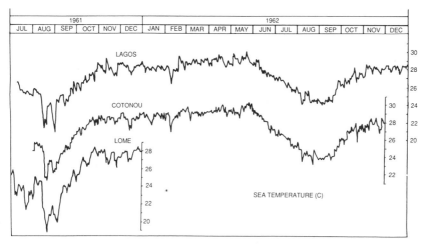

Fig. 3.8. Seasonal evolution of coastal sea surface temperature in the eastern Gulf of Guinea, from observations at coastal stations in West Africa. (From Longhurst, 1964a.)

wind stress off South America causes a series of eastward-propagating, equatorially trapped Rossby waves which are transformed at the eastern margin of the ocean into poleward-propagating Kelvin waves, a northern series trapped by the coast of the Gulf of Guinea and a southern series on the coast of Angola. Appearing as a westward-moving doming of the isotherms above 300–400 m, and a consequent shoaling of the mixed-layer depth, this is sufficient explanation for the major features of the sea-surface temperature cycle in the Gulf of Guinea.

However, imposed on this seasonal temperature cycle is a series of discrete, rapid drops in surface temperature of from 4 to 6°C (Fig. 3.8) that appear throughout the year, but more strongly in the wet season when surface temperatures are already at their coolest. These events have a frequency of about 14.3 days, and are caused by the dissipation of the fortnightly tidal energy (Houghton and Beer, 1984) through continental shelf waves as described by Gill and Clarke (1974). These propagate westwards—with the coast on their right, as they must in the northern hemisphere—against prevailing wind and current regimes. Sea-surface temperatures are coldest just east of the two most prominent capes—Cape Palmas and Cape Three Points—probably because topographic intensification of vertical turbulence is caused by the eastward flow of the Guinea Current past these capes.

The Gulf of Guinea upwelling serves as an excellent demonstration of what is frequently forgotten or ignored: that processes important for understanding coastal fishery dynamics may be driven by events a whole ocean away. Parochialism is a dangerous attitude of mind for fisheries oceanographers at any latitude. As will be shown later, an understanding of the fluctuations of the Ghanaian sardine fishery can be greatly improved by reference to wind stress on the ocean off northeast Brazil, thousands of kilometers to the west.

The more important upwelling along the Somali coast in the Indian Ocean is likewise only to be understood in relation to ocean-wide processes. In fact, mixed-layer cooling far to the east of the Somali upwelling area is now thought to be caused by midocean upwelling in response to the curl of the wind stress between 10 and 15°N (Krishnamurty, 1981). Previously, this was attributed to reduced solar radiation, or to the shedding eastward of eddies from the Somali upwelling. The transition from northeast (winter) to the much stronger southwest (summer) monsoon wind system in early May over the Indian Ocean brings a number of important changes in the Arabian Sea. At this same time, the relatively slow (50–100 cm/sec) southerly surface flow of the Somali Current is reversed, and by mid-May northerly flow of 200 cm/sec can be expected, rising to 350 cm/sec during June. However, reversal of deeper (100–450 m) flow toward the northeast begins much earlier, by mid-April, apparently fed by westward flow at the equator (Leetma et al., 1981). While the reversal of the surface flow seems to be driven by local wind stress rather than by a northward extension of the southern hemisphere African coastal current as some have thought, the deeper reversal does appear to respond to remote forcing propagating westward along the equator. After the reversal, a very strong coastal upwelling (Fig. 3.9) occurs along the Somali coast with intensification at about 5° and 7°N, and especially downstream of the northern peninsula at Ras Hafun inshore of a permanent clockwise eddy, the Great Whorl of oceanographers. While offshore mixed-layer temperatures remain at about 25°C, upwelling water is as cool as 13–15°C. The upwelling water leaves the coast in a plume around the northern side of the Great Whorl and then follows the line of the continental edge as it passes south and east of Socotra Island. This is a dynamic response of the Arabian Sea gyre to increasing Coriolis force with increasing distance from the equator. Geostrophic balance is maintained by upsloping of the isopycnals toward the coast that is sufficiently strong that the thermocline surfaces. The biological and fisheries consequences of this phenomenon will be discussed in a later section.

Elsewhere in the Indian Ocean, seasonal upwelling in many locations is induced by the dynamics of the monsoon-driven circulation. The western coast of India, from Cochin in the south to Karachi in the north, is influenced by an uplift of deep water onto the shelf, the 20°C isotherm rising

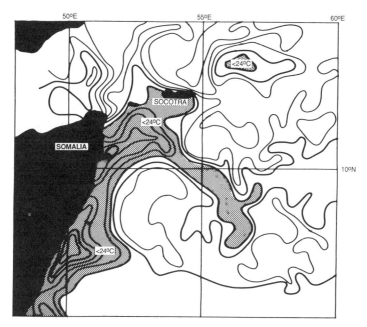

Fig. 3.9. Sea-surface temperature east of the coast of Somalia during a strong upwelling season. (Redrawn from Dietrich *et al.*, 1980. From "General Oceanography, An Introduction," Copyright © 1980 John Wiley & Sons, Inc.)

by 90–100 m (Banse, 1968); there is a reversal of the coastal current, though not so striking as off Somalia. Flow is southwards from February to October, being strongest (but still only about 25 cm/sec) in July–August, during the height of the southerly monsoon. During this period, upwelling is indicated by the lowering of sea level the length of the coastline of western India, and cool subsurface water occupies much of the shelf below a thin, warm mixed layer. The upslope of isopycnals is probably related more to the geostrophy of the Arabian Sea gyre than to any local wind-forced divergence at the coast, though this may be a factor off southwest India. As will be discussed below, the biological importance of the upsloping of isopycnals in this region is perhaps not so much the surfaceward transport of nutrients as the movement of water of extremely low oxygen content from the oxygen minimum layer of the Arabian Sea onto the shelf, and even very close to the coast.

It is to be expected that the East Madagascar Current, resulting from the deflection of the South Equatorial Current by the eastern coastline of Madagascar, and which has the characteristics of a western boundary current (Lutjeharms *et al.*, 1981), will induce upwelling and biological enrichment off the southern tip of Madagascar, where the strong narrow current is retroflected eastward in a series of mesoscale eddies, separated from the more gentle westward drift of some of the transported water into the southern Mozambique Channel. Along the northern coast of Mozambique (10–16°S) the coastal jet, which is one of the origins of the Agulhas Current to the south, includes an upwelling zone during the northeast monsoon (November–April) about 30–50 km wide, which is in agreement with the apparent radius of deformation (Saetre and da Silva, 1982). There is a monsoonal reversal in this region; the South Equatorial Current is most intense during the southwest monsoon. The East Madagascar Current, and the coastal Madagascar current running southerly along the coast of Africa, run strongest during the season of the northeast monsoon (Saetre, 1985).

In the archipelagic region of the Indo-West Pacific, monsoonal current reversal is also very marked. This is a very complex region and many important details remain to be worked out, though the general outline has been clear for some time in the very comprehensive description by Wyrtki (1961). Figure 3.10 shows how during the northerly (winter) monsoon the currents set southerly along the whole western coast of the South China Sea. Water from the South Equatorial Current of the western Pacific feeds into this circulation through all the east–west passages in the Indonesian archipelago, eventually finding its way southeast into the Banda Sea. Currents flow along the southern coast of Sumatra and Java in a southerly direction until they turn seaward in the region of Lombok and Sumbawa to join the origin of the South Equatorial Current. Figure 3.10 also shows how, during the southerly (summer) monsoon, flow is reversed and water from the Pacific South Equatorial Current flows westward through the passages between the Moluccas and western New Guinea, feeding the northerly coastal monsoon current of the South China Sea and the South Equatorial Current of the Indian Ocean.

This system of reversing currents is known to result in several significant regions of upwelling. In August, off the northwest coast of Australia and especially along the southern coast of Java and Sumbawa, the southerly monsoon induces strong upwelling. This is an important source of water for the South Equatorial Current of the Indian Ocean (Wyrtki, 1961), and originates at depths sufficient to bring up nutrient-rich water along the coast of Java and in the Bali Strait, which supports an isolated population of the Indian oil sardine *Sardinella longiceps* (Dwiponggo, 1972). At this season, also, there is significant upwelling along the northern coast of Papua New Guinea.

During the same season but for a shorter period—just May to August—a less strong upwelling occurs on the eastern side of the Banda Sea, along

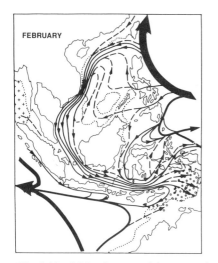
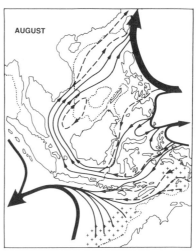

Fig. 3.10. Major features of the reversal of currents in the western Pacific archipelagos under the influence of the reversing monsoon winds; crosses indicate divergence, circles are convergence areas. (Redrawn from Wyrtki, 1961.)

the shelf edge of the Arafura Sea, and surface temperatures drop about 2°C. Farther to the west, off Macassar, water from the Flores Strait and the Macassar Strait meet, to flow west into the Java Sea, and high surface salinities in June and July suggest upwelling in consequence. There seems to be some divergence of surface water in the northward monsoon current along the coast of Vietnam, which may not alone be sufficient to cause upwelling but which may precondition the water mass to local wind-induced upwelling, that Wyrtki thinks may have been observed. It seems unlikely, as Wyrtki suggests, that significant upwelling occurs anywhere in the Philippines, since sea levels are permanently high throughout this archipelago.

Wyrtki (1961) suggested that during the northern monsoon there may be some coastal upwelling south of Hong Kong, and perhaps also along the coast of Sarawak. Lim (1975) has described topographically induced upwelling along the edge of the shelf in the central South China Sea, at 6°N 109°E, north of the island of Natuna. This occurs during the southeast monsoon and brings water from about 90 m to the surface.

In the Andaman Sea, on the west side of Peninsular Malaysia, rising salinities during the offshore northerly monsoon suggests some upwelling along the inner, eastern part of the sea. The complexity of a seasonal current reversal passing through an archipelago of many hundreds of islands, and including major deep and shallow regions, is best examined for undescribed processes by the use of satellite imagery: sea-surface temperatures, sea-state climatology, and the existence of trains of internal waves and other observables are all recoverable from satellite data and give direct information on the occurrence of upwelling phenomena. The processes that cause the upwelling may then be deduced from a knowledge of topography, tidal regimes, depths, and currents.

Finally, to the south of the monsoon region, along the northeast coast of Australia, Andrews and Gentien (1982) have shown how the East Australia Current pulsates in strength with a period of about 90 days; where the continental slope is especially steep, this leads to intrusions of cold bottom water onto the outer shelf in the same manner as occurs elsewhere in poleward western boundary currents. Subsequently, the upwelled water is pumped inshore as the bottom Ekman layer responds to quasi-periodic (25–30 days) reversals in the longshore winds. As we shall discuss later, this process results in plumes of nitrate-rich water passing inshore across the Barrier Reef and silicate-rich compensatory offshore flow.

These, then, are some of the smaller-scale upwellings along tropical coasts that have regional importance but are smaller than the classical upwellings of eastern boundary currents where great fisheries of global importance occur. We have indicated some of the diversity of the mechanisms that drive regional coastal upwelling. Unfortunately, oversimplification of these systems in the fisheries literature by simply attributing the upwelling process to wind-forced divergence of the Ekman layer is often done but conceals important detail and many anomalies. Consider the remarkable pattern of many small upwelling plumes, each quite distinct and related to topographical detail on the coast, shown in Fig. 3.11, and compare it with the diagram illustrating the Peruvian upwelling region in the textbook closest to your hand. It is at once obvious that oversimplification will yield a quantification of the upwelling process that can be quite incorrect. Fortunately, satellite data is now capable of showing even greater detail for sea-surface temperature and chlorophyll, to yield even more accurate estimates of regional productivity than the Peruvian example, which was based on data collected synoptically by about 100 chartered bolicheras.

These tropical upwelling regions merit careful attention for, as Cushing (1971a) pointed out, they have some similarities with high-latitude regions where winter overturn of the water column occurs. One would not want to take the comparison too far, but it is true that tropical upwelling ecosystems resemble those of high latitudes in their intermittent production cycle, and in the relatively short, simple, and "inefficient" food chain that they support. In these characteristics, they differ greatly from other tropical ecosystems.

Fig. 3.11. Coastal sea surface temperature plots to show upwelling centers, obtained by simultaneous observations by very many chartered fishing boats over a 2-day period. (Redrawn from Instituto del Mar del Peru, Eureka expeditions, for November, 1973.)

Turning now to the major upwelling regions in the two great pairs of eastern boundary currents—Canary and Benguela, California and Peru—the process does indeed appear to be dominated by classical wind-driven divergence of the Ekman layer. As Barber and Smith (1981) point out, the dimension of the upwelling is much less than the area over which the appropriate wind stress is applied. The radius of derformation of the density field is determined by HN/f (f = Coriolis value, N = Brunt–Väisälä frequency, and H = water depth). The limited depth of water available over a continental shelf usually restricts upwelling to a strip of only 10–20 km, though at a midocean Ekman divergence it may be four or five times as wide. The period during which favorable wind-stress must be

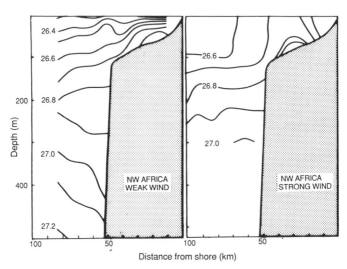

Fig. 3.12.   Temperature sections during upwelling and nonupwelling periods, Mauretania. (Redrawn from Barber and Smith, 1981.)

applied before upwelling is induced (Fig. 3.12) is related to the value of
the inertial period f^{-1}; at higher latitudes, where Coriolis' force is greater,
f^{-1} may be of the order of 1 day, while at 5° from the equator it is as long
as 5 days.

The characteristics of a coastal upwelling system that determine its
biological and fisheries impacts are rather complex. First, the depth of
water available on the continental shelf over which upwelling occurs will
determine the width of the upwelling zone. Second, the nutrient content
of the water at the depths at which upwelling water originates will determine its productivity after being upwelled. Third, the seasonal intermit-
tency and changes in direction of the wind stress may determine strength
and duration of the upwelling itself. These variables drive the principal
upwellings of eastern boundary currents at various latitudes along each of
the coasts concerned and induce differences between them. For instance,
upwelling is of longer duration, and is less intermittent, in the Canary
Current off Mauretania than it is in the geographically similar southern
California Current off Mexico. Upwelling is more continuous on the Peruvian coast than in the other major upwelling regions.

It must not be assumed that similar conditions occur all along each
upwelling coast. As was evident from Fig. 3.11, off Peru many upwelling
centers, each with an individual upwelling plume, can be identified; off
Baja California, in Mexico, the same thing is evident, with major upwell-

ing occurring on the downstream side of Punta Baja, Cabo San Eugenio, and Cabo San Lazaro. The same is true also in the Canary Current, where major upwelling plumes develop behind Cap Blanc and Cap Timiris. As Herbland (1978) has shown, at the latter cape upwelling is intensified topographically at Cap Timiris by a submarine canyon that approaches very close to the coast.

Nor must it be assumed that a simple generalization of upsloping isotherms toward the coast, perhaps derived from such sections as shown in Fig. 3.12, describes all the biologically significant features of an upwelling coast. For instance, as Barber and Smith (1981) and Smetacek (1985) point out, if the thickness of the offshore-flowing plume is shallower than the euphotic zone, the dormant phytoplankton cells being carried shorewards below it to be upwelled at the coast are preconditioned to light and are activated more rapidly when they reach the surface than if the plume were deeper. Then, the strength of the longshore currents may determine whether or not there is a much accumulation of organic material on the deeper parts of the shelf below the upwelling plumes; the shelf below the Canary Current does not accumulate material as rapidly as below the Peru Current, where so much accumulates as to produce local anoxia of the bottom water. Below the California Current, the poleward countercurrent often moves so far onto the shelf as to surface intermittently at the coast; off northwest Africa, on the other hand, the countercurrent is purely a slope phenomenon. Finally, one may expect to find a second frontal system some distance from the coast that may itself cause secondary biological enrichment.

VARIABILITY IN THE TROPICAL OCEAN

In high latitudes, seasonal change in the ocean cannot be ignored, for winter and summer bring very different conditions. For the tropical oceans, it sometimes appears possible to describe the principal environmental features as if they were steady-state. But a sojourn of more than a few months on any tropical coast should convince even a myopic northerner that seasonality exists and is important also in tropical oceans. We have already noted the inherent ability of the tropical ocean to exhibit seasonal changes in current patterns and the rapidity of its adjustment to changes in wind stress. However, as we shall see, changes on time scales much longer than seasonal may be the most important variability of tropical oceans and may have importance on the planetary scale. In discussing tropical variability, we will proceed from the shortest scales to the longest.

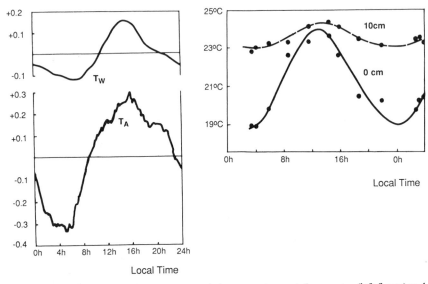

Fig. 3.13. Diel changes in sea water and air temperatures at the equator (left figure) and in the Black Sea (right figure), T_w = sea water temperature at 30 cm, T_A is air temperature 8 m above the sea surface. (From Dietrich *et al.,* 1980. From "General Oceanography, An Introduction," Copyright © 1980 John Wiley & Sons, Inc.)

Within-Year Variability

The daily cycle of solar radiation warms and cools the lower atmosphere just above the sea surface by 4–5°C in the tropics, but even under calm sea conditions there is a much smaller daily cycle in water temperature just below the sea surface (Fig. 3.13). In fact, it is only in very shallow water, and in much higher latitudes, that significant daily changes occur: In the upper Baltic, the temperature of the whole 10-m water column may change between day and night by almost 2.0°C. Daily changes of this magnitude probably occur in the tropics only where cold upwelled water happens to occur right inshore over a shallow sandy beach. Measurement problems still make it quite difficult to be certain what is the exact nature of the daily temperature cycle in the upper centimeter or so of the water column at sea, which is the habitat of the oceanic neuston: Theoretically, much greater changes and much higher temperatures in daytime are possible than seem to occur at even 10 cm depth in the tropical ocean.

On a slightly longer time scale, the passage of tropical cyclonic storms bring important consequences, not only if the storm strikes a coast, but also in the open ocean. The energy to maintain the cyclonic circulation of a tropical storm depends on heat extracted from the sea surface, associated with the imposition of wind stress at the surface. The mixed layer is deepened, and mixed-layer temperatures significantly lowered, by the passage of a cyclonic storm over a warm ocean. Of course, in some regions, where topography funnels the energy of a storm over shallow water on a low-lying tropical coast, devastating storm surges may result with important social consequences. This is especially common in the northern Bay of Bengal among the islands of the Ganges and Hooghly deltas, and also along the Pacific coast of the Philippines. Indeed, as we write, deltaic Bangladesh is even now recovering from storm surges associated with the onset of the 1985 monsoon season; as well as the direct human suffering caused by such events, the probability of their recurrence renders impracticable many normal fisheries development schemes, especially in coastal aquaculture and the building of port facilities; the fishing port of San Felipe at the head of the Gulf of California, and its fleet of shrimp trawlers, was almost totally destroyed by the passage of a chebasco up the Gulf in the early 1970s. Fortunately, progress is now being made in the prediction of storm surges by physical oceanographers, especially for the Bay of Bengal region; a successful numerical simulation has been made of the 5-m storm surge that overwhelmed the coast of Andra Pradesh in 1977, for instance (Johns *et al.*, 1981).

On a time scale of weeks and months, rather than days, the variability of the ocean is expressed principally as mesoscale eddies, most of which originate by the cutting-off of the meanders that form at current boundaries where the flows are in opposite directions. Remarkable progress has been made in recent years in mapping and understanding these features, which occur almost everywhere in the ocean. Not only sensors on satellites, but also satellite-tracked drifting buoys and arrays of moored buoys with vertical strings of current meters, have supplemented voyages of oceanographic ships dedicated to investigating the physics, chemistry, and biology of eddies. Because they are especially prominent in the vicinity of the jet-like western boundary currents (especially the Kuroshio and Gulf Stream), it is in these regions that they are best mapped. Oceanographic experiments such as MODE, the Mid-ocean Dynamics Experiment, or GATE, the Tropical Atlantic Experiment of the Global Atmospheric Research Programme, have probed the physics of eddies with arrays of current meter moorings in midocean. As reviewed by MacLeish (1976) and Robinson (1983) we now know that mesoscale eddies are usually large (100–1000 km), deep (200–1000 m), and long-lived (1 month–1 year) and that biological conditions within them evolve in a predictable manner. Mesoscale eddies therefore have very significant implications both for oceanic tuna fisheries, and also for the fisheries of the continental shelves against which they impact. It is now also becoming clear that small anticylonic eddies of order 20–100 km in diameter are a common feature not only at the surface, but also at all depths. In the southern Sargasso Sea, Kerr (1985) notes that eddies have been observed which originated 4500 km eastward at the Mediterranean outflow, 800 km southward in the central oceanic gyre, and 500 km to the west in the Gulf Stream.

We are now able to map the occurrence of mesoscale eddies in the tropical oceans (e.g., Bruce *et al.*, 1985, for the Brazil and Somali currents) and to identify regions globally where mesoscale variability of this kind is particularly common. Two techniques for locating mesoscale variability are available: the first is to measure the height of the sea surface by radar altimeter from a satellite whose orbit enables repeated passes to be made along the same track. The variability of the height of the sea surface in relation to the geoid measures the variability of flow, and hence, in part, of eddying. The second, which supports the more direct technique remarkably well, is to use very many observations of ships' drift to contour the kinetic energy of mean flow and its variability to derive a field for eddy kinetic energy per unit mass that can then be applied to 5° squares in all oceans (Fig. 3.14). As might be anticipated, both techniques show that eddying occurs most frequently where jet-like currents occur in the tropics, as well as along the equatorial divergence, presumably related to "snaking" of the equatorial undercurrent.

Seasonality in the tropical ocean is driven by the twice-yearly passage of the atmospheric intertropical convergence zone (ITCZ) across the equator. On a tropical coast this passage is associated with a wet season–dry season alternation; dry when the seasonal trade winds blow offshore from a continental mass (as in the West African Harmattan season), wet when they bring moisture-laden oceanic air onto the landmass. At the coast, therefore, seasonality is most often expressed by sea state and the amount of fresh water discharged by rivers and lagoons. In the ocean, it is not only the reversing monsoon currents of the western Indo-Pacific which show seasonality, for in the eastern parts of the tropical ocean basins the Equatorial Counter Currents are seasonal in their velocity and extent. The North Equatorial Current of the eastern tropical Pacific does not extend father east than about 120°W during the winter months but for May to November extends clear to the coastline of the American continent, where it is deflected north and south. Its Atlantic analog, the Guinea Current, extends the whole length of the coast of the Gulf of Guinea as far as the Bight of Biafra year-round, but has much higher velocities during the season of the southwesterly winds, June–October.

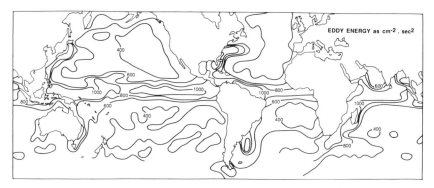

Fig. 3.14. Distribution of eddy energy in the oceans to show regions of high probability of mesoscale eddy occurrence. (Redrawn from Dickson, 1983.)

Somewhat analogous to the monsoon reversal in the western Indo-Pacific, whose upwelling consequences we have already discussed, is the seasonal reversal of circulation in the Caribbean region. The North Equatorial Counter Current is driven by westward wind stress, and when this relaxes during the period February–April (Garzoli and Katz, 1983; Richardson, 1984), flow of the Counter Current is reversed. This reversal reduces the influence of the low-salinity Amazon/Orinoco plume in the southern Carribean, which is strongest when the Counter Current is flowing most strongly (Merle and Arnault, 1985); Mahon *et al.* (1986) suggest that this is due to the increased strength, and hence influence in the southern Caribbean, of the Guiana Current at this season. Similarly, the seasonal changes in strength of the Guinea Current mentioned above are related to the same reversal phenomenon; it is likely that the Guinea Current is more influenced by Canary Current water at this time than when the North Equatorial Counter Current is flowing strongly.

Except where there is a seasonal change in the water mass present at any location, or where upwelling is seasonal, local changes in biological communities are usually rather small in the tropical ocean. Standing stocks of organisms—represented by chlorophyll, plankton, and nekton biomass—may change by a factor of about 2 or 3 (e.g., Blackburn *et al.*, 1970) in the open ocean, and perhaps 3–5 (e.g., Longhurst, 1984) in the coastal zone. Such differences are, of course, significantly smaller than normally occur in higher latitudes.

Between-Year Variability and Its Biological Consequences

It is variability on the interannual scale that is presently drawing the attention of so many oceanographers toward the dynamics of the tropical ocean; we now know that every few years an event occurs in the tropical ocean that has global consequences that touch almost everybody. Such events, that have come to be called El Niño from the Peruvian name by which they were first described, can be hindcast from rainfall records for the west coast of South America, and are found to coincide with almost every recorded famine on the Indian subcontinent during the late eighteenth and throughout the nineteenth century. For those few recorded famines that did not occur at times of El Niño, a major dust veil in the stratosphere is likely to have occurred from a major volcanic eruption. When the tropical ocean changes from its accustomed pattern, people starve.

Between-year variability expresses itself in two principal ways in the marine biosphere. First, Niño events mediate environmental changes on a global scale, which are self-evidently driven directly by the event itself. Second, there are local changes in the biological environment from year to year, especially in the coastal zone, which may be driven by regional rainfall patterns or by stochastic events in ocean circulation, especially at the mesoscale. There is also evidence that similar biological changes may also be mediated from within the biological components of the ecosystem by interspecific interactions; almost every philosopher of tropical ecology points out that biotic, rather than abiotic, interactions may dominate tropical multispecies ecosystems.

Looking first at the regional rather than global variability, it has been suggested that between-year changes in species composition and relative abundance within tropical ecosystems are more important than seasonal changes; we would not go so far as to support this concept, but certainly between-year changes, apparently stochastic in nature, are characteristic of tropical coastal marine ecosystems. There is evidence for between-year variabity in the settlement of benthic invertebrates in the estuaries of the Gulf of Guinea (Longhurst, 1957b); over the 3-y period 1952–1954, the occurrence of *Lingula* in the Sierra Leone estuary (as indicated by the stomach contents of the sole, *Cynoglossus,* routinely sampled every 5 days) was concentrated in an 8-month period of 1953. In the period of peak abundance, in May–June 1953, *Lingula* crammed the stomachs of soles, while outside the period of anomalous settlement, almost none occurred. From regular 2-weekly trawl hauls in an estuary in Papua New Guinea, Quinn and Kojis (1986) found that the annual cycles of abundance of 40 species of fish and crustacea were closely linked to river discharge, but that interannual variability was greater than the seasonal signal. In the first of these examples, the variability in community composition is thought to be driven by changes in the estuary by changing river discharge; in the second case, where no major river enters the sampled system directly, the situation is not so clear. Under conditions of high flow rates, estuaries and the inner shelf in many parts of the tropical ocean are strongly influenced by high turbidity, lower plankton production rates because of reduced solar radiation, and shifted distributions of estuarine conditions. Such as probably the principal determinants of between-year ecological variability in coastal regions.

On the oceanic scale, it has taken more than half a century of study, and the availability of modern equipment, fully to understand the global consequences of El Niño: It was described originally as an incursion of warm tropical water over the Peruvian upwelling area, often near Christmastime, and was thought to be of purely local significance. Not until the 1957–1958 event did oceanographers begin to suspect that El Niño might also affect areas of the eastern Pacific as far away as California, apparently (as was then assumed) inducing a cessation of upwelling by a reduction in wind stress at the coast. It was only study of the 1972–1973 and later events that brought a realization that this was a phenomenon on an altogether grander scale. The most recent event, of 1982–1983, which was the most powerful global ocean and atmosphere perturbation of the present century, finally convinced the global oceanographic community that these events can bring about a global change in weather patterns that can be expected to recur every few years at some level of intensity.

It is still, for the moment, customary to describe the El Niño phenomenon as if it originated and developed purely in the Pacific Ocean; however, the 1982–1983 El Niño data now show quite clearly that parallel changes occurred concurrently in the Indian and Atlantic Oceans. The western Pacific, it is thought, does indeed hold the key to unleashing El Niño, but it seems certain that future events (statistically, one is quite likely to occur during the writing of this book and its publishing)* will come to be studied as a global response of the ocean to global changes in atmospheric circulation.

El Niño is now seen to be a complex and variable event for which a variety of models is now available, each with its proponent (Ramage, 1986). However that may be, what is now known as the "canonical El Niño" model to distinguish it from its many competitors, may be reviewed briefly if only to sensitize tropical biologists to the full implications of bizarre oceanographic events they may encounter on their own coasts, which may be far from Peru. Though each El Niño event that has been well studied has had its own peculiarities, the general pattern of how one develops, and why, is now fairly clear. The occurrence of El Niño is related in the first place to the state of a periodic (2–7 y) oscillation of atmospheric pressure between the Indian and Pacific Oceans, the so-called Southern Oscillation. A weakening or reversal of the normal east to

* Statistics were vindicated: as we correct proofs, the 1986–1987 Niño is winding down.

west pressure gradient along the equator weakens the easterly trade winds, replaces them with westerlies at the equator, and may lead on to an El Niño event. To recognize this connection, oceanographers now refer to ENSO (El Niño–Southern Oscillation) events. However, even this may be a somewhat parochial view for there is an association also between the strength of the Indian monsoon and a Northern Oscillation of pressure anomalies across the northern Asian continental land mass, and as Birch and Marko (1986) have pointed out a severe ENSO event is a prelude to several years of heavy iceberg frequency on the Grand Banks of Newfoundland.

As the Southern Oscillation index (representing the pressure gradient) weakens, the Indonesian–Australian low pressure shifts eastward to the Polynesian islands. This movement (indicated by a pressure difference index for Darwin and Tahiti) is an indicator that an ENSO event may be imminent. Elevated sea levels, high sea-surface temperatures, and a deep mixed layer in the western Pacific are other predictors, and are all associated with a long period of strong wind-stress from easterly trades.

The trades are normally progressively weaker westwards, their dry regime being replaced by a rainy regime of more variable winds in the western Pacific. The first major impact of the onset of an ENSO episode is that the trades in the western Pacific become even weaker, and the rainy, variable westerlies occur progressively further to the east. Reduction of westward wind stress at the sea surface, and imposition of eastward stress, then initiates eastward-propagating Kelvin waves that travel toward the American continent at almost 3 m sec^{-1}, or about 5 knots. In this process, the normal topography of the thermocline becomes strongly modified. The mixed layer shallows in the west, deepens toward the east, and the sea level rises there. Even though it is now known that upwelling winds, and upwelling itself, continue on the Peruvian coast during most ENSO events, what then reaches the surface is warm water from above the deepened thermocline, so that nutrient enrichment at the surface is much reduced.

The 1982–1983 ENSO followed the above pattern, but was an extremely strong event. In September 1982, values for the Southern Oscillation Index were extremely low, and the Darwin–Tahiti pressure difference was higher than at any time for 50 years. Meterologists then became aware of very strong, sustained low-level westerly winds and heavy rains over the western equatorial Pacific that progressively extended eastward as far as the dateline during the next couple of months. It is now known that unusually sustained southwesterly winds in the northern Tasman Sea preceded development of the equatorial westerlies by some months. In the same period, anomalous eastward currents were observed in the west-

ern Pacific, and there was a rapid shoaling by almost 50 m of the thermocline. By May and June, 1983, with the ITCZ in an anomalously southerly
position at the South American coastline, the eastward extension of rainy
conditions had reached Peru and California: coastal Ecuador received 580
mm of rain in June, compared with a norm of 14 mm; Piura in Peru
received 2400 mm compared with a seasonal norm of 45 mm. On the other
hand, searing drought conditions were already evident in China, India,
Africa, and Australia. Major changes occurred in both Pacific and Atlantic tropical current systems: In both oceans, the equatorial undercurrent
decayed or reversed in mid-1982 and did not reappear until early 1983.
The South Pacific Subtropical Gyre was centered anomalously far to the
south of its normal location.

Anomalous sea-surface temperature patterns were present throughout
all tropical oceans and as far north as the Gulf of Alaska (Fig. 3.15). These
were associated with anomalous winds and global cloud cover from at
least 45°N to 45°S. Tropical hurricanes were unusually few in the Atlantic,
unusually late but very severe in the China Sea. In the Atlantic, the Inter-
Tropical Convergence Zone was far to the south of its usual position, and
upwelling brought warm, nutrient-poor water to the surface in the Canary
and Benguela Currents because of the deepening of the thermocline in the
eastern part of the ocean relative to the west; generally, there appears to

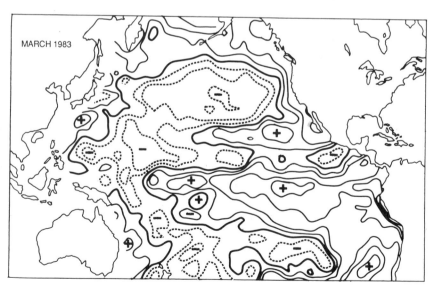

Fig. 3.15. Pacific sea surface temperature anomalies, positive and negative, during the
1982–1983 ENSO event. (Redrawn from Mysak, 1986.)

have been a 6-month lag between Pacific and Atlantic processes. Relaxation to normal conditions did not occur until the winter of 1983–1984, almost 18 months after the onset of the ENSO.

Picking up the pieces after the event, almost every available biological data series shows major changes in the environment or to fish stocks that must have had profound consequences for fisheries. Until the 1982–1983 ENSO, it was thought that such changes were restricted, more or less, to affecting the distribution and recruitment of the Peruvian anchovy stock, and to some transport of tropical organisms poleward off California and Peru. It is now known that the effects are very much wider. From the central equatorial Pacific to the South American coastline, surface chlorophyll concentration was abnormally low, and plumes of high chlorophyll lay to the northeast rather than westward from the Galapago in response to changed circulation patterns. Net primary production of plant material showed a deficit of the order of 1.1×10^{15} tons of carbon for the eastern tropical Pacific alone, representing 1×10^7 tons of fish production, assuming a food chain of three trophic levels. The deep chlorophyll maximum was twice as deep as usual off California, and off Peru oceanic and equatorial dinoflagellates *(Ceratium, Pyrophycus)* and oceanic diatoms *(Planktoniella, Ethmodiscus)* replaced the normal coastal diatoms (e.g., *Nitzschia*). Coastal red tides of dinoflagellates and *Mesodinium rubum* were widespread off Peru. Biomass of small epiplanktonic copepods was low throughout the whole eastern Pacific, and oceanic copepods *(Acrocalanus, Ischnocalanus, Euchaeta)* occurred at the coast off Peru.

Catastrophic mortality occurred in all coral reefs in the eastern tropical Pacific Ocean (Glynn, 1985), especially among the reef-building scleractinian genus *Pocillopora*; large tracts of totally dead reef now exist from Costa Rica to Columbia and in the Galapagos Islands. Death of the protective *Pocillopora* thickets is now permitting the coral predator *Acanthaster planci* to enter reef areas and feed on previously protected massive corals. Analysis of coral growth patterns as recorded in the skeletal material suggests that a disturbance of this importance had not occurred in this region during the previous 200 years.

Off Ecuador, Peru, and at the Galapagos, major changes in the pelagic fish faunas occurred and among other events, 61 species of unusual Panamanian species of fish were recorded off Peru. Tropical shrimp fisheries boomed off Ecuador and Peru, south of their normal latitudes. A warmwater scallop fishery developed off Peru, while demersal fish catches were about half of their normal level off Ecuador. Peruvian anchovy withdrew first to deep water, where most apparently died of starvation and perhaps low oxygen levels, and then disappeared completely from the Peruvian coast. Fish that later returned to the Peruvian coast were probably the survivors of part of the population that had migrated south early in the event. Sardines, which had in recent years occupied the old anchovy region off Peru, migrated southward and almost entirely left the Peruvian coast. Both anchovies and sardines suffered unusual mortality offshore from pelagic predatory fish; the 1982 reproductive season for both species was very brief.

On the other hand, jack mackerel and mackerel had excellent spawning years off Peru. Spotted porpoise schools were anomalously distributed throughout the eastern tropical Pacific, and albacore tuna changed their migration routes over the whole of the North Pacific Ocean. At the Galapagos, and on the South American coast, seabirds failed to breed, as they did at equatorial islands far to the west. Populations of adults pelicans and cormorants moved south and suffered very heavy mortality. Fur-seal pups were aborted or did not survive normally, and many pups and adults starved. Marine iguana reproduction failed due to replacement of their intertidal algal food with unpalatable species.

Such are some of the observations already to hand for the eastern part of the Pacific basin, mostly as reported in the pages of the "Tropical Ocean-Atmosphere Newsletter" up to late 1985, and in the comprehensive reports by Arntz et al. (1985) and Arntz (1986) of the effects at the Peruvian coast. Undoubtedly, as wider studies are completed, consequences will be noted far from the eastern Pacific tropical region. Some indications of major changes in the northwest Pacific are already available from a report by Yamanaka (1984). Describing the cold-water anomalies in this region (of from $-2°C$ to $-7°C$ colder than normal), Yamanaka relates them to changed distribution patterns of fish and marine mammals around the Japanese islands (more southern species on the west coast, more northern species in the east coast), a changed migration route of oceanic bluefin tuna in the western Pacific (unusually large proportion of stock went to the southern hemisphere), and lowered reproductive success for the Japanese sardine, already probably approaching peak population abundance.

Chapter 4

Biological Communities
of Tropical Oceans

In this chapter we review the nature of the tropical marine fauna, the manner in which it is organized into discrete ecosystems—or at least into entities that we can conveniently discuss in such terms—and how these ecosystems produce and transform organic material. Though it is not our principal intention, we shall sometimes be forced into a comparative discussion of how tropical biota differ from those of higher latitudes. In the earlier literature, such comparisons were made in many papers describing the tropical fauna and flora, perhaps because in many cases the authors were expatriates from colder countries to a greater extent than today. In fact, of course, the utility of such comparisons is somewhat limited: Their greatest use is to help us transfer assumptions concerning the rates of physiological processes from studies done in high latitudes to the tropical biota.

ORIGINS OF THE TROPICAL MARINE FAUNA

Though some of the factors determining the nature of the tropical fauna are valid for both pelagic and benthic biota, there are sufficient differences that they are best discussed separately. Fish of continental shelves tend to have the same zoogeographic patterns as shallow water benthic invertebrates, while species of plankton and oceanic pelagic fish tend to be distributed similarly. Because of their paramount importance for fisheries, we shall first discuss the zoogeography of continental shelf biota, including both fish and invertebrates.

It is easily overlooked that there are real discontinuities in the north–

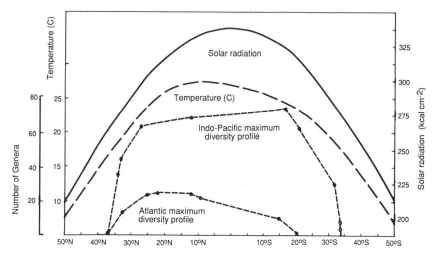

Fig. 4.1. Diversity gradients with latitude for coral genera, Atlantic and Pacific Oceans, with solar radiation and sea temperature. (Redrawn from Rosen, 1984. From "Fossils and Climate" (P. Benchley, ed.), Copyright © 1984 John Wiley & Sons, Ltd.)

south distribution of marine benthic biota. As Ekman (1953) showed, there is a relatively homogenous benthic and demersal "warm water fauna" that occurs where minimal seasonal sea-surface temperatures are at least 16–18°C; within this zone of warm water, the division between subtropical and tropical faunas *sensu stricto* occurs along a line defining mixed-layer temperatures above 20°C. There is greater similarity between the tropical and subtropical faunas than between the subtropical fauna and the temperate fauna lying poleward of it. The warm water fauna is increasingly diverse equatorward, though Fig. 4.1 shows that diversity and latitude (or other attributes of latitude such as temperature) are not linearly related. Indeed, since sea-surface temperature does not form a simple poleward gradient, there is no reason to suppose that a simple latitudinal gradient in diversity should exist, as seems to have been assumed in the past.

So many clear examples of diversity gradients have been demonstrated that it has come to be accepted as a truism that biological diversity is a general and negative function of latitude in most habitats. This has been related to several independent hypotheses (e.g., Pianka, 1966) or to a general hypothesis (e.g., Sanders, 1968; Huston, 1979). However, as we hope to clarify, this is far from a universal rule, and important cases are now known where the relationship is weak or nonexistent. One of the most striking such examples are the benthic communities of level sea bottoms, as we shall see below.

Within the benthic warm water fauna, many circumtropical species occur in all ocean basins, so that there is a considerable degree of longitudinal homogeneity in this fauna. Many authors have commented on this homogeneity, and related the general distribution of tropical fauna to the extremely long existence of the Tethys Sea, which occupied a circumtropical basin from the Palaeozoic until the late Tertiary, at times joining the Mediterranean to the western Pacific; during the same period, the present Isthmus of Panama was periodically represented only by isolated islands.

The general composition of the marine faunas of fish and benthic invertebrates that progressively occupied the Tethyan basin are relatively well known; they were rich in species and tropical in character. By the early Tertiary the fauna of the western "Atlantean–Mediterranean" portions of the Tethys Sea strongly resembled the present fauna of the western Indo-Pacific. Ekman (1953) lists many genera of coelenterates, including corals, crustacea, echinoderms, and fish that occurred in the eastern Atlantic region, and now occur only in the western Indo-Pacific. As an example, of 154 species of fish in Italian fossil beds of tropical facies, half now occur only in the western Indo-Pacific, 15% are ancestral to modern Mediterranean species, and the remaining 35% have modern descendants among tropical Atlantic fish species. Such fossil evidence is valuable for understanding the distribution of modern fishery resources because it tells us (for instance) that Leiognathidae, which are an important constituent of western Pacific fish catches but are now absent from the Atlantic, occurred in both oceans during the Miocene (Danil'chenko, 1960).

The richness of the Atlantic–Mediterranean fauna of the Mesozoic and Tertiary periods progressively dissipated during the last 20 million years of the Tertiary, during and after the Miocene. As Ekman and others point out, this was a global phenomenon, but it affected the Atlantic fauna to a much greater extent than the Pacific. This was partly due to the general cooling of the global climate during this period and the withdrawal of the tropical fauna to a much smaller latitudinal extent than formerly, especially in the Atlantic. However, as we shall see, other explanations of the present distribution have also been suggested, and great care must be taken in interpreting faunistic evidence, for example in relation to the Leiognathidae. The great ichthyologists L.S. Berg and J.R. Norman, for instance, both considered the Gerreidae to be properly placed within the Leiognathidae. Their interpetation was of a single family of small, protrusible-mouthed fish would then still occupy both ocean basins; however, more modern taxonomy suggests that while Gerreidae occur in all tropical oceans, Leiognathidae (*sensu stricto*) occur only in the Indo-Pacific (Fig.

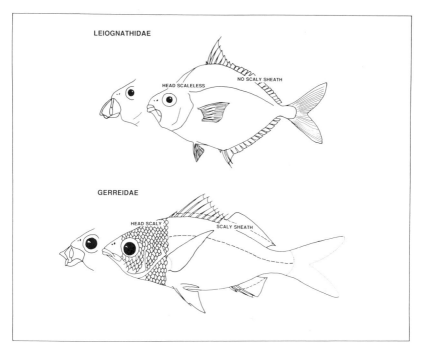

Fig. 4.2. General morphology of two sibling families of tropical fish: Leiognathidae and Gerridae. Redrawn from FAO species identification sheets (Fischer and Whitehead, 1974).

4.2). The only exception is *L. klunzingeri,* which has recently invaded the Mediterranean through the Suez Canal (Por, 1978).

The zoogeography of the benthic warm-water fauna is dominated by the existence of two principal elements: an Indo–West Pacific and an Atlanto–East Pacific fauna, to which the best introduction still remains the text by Ekman (1953). Zoogeographical barriers exist in the eastern Pacific, in form of the great expanse of empty ocean lying between the American continent and the most easterly of the islands of Polynesia, and also in the South African region between the Indian and Atlantic Oceans. Africa and the cold waters of the Benguela Current are a more permanent barrier to the distribution of tropical marine organisms than the central American isthmus, which has been breached many times in recent geological history.

Of these two regions, the Indo-West Pacific is faunistically by far the more diverse: Sponges, corals, medusae, molluscs, crustacea, echinoderms, and fish all have more families, genera, and species there than in the Atlanto–East Pacific region (Fig. 4.3). The Indo–West Pacific not only

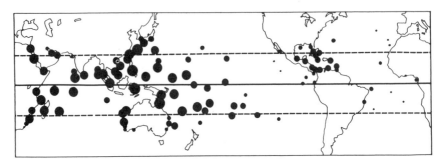

Fig. 4.3. Coral generic diversity, showing the high diversity in the western Indo-Pacific compared with the remainder of the tropical oceans; size of symbol represents relative degree of generic diversity at each location. (Redrawn from Rosen, 1984. From "Fossils and Climate" (P. Benchley, ed.), Copyright © 1984 John Wiley & Sons, Ltd.)

has a more diverse fauna, but that fauna includes (as it logically must) a high proportion of endemic genera and species (Springer, 1982); in the eastern Atlantic, on the contrary, very few endemic species occur. Ekman goes so far as to say that the Indo–West Pacific preserves to our day the richness of the Tertiary marine fauna, and certainly some very ancient genera still occur there: *Heliopora coerulea,* the blue coral, still exists today in abundance in the central Indo-Pacific, and occurs in the fossil record back to the Cretaceous in the Tethyan ocean, as well as in the Indo-Pacific (Zann and Bolton, 1985). Within each region, however, there is a somewhat parallel subregionalization of diversity. Leaving aside for the moment the question of the eastern Pacific barrier, there is similarity in the restricted north–south extent of the tropical region on the eastern sides of the Atlantic and Pacific Oceans and in the concentration of island archipelagoes in their western basins.

Do these facts alone, together with the relative sizes of Pacific and Atlantic Oceans, sufficiently explain the greater diversity of the Indo–West Pacific? The diversity of the Tethyan Atlantic fauna suggests that the explanation may not be as simple as this. The classical suggestion that late Tertiary and Quaternary climatic regressions left extensive refuges for the Tethyan fauna in the Indo–West Pacific while the Atlantic tropical fauna was largely destroyed has been challenged by several other hypotheses: The stability-time hypothesis of Sanders (1969) has been invoked, but Abele and Walters (1979) question the generality of Sanders proposal. Rosen (1981) claims that the disturbance hypothesis of Paine (1966) and Connell (1978) cannot apply to these biogeographical regions, neither of which is demonstrably the more stressed, anyway (Rosen, 1984). The area effect (Connor and McCoy, 1979) applied to the two regions does not show that either is the more heterogeneous, though the Indo–West Pacific is larger and may have lower extinction rates (Rosen, 1984). The hypothesis that has attracted most attention is that the Indo–West Pacific is a focus for the evolutionary radiation of the tropical fauna from which subsequent dispersal occurred, though the general theory of dispersive centers in zoogeography and evolution is now considered by many to be logically fallacious. However, as Rosen (1984) points out, a residual problem in rejecting this hypothesis is the observation of Stehli and Wells (1971) that the youngest genera of reef corals are confined to the Indo–West Pacific, while older genera are more widespread.

The most attractive hypothesis is that proposed by Rosen (1984) himself. He suggests a mechanism based on vicariance, or allopatric speciation, resulting from the changing pattern imposed on a region by the arising of new geographical barriers to the distribution of organisms. According to this suggestion, it is a changing critical pattern of islands, biological barriers, and hence vicariance in each ocean brought about by eustatic changes in sea level due to glaciation in high latitudes that can best explain the principal east–west features of the present-day distribution of diversity in the warm-water fauna. According to this suggestion and more recent data (McManus, 1985a), it is the present-day diversity of geographical features, with associated barriers to distribution of pelagic larvae, that has caused the present-day diversity of the Indo–West Pacific. An extension of this argument is that the east Pacific barrier is simply maintained by the unlikelihood that viable larvae of benthic organisms and fish will be transported between Polynesia and the American continent.

Whatever becomes the generally accepted view, it is sure to be more complex than provided for by any single hypothesis (Potts, 1985). For all its age, the splendid review by Ekman should be consulted by those needing detailed information on such regional problems, and on the sub-distribution of shallow water marine organisms within the Indo–West Pacific and Atlanto–East Pacific regions.

Beyond the break of slope at the edge of the continental shelf, the biota of the continental slope have been very much neglected, and we know less about their general distribution than we do about the animals of the deep ocean floor, yet in many places fisheries extend down into this region and we should know something of their biological environment. For the eastern Atlantic, some information is available that may have wider relevance. Along the tropical coast, benthic surveys off Senegal, Guinea and Sierra Leone (Longhurst, 1957a), and Ghana (Buchanan, 1957) have revealed the existence on the deep shelf and upper continental slope of a benthic epifauna remarkably similar to the slope fauna of the Mediterranean and the western coast of Europe as far north as the Celtic Sea at similar depths. Of 23 species chosen to characterize the deep continental shelf fauna off Sierra Leone at 8°N, 13 also occur among characterizing species off western Europe from 43 to 54°N at comparable depths.

In fact, this similarity could have been expected, for the temperature at such depths shows very little change over this wide range of latitudes; other environmental variables—light, salinity, and the subtrate—are rather similar, and there are no geographical discontinuities to break the faunistic continuum. As we shall discuss later, it is most likely that a similar situation occurs elsewhere between tropical slope fauna and the biota below colder temperate seas. To what extent this represents the same phenomenon as the tropical submergence of plankton awaits analysis of the extent to which the slope fauna resembles the shallow-water fauna of even higher latitudes.

Turning now to the zoogeography of the plankton of tropical water masses, we note at once a much lower level of endemism within ocean basins than for benthic and demersal biota, and negligible differences in numbers of species between Atlantic, Pacific, and Indian Oceans; therefore, we are not led so compellingly into the analysis of Tethyan geography as we are to explain the distribution of benthos and continental shelf fish.

For plankton organisms and oceanic pelagic fish, endemism between ocean basin is rare for species, and extremely rare for genera; whether the reason for this lack of endemism is to be sought in the greater probability that plankton organisms will be advected between ocean basins, or to a greater uniformity already in existence in Tethyan times, is not clear. In any event, the copepods that dominate modern plankton are a relatively recent group with an extremely sparse fossil record (Longhurst, 1985a), and it would be an unjustified extrapolation to attempt to explain their present distribution in Tethyan terms. Similar numbers of species of planktonic euphausiids have been recorded in all oceans at comparable latitudes ranging from 25 to 30 in the tropics to 5 to 10 in boreal oceans (Reid et al., 1978). Numbers of species within the tropical regions are lower only in the eastern tropical Pacific and the northern Arabian Sea in response, it is suggested, to the areas of very low subsurface oxygen levels (Longhurst, 1967a); one suspects that the same phenomenon awaits observation in the Atlantic Ocean.

Expatriation, or the existence of "dead-end" populations of old, nonreproducing individuals of species far from their core regions of distribution, is normally seen in the data from any regional plankton survey: Such expatriate populations are not always easy to distinguish from cosmopolitan species that occur at increasing depths toward the tropics so as to remain within their range of temperature tolerance. We should perhaps also note here in passing that it is not exclusively among plankton that expatriation occurs: as we have already described, during El Niño–Southern Oscillation (ENSO) events there is a massive movement of many species of warm-water fish polewards in the eastern boundary currents, with the potential for establishing new populations in response to decadal changes in ocean conditions. Vagrant expatriates also occur among demersal fish, like the apparently healthy European sea bass (*Morone saxatilis*) that one of us took in a commercial trawl off the coast of Nigeria, or the first known specimen of *Latimeria chalumnae* taken off Durban, far to the south of its normal Comoran habitat.

North–south distribution limits for plankton species invite comparison with the limits of the warm-water fauna of benthos: Fig. 4.4 shows the distributions for the Pacific Ocean of the main elements of the zoogeography of warm-water plankton and illustrates the difference between equatorial (or tropical) and central (or subtropical) species distributions. This figure also shows how some equatorial species are specialized to the eastern part of the ocean. It is obvious that the patterns exhibited by this model of plankton zoogeography correspond broadly with the oceanographic features of the Pacific, as do benthic distributions along the coasts. The tropical plankton species have their core distributions in the North and South Equatorial Curents, extending into the warmer parts of the western boundary current, the southern Kuroshio; the subtropical species occur principally in the two great midlatitude, midocean gyral regions which have greater seasonality than the equatorial zone.

Genera of plankton are mostly cosmopolitan, and many, perhaps most, tropical species have been thought to occur in each ocean, so that about 60% of the epipelagic copepods of the northern Indian Ocean occur also in the Atlantic and 91% in the Pacific at comparable latitudes (Madhupratap and Harides, 1986). It is mostly in neritic and estuarine species that regional speciation occurs importantly. However, critical modern study of some circumtropical plankton species has revealed that they are in reality better considered as groups of cognate species inhabiting different ocean basins. Frequently, such distinctions are described as a subspecific level of difference and are referred to as "forms" of a single species. Actually, the species concept itself is sufficiently imprecise in planktonic invertebrate taxonomy as to suggest that no formal distinction is possible between morphs described as forms of a single cosmopolitan species, or a group of closely related species having representatives in two or more ocean basins.

As Shih (1979) has suggested, perhaps the presence of the central Americal isthmus separating the Atlantic and eastern Pacific faunas has

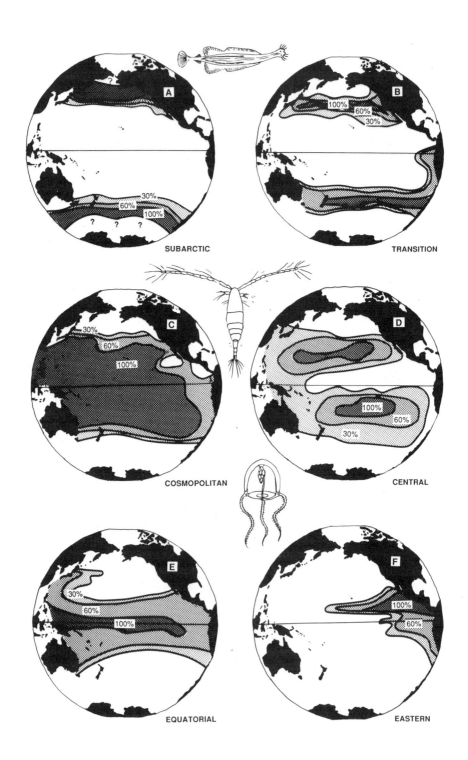

| A | B |
| SUBARCTIC | TRANSITION |

| C | D |
| COSMOPOLITAN | CENTRAL |

| E | F |
| EQUATORIAL | EASTERN |

given taxonomists greater confidence in describing Atlantic morphs as distinct species than in the case of Indian and Pacific Oceans, between which the physical barrier to mingling is less complete. However that may be, in the taxonomic literature for tropical plankton, there is a very clear pattern, as Reid *et al.* (1978) point out: Noncosmopolitan species or morphs tend to occur within Atlantic or Indo-Pacific regions, with a tendency for an eastern tropical Pacific subregion to be distinguishable from the rest of the Indo-Pacific. For plankton, this may be an equivalent of the east Pacific barrier that is so clear in benthic distributions, or may be related to the rather special oceanographic conditions characteristic of the eastern part of tropical ocean basins: Planktonic distributions in the Atlantic are much less well explored than the Pacific, and a similar pattern may occur also in that ocean. For those planktonic groups whose specific distributions have been properly explored, patterns similar to those shown in Fig. 4.4 tend to be repeated in each tropical ocean basin.

Below the tropical water masses the composition of the deep plankton is highly cosmopolitan. Grice and Hulseman (1965) found that more than 80% of the plankton species of the bathypelagic fauna in the northern Indian Ocean occurred also in the North Atlantic Ocean. This assemblage must be relatively ancient, and probably has been little modified by the glacial episodes of the Tertiary era.

TROPICAL BENTHIC COMMUNITIES

In this section and the next, we discuss the nature of the benthic and plankton communities of tropical oceans that are the ecological expression of the zoogeographical patterns described above. For neither benthic nor planktonic communities are there yet sufficient data from tropical regions to attempt a synthesis of the kinds and numbers of small meiobenthos and the microplankton that accompany the species assemblages of larger, macroscopic organisms that we shall discuss.

To begin with the shallowest area first: The emergent benthos communities of the littoral zone are best regarded as a special case, though on sandy and muddy shores there is continuity between the emergent littoral or intertidal and the always-submerged sublittoral benthos.

Few studies of tropical shores have been published, but these show that though tropical rocky shore zonation conforms generally to the formalism

---

Fig. 4.4. Generalized distribution patterns for species of zooplankton in the Pacific Ocean; shows percentage of species having each distribution type among all species present in each area. (Redrawn from McGowan, 1972.)

of Pérès' classification for the Mediterranean (Pérès and Picard, 1958) and
the wider studies of Stephenson and Stephenson (1949), it is unprofitable
to try to press this formality too far. Lawson (1966), reviewing studies of
rocky shore zonation in Senegal, Sierra Leone, and Ghana found himself
unable to locate some of the formal identifiers for the infralittoral fringe,
eulittoral zone, and supralittoral fringe that are appropriate to higher latitudes. Nevertheless, the general appearance of a tropical rocky shore
follows the same pattern so well described for higher latitudes. At the
lowest exposed levels, there is a zone of small macroalgae (*Sargassum,*
Caulerpa, and *Dictyopteris* in West Africa, *Sargassum* and *Bifurcaria* at
the Galapagos, and *Turbinaria* at Madagascar), which are themselves
zoned and occur with various echinoids, anemones, and zoanthids.
Above is the eulittoral or midlittoral zone, dominated at the top by barnacles (*Chthamalus*) with *Siphonaria, Nerita,* and even *Ostrea* in sheltered
places, and lower down by encrusting coralline algae and various molluscs (*Fissurella, Mytilus, Patella, Thais*) and crabs. Above the *Chthama-*
lus zone of the midlittoral, a splash-zone occurs with littorinid molluscs
and higher still with encrusting lichens. Such a general zonation, driven
principally by tidal exposure and wave action, occurs the length of the
Gulf of Guinea (Lawson, 1966), on the coast of Madagascar (Pérès and
Pichon, 1962), and at the Galápagos, though without littorinids or *Nerita*
(Hedgepeth, 1969), and presumably throughout the tropics.

It is to be expected that this zonation would extend onto wharf pilings
and to mangrove stems, and so it does, as appropriate to other environmental factors: Such sites in estuaries, for instance, have a reduced fauna.
On mangrove stems, especially near the outer side of the mangal, two of
the three principal rocky shore zones can be recognized, though with
some major differences in emphasis (Lawson, 1966; Longhurst, 1957a). A
littorinid zone is dominated by species of *Littorina* which specialize in this
particular habitat; the eulittoral zone is dominated above by barnacles,
sometimes by species that are mangrove specialists, and by dense growth
of *Ostrea* somewhat lower on the stems with the gasteropod *Thais.* Below
this zone, however, things are very different from a rocky shore and in the
place of the sublittoral bunches of brown and red algae, there is a mud
fauna at the foot of the mangrove stems and the pneumatophores, including gasteropods, hermit crabs, and the mud skipper *Periophthalmus.*

Sandy beaches in the tropics have a much more varied fauna than in
temperate regions, though this does not necessarily mean that formal
diversity is higher. Two principal zones occur: above high water a population of several species of ghost crabs (*Ocypode*) have their semipermanent burrows, the crabs themselves feeding at the surf line (especially at
night and on cloudy days) on flotsam, including the nutritive material

contained in the spume blown onto the beaches, and even dead stranded fish; the highly colored *O. gaudichaudi* of more sheltered beaches manipulates sand into little pellets from which benthic diatoms are extracted; fiddler crabs (*Uca*) feed in the same way on estuarine sandy beaches. Also feeding at the surf line, by filtration of the backwash of each wave, are several species of small bivalves (*Donax*) that accompany the surf zone in its tidal migration up and down the slope of the beach. Associated with *Donax* are some small gastropods, such as *Terebra,* and small polychaetes (*Sigalion, Urothoe, Spio,* etc.). Above the beach, in the vegetation that often includes coconut palms, large land crabs of various genera (*Cardisoma, Gecarcinus, Coenobita*) are encountered frequently.

For benthic communities of continental shelves, Tethyan zoogeography might be expected to impose significant constraints on the kinds of communities that occur in comparable environments in each ocean. However, both in the faunistically poor Atlantic and in the faunistically rich Indo–West Pacific there are several benthic communities whose species composition closely parallels communities in similar habitats in temperate latitudes. Comparisons have been made between the phyletic composition of the benthic faunas between high and low latitudes (e.g., Longhurst, 1959; Thorson, 1957) that seem to show that there are more epifaunal organisms in tropical benthic communities than in higher latitudes: there are more Brachyura, Gastropoda, Nudibranchiata, and Ascidiacea in the tropical communities, but similar numbers of Ophiuroidea, Holothuroidea, and Polychaeta. However, this does not always apply, and, in fact, Gallardo *et al.* (1977) have shown that Chile Bay, Antarctica, was richer in species and had higher benthic biomass than Nha Trang Bay on the southern coast of Vietnam. The relative importance of benthos in the tropics may thus be principally an expression of the fact that some kinds of communities—especially those containing hermatypic corals— occur only in the tropics and are associated with an abundant epifauna of many phyla. It is also secondarily an expression of the fact that there are more large invertebrates as components of the demersal fish communities in the tropics: swimming crabs, penaeid shrimps, cephalopod molluscs, and so on. In fact, when similar habitats are compared, it seems probably that species/biomass curves are similar for epifauna at all latitudes (Golikov and Scarlato, 1973).

As Table 4.1 shows, polychaete worms, crustaceans, and gastropod and bivalve molluscs comprise about 85% of all organisms in tropical level sea benthic communities: as we shall discuss later, these form the basis for energy flow from the benthos to many species in the demersal fish communities. Polychaetes are one of the principal food items taken by demersal fish from the benthic infauna everywhere, including the tropical seas, and polychaete numbers range from very low values in some areas dominated by thick-shelled bivalves clustered above black mud to very high values in more typical situations. As Harkantra *et al.* (1982) and Table 4.1 show, polychaetes totally dominate the benthos numerically over the whole continental shelf of the western Bay of Bengal at all depths; only on some clay-like deposits are bivalve molluscs the dominant organism.

Without accepting or debating the reality of benthic communities identified according to highly subjective, but apparently rational, rules of thumb following the biocoenosis method of Petersen (1918) or the biotope methods of Lindroth (1950), of Jones (1950) or of Pérès and Picard (1958), a survey of where benthic ecologists have encountered animal associations having the appearance of such communities is a useful starting point for understanding tropical benthic ecology. Petersen's approach ("name the community after two or three dominant organisms") has captured the attention of benthic ecologists more than Lindroth's ("name the community after the substrate in which it sits"), and already by 1957, Thorson was able to list many "parallel" Petersen communities worldwide, several from the tropical oceans. Thorson was able to extend the concept to what he termed iso-communities, in which the dominant genera were different taxonomically, but similar morphologically and (presumably) ecologically.

Parallel and iso-communities of the main benthic communities of temperate latitudes have been described as follows. *Macoma* communities are the biota of soft muddy deposits inshore or in estuaries dominated by small bivalves and many polychaetes; in addition to five parallel communities in temperate seas, the community has been noted off Kenya, off Senegal and Ghana, in the Mediterranean (Thorson, 1957, 1966; Buchanan, 1957), and off Java (Warwick and Ruswahyuni, 1987). *Amphiura, Amphioda,* or *Amphioplus* communities occur in slightly deeper muddy sand and comprise many small ophiuroids, polychaetes, and burrowing crustaceans with some bivalves. Thorson (1957) lists 10 high-latitude examples, as well as two from the Persian Gulf and Florida. Subsequently, two were described from Senegal, Guinea, and Sierra Leone (Longhurst, 1957a), one each from Sek Harbour, Papua New Guinea, and Moreton Bay, Australia (Stephenson and Williams, 1970; Stephenson *et al.,* 1971a); others are inherent in descriptions of biota from Ghana (Buchanan, 1958), Jamaica (Wade, 1972), off the Malabar coast of India (Seshappa, 1953), and off Java (Warwick and Ruswahyuni, 1987). *Tellina* communities lie in clean sand beyond the surf zone of beaches, compris-

Taxonomic Composition of Tropical Benthos, as Relative Numerical Abundance of Taxa[a]

	Continental shelf						Coastal and neritic						Estuaries and lagoons				
Reference[b]	4	7	9	10	12	15	2	3	5	8	11	16	1	6	13	14	17
Latitude	20N	11N	16–21N	16–21N	8–10N	5N	27S	26S	16N	10N	10S	5S	30N	17N	8N	6N	3N
Number of Stations	200	180	57	93	175	187	400	Many	6	41	11	62	240	8	104	390	2
Sampler	grab	grab	grab	grab	grab	grab	dredge	hand	grid	grab	grab	grab	corer	grab	grab	grab	dredge
Temperature °C	15–30	26–28	25–28		27–28	21–28	22–27	18–29	26–28	26–28	24–28	26–28	24–29	27–31	26–28	26–28	26–28
Sediments	M	MS	S	M	MS	MS	MSH	Var.	S	MS	MS	MS	S	S	MS	MS	M
Depth (m)	13–23	5–25	20–200	20–200	20–200	2–100	3.5–25	<3.7	1–4	0–65	<50	2–20	0–8	5–25	0–25	<25	<5
Foraminifera, Bryozoa							3.16										
Porifera		0.19					1.58	0.27	1.08		2.73	1.41	2.36		1.51		
Hydrozoa			0.26	0.64			2.90	2.13			3.64	2.82	4.72		2.01		
Anthozoa	2.30		0.69		0.76		5.28	1.60	3.23	0.49		2.85		1.39	1.01		
'Coelenterata'					2.47	5.02	1.85								0.50		
Platyhelminthes, Nemerteans	1.52			0.22		5.02	1.32	0.80	3.23		0.91	1.41		2.78	1.51	1.56	
Errant polychaeta	54.55				12.38		7.39						7.09		8.54		
Sedentary polychaeta	4.55				4.38		8.18						2.36		3.52		
'Polychaetes'		24.10	63.75	81.37		18.26		21.86	50.35	58.33	39.09	31.42		43.75		37.50	
Cirripedia					0.38		0.79								2.51		
Isopoda			8.10	8.39	0.95		0.26						4.72		2.01		
Stomatopoda					1.52		0.79							0.69	1.01		
Brachyura	9.85				14.86		11.87		2.15				2.36	5.56	11.56		
Other decapoda				0.32	10.10				3.23				1.57		8.04		
'Crustacea'	5.34		1.22			28.31	3.69		4.30			28.57	13.38	19.44	4.50	28.12	8.68
Polyplacophora					0.38		0.26						0.79				
Gastropoda	6.82	10.34	0.16	0.02	14.86	9.13	10.55	15.73	3.23	5.37	9.09	7.04	41.73	4.17	20.60	20.60	56.52
Bivalvia	6.06	27.59	23.99	7.45	11.62	12.33	27.70	12.27	8.60	5.37	7.27	9.86	8.66	9.72	20.60	14.06	34.78
Echiuroidea	0.76					0.46	0.26	0.27					0.79	4.17			
Sipunculoidea	2.27						0.26			2.44			0.79			3.13	
Asteroidea	0.76				3.05	2.74	2.64	3.47	4.30	0.49	2.73		0.79	0.00	2.51		
Ophiuroidea	1.52		1.09	1.58	7.24	7.31	1.32	2.67	1.08	0.49	2.73	8.45	0.79	3.47	2.51		
Echinoidea	1.52				2.29	3.20	1.32	2.67	2.15	0.49		2.82	0.79	0.69	0.50		
Holothuroidea					2.86		1.32	2.67	3.23	0.49		1.41	0.79	0.69	2.01		
Tunicata					2.10		4.22						1.57	0.69	1.01		
Cephalochordata	0.76				0.57	0.91	0.79				0.91			1.39			
Percentage coelenterates	2.30	0.19	0.95	0.64	3.23	5.02	10.03	3.73	3.23	0.49	0.91	2.85	4.72	1.39	3.52		
Percentage polychaetes	59.09	24.10	63.75	81.37	16.76	18.26	15.57	21.86	50.35	58.33	39.09	31.42	11.81	43.75	12.06	37.50	
Percentage crustacea	15.19	27.58	9.50	8.71	27.81	28.31	17.41	33.33	19.35	22.43	30.00	28.57	22.04	25.69	29.63	28.12	8.68
Percentage gastropods	6.82	10.34	0.16	0.02	14.86	9.13	10.55	15.73	3.23	5.37	9.09	7.04	41.73	4.17	20.60	20.60	56.52
Percentage bivalves	6.06	27.59	23.99	7.45	11.62	12.33	27.70	12.27	8.60	5.37	7.27	9.86	8.66	9.72	20.60	14.06	34.78

a Compiled from the sources indicated.

b 1. Holm, 1978. Florida. 2. Stephenson et al., 1971a. Queensland. 3. Macnae and Kalk, 1962. Queensland. 4. Shin and Thompson, 1982. New Guinea. 5. Aller and Dodge, 1974. Jamaica. 6. Wade, 1972. Jamaica. 7. Seshappa, 1953. Malabar coast. 8. Maurer and Vargas, 1984. Gulf of Nicoya. 9, 10. Harkantra et al., 1982. NE India. 11. Plante, 1967. Nosy Bé. Madagascar. 12, 13. Longhurst, 1957a. West Gulf of Guinea. 14. Zabi, 1982. Ivory Coast. 15. Buchanan, 1957. Ghana. 16. Stephenson et al., 1971b. New Guinea. 17. Broom, 1982. W. coast, Malaysia.

ing a few bivalve molluscs, urchins, and sand dollars, and specialized gastropods and polychaetes; Thorson (1957) located five high-latitude examples while Longhurst (1957a) described another off Sierra Leone sandy beaches, Pichon (1962) a sixth off Madagascar, and Warwick and Ruswahyuni (1987) a seventh off Java. Finally, *Venus* communities lie deeper on shelly sand and shell gravel at least partly of their own making, and comprise many species of bivalve molluscs, urchins, gastropods, small crabs, *Branchiostoma,* and specialized polychaetes. Thorson (1957, 1966) lists six high-latitude examples, and briefly notes others from the Persian Gulf, the Indian coast, and Florida, while Longhurst (1957a) formally described a parallel community from tropical West Africa between Senegal and Liberia.

For the three parallel communities that they describe off Java, Warwick and Ruswahyuni (1987) show conclusively that their diversity, measured as k-dominance (Lambshead *et al.,* 1983) is similar to that of counterpart communities off northern Europe. The comparisons are between the Javan *Amphioplus/Lowenia* and European *Amphiura/Brissopsis* communities, the Javan *Tellinoides* with a European *Tellina* community, and the Javan *Laciolina* with a European *Macoma* community. In each pair, the k-dominance was similar, even though biomass and numbers of organisms per unit area were both lower off Java.

Whether or not these communities have any objective reality is not relevant to the present discussion: We should simply note that in many tropical regions benthic ecologists have recognized similar associations of organisms, with approximately the same diversity, as temperate benthic communities in comparable environments. This observation is of great importance in interpeting the ecology of the tropical demersal fish communities that we will discuss later. For a discussion, using tropical examples, of the reality of Petersen communities, the reader may consult Stephenson *et al.* (1971a,b) who believe that the computer techniques they employed do not, in fact, fully confirm Petersen's concepts.

Obviously, the surveys at which tropical Petersen communities were recognized do not comprise the totality of benthic synecology in the tropics. Numerous benthic ecologists have remarked on the high diversity within their samples and the impossibility of applying the Petersen concept to their data. This must imply that at least in parts of the tropical continental shelves the benthic environment of demersal fish is quite differently ordered from temperate seas. But the important point is that whether the benthic ecosystem is, or is not, too diverse to support Petersen's concept, the general nature of the benthic fauna is predictable throughout the tropics over the warm-water depth zones of continental shelves. On sands and shelly sands, a more diverse community characterized by poriferans, nemerteans, small errant polychaetes, small vagile crustaceans, suspension-feeding bivalve molluscs, asteroids, echinoids, and ascidians; on muds and sandy muds, the fauna will be found to be less diverse and dominated by burrowing gephyreans, thallasanid crustaceans, polychaetes, and deposit-feeding bivalve molluscs. As Warwick (1982) has shown for the Bristol Channel, the basic determinants of both substrate and benthic biota are regional supplies of mineral and organic material and bottom stress imposed by tidal and residual currents: undoubtedly, this is true also for the tropical shelves. As Warwick's studies demonstrate, when Petersen communities are plotted as production–rank curves for all constituent species, the slopes of the curves are independent of the actual species comprising the communities but are closely correlated with bottom currents and quantitative bed stress.

Thorson, in 1961, was one of the first to comment that some tropical surveys produced data that indicated a much more diverse benthic ecosystem which could not be described within the framework of his parallel communitites; he was referring to almost a thousand grab hauls taken by him and Gallardo in the Gulf of Thailand and off Vietnam. However, the surveys by Shin and Thompson (1982) around Hong Kong produced data that were probably rather similar; though the number of species encountered was not large ($N = 109$), there was little fidelity between stations, and Shin and Thompson concluded that their data showed species distributions reacting independently to environmental gradients, so that there was no reason to suppose that the species associations given by their Bray–Curtis grouped stations would recur elsewhere in the same arrangements. In fact, we should note that few published surveys of tropical benthos communities have actually demonstrated very large numbers of species, and more authors have commented on this fact than to the contrary. Recently, Maurer and Vargas (1979) made the same comment for their polychaete-rich ecosystem in the Gulf of Nicoya, Costa Rica, as did Wade (1972) for Kingston Harbour, Jamaica.

Very little can yet be said of the smaller (meiobenthic) organisms that are too small to be included in normal benthos samples in the tropics. Dominated by small copepods, especially harpacticoids, and a wide range of ecological types of nematodes, we know that they must occur everywhere and probably comprise 10–20% of the biomass of the total benthos. Rudnick *et al.* (1985) have assembled a few data for nematode and harpacticoid abundance in the Arabian Sea and in the Bay of Bengal from 26 grab stations: total abundances of meiofauna thus indicated are in the range $0.03–0.40 \times 10^6 \, m^{-1}$, which is somewhat lower than abundances indicated for similar organisms on the coasts of the North Atlantic by a factor of about 5 if one anomalous station is excluded. As Schwinghamer (1981) and Warwick (1982) have shown, meiofaunal organisms do not form part of a continuous size spectrum with the macrobenthos (as seems to occur in the plankton); rather, there is a discontinuity in the size spectrum corresponding to the dimensional constraints of living in the interstitial water between sand grains. Apparently, benthic organisms form two non-overlapping size spectra—those small enough to live between sand grains and those large enough to burrow through sediments. A third peak in the size-frequency spectrum represents microbiota, mostly bacteria, living on the surface of sand grains.

Some demersal fish habitats in the tropics are rather special and should be noted as end members of a global series of shallow-water habitats. In many regions, as we have already seen, coastal banks of extremely soft, mobile muds occur. Seshappa (1953) and Joseph (1985) describe how the benthic animals inhabiting the famous mud banks or chakara of the Malabar coast of western India are destroyed when the southwest monsoon in July to September resuspends the mud banks as "liquid mud," calming the sea by thixotropic damping of the onshore swell. Populations of polychaetes, ophiuroids, and molluscs disappear completely and there are accounts of mass mortalities due to the occurrence of up to 30% suspended organic mud in the inshore water. Benthic fish and invertebrates do not fully recolonize the reestablished mud banks until about the end of the year. The liquid mud often forms temporary littoral mud banks which persist for only a matter of weeks before migrating back offshore again. During this incursion, the littoral fauna of the sandy beaches is in its turn destroyed. This period of calm water inshore of the mud banks, and of abundantly available benthic food for fish, is of great significance for local fisheries and may be good model for biological events in the other tropical regions that also have soft, migrating mud banks on the inner continental shelf. In the Bight of Biafra, to the east of the Niger delta, a large area of reduced muds occupies much of the shelf between the Cameroon coast and the large island of Fernando Po and is almost without benthic fauna or (as we shall see later) demersal fish. This almost certainly represents the most extreme case of impoverishment of the benthic fauna in soft, reduced mud in shallow water. The West African *Amphioplus* community becomes progressively less diverse on deposits with higher silt content until only a few specialists survive in the blue mud of the inner continental shelf (Longhurst, 1957a): *Cerebratulus, Diopatra, Mactra, Amphioplus,* and *Amphipholis*. The Bight of Biafra is an end member of this series, and other examples probably occur in the Bay of Bengal, off the Ganges mouths (Parekular *et al.,* 1982).

In tropical regions of strong coastal upwelling and biological production, a somewhat similar situation may occur in deep water over the upper continental slope where reduced muds support a sulfide-driven benthic community fueled by a rain of diatoms and fecal pellets from the rich near-surface production. Off Peru and northern Chile, from about 50 to about 400 m, the shelf benthos is highly impoverished, the only invertebrates being a few polychaetes (*Nepthys, Magelona*) and the red galatheid crab *Pleuroncodes*. However, a dense mat of filamentous prokaryote sulfur bacteria (*Thioploca, Beggiotia, Achroonema*) and cyanophyta (*Schizothrix* and *Oscillatoria*) associated with nematodes inhabits the zone of reduced oxygen levels below the eastern boundary current (Gallardo, 1975).

The opposite end member of the benthic series is the clean, white carbonate-rich sand of the lagoons within coral atolls and other coral lagoon systems. While there are often small pockets of organic mud within such lagoons that support, as we have already noted, small groves of mangroves, the benthos of such lagoons is probably determined, as Aller and Dodge (1974) found for a backreef lagoon in Jamaica, by the gradient in physical stability of the sand. Resuspension or mobility of sand grains is controlled by wave and tidal stress, by the nature of the substrate, and by the presence or absence of algal mats that bind the limestone fragments together. Suspension-feeding benthos dominate where resuspension is limited, and here biological reworking rates are low. Where resuspension rates are high, deposit-feeding benthos are common, and biological reworking rates are also high, so that pits and mounds dot the lagoon floor.

We can perhaps most appropriately take notice here of the benthic plankton (or hyperbenthos) which is now known to occur in substrates such as these large-particle limestone sands, and also in the clean sands in tropical sea-grass meadows. This is a special assemblage of copepods and other organisms that emerge at night from daytime refuges in the interstitial spaces between the granular sedimentary material of the lagoon bottoms. These benthic zooplankton join at night with the other zooplankton dispersed through the water column but which at daybreak form remarkable aggregations (Hammer and Carleton, 1979). Monospecific swarms of *Acartia, Oithona,* and *Centropages* occur at densities of $<1,500,000/m^3$, and are usually 1 or 2, in diameter. These swarms occur in sea-grass habitats and over the white sand of reef lagoons, and orient themselves with isolated coral heads or, as at Palau, in a linear band within the solution notch that undercuts limestone islands. A proper evaluation of the food available to coral reef carnivores must be based on a proper quantification of these swarms of copepods. However, it is to be noted that the individual copepods are able to maintain distance and orientation even against the effects of current swirls around coral heads: Zooplankters are not all the passively drifting prey of corals that we have come to believe in. Perhaps only those individuals of oceanic plankton that are brought inadvertently across the reef from deep water beyond the reef face can be expected to be disoriented and to be prey for corals.

Though the benthos of tropical coastal lagoons has been investigated in only a few areas, notably western Mexico, the Gulf of Guinea, and the Cochin backwaters, there are indications (e.g., Zabi, 1982, for the Abidjan lagoon) that polychaete-rich communities occur in some coastal lagoons, while in others (e.g., Yáñez-Arancibia and Day, 1982, for the Terminos lagoon in southeast Mexico) benthic communities are varied and appropriate to a range of salinity in sandy and sandy mud deposits.

So far, we have discussed mainly those benthic communities that are most important for tropical demersal fish; we must remember, however, that below the warm tropical surface water masses, coldwater benthic communities extend throughout the tropical latitudes. On continental shelves in the eastern parts of oceans, the tropical surface water may be extremely shallow. Off Sierra Leone and Guinea, almost half of the width of the continental shelf is covered with water of <17°C. Below the thermocline, on muddy sand and shelly sand deposits, a benthic community occurs that is quite unlike any of Petersen's series, yet is not unusually diverse; it comprises a number of pennatulids, holothurians, crinoids, ophiuroids, and echinoids together with a range of small brachyuran and galatheid crabs and polychaetes, especially the quill worm *Hyalonoecia,* which very frequently occurs in the stomachs of demersal fish. The relative distribution of the individual species of this deep benthos followed the nature of the substrate rather than depth, down to the shelf break at 200 m.

At the start of this chapter we noted how similar the species lists are at such depths off West Africa with the benthos at similar depths off western Europe 2000–3000 miles to the north; it is clear that, in principle, the ecosystems discussed by LeDanois (1948) on the deep continental shelf and upper slope off Europe will serve as excellent models for those of the eastern tropical Atlantic. Indeed, it is most likely, as a first approximation and in the absence of direct evidence, that benthic communities in the tropics below the thermocline will be found to be rather homogeneous globally, and not very different from LeDanois' descriptions. We should perhaps note the existence of the oceanwide surveys of the benthos of the deep ocean floor carried out principally by Soviet oceanographers aboard *Vityaz* with their famous Okean grab. These surveys, of which an excellent example is that of Neyman *et al.* (1973), show how depth, water movement near the bottom, and the overlying pelagic production regime influence the nature of the benthic communities of the deep ocean floor.

One can trace in these benthic distributions the great zonal bands of different ecological conditions related to the equatorial current systems right across the Pacific Ocean. There is very little comprehensive information on seasonal variability in benthic communities for any part of the ocean, and especially not for the tropics. Perhaps the North Sea data of Buchanan *et al.* (1978) best serves as an interim general model. Here the benthos community builds biomass and numbers during the summer and suffers heavy mortality in the winter. The winter–summer range for both biomass and numbers is a factor of about 5. A study of the benthos at nine coastal stations in the Philippines (Tiews, 1962) seem to show a similar change, of similar or slightly higher magnitude, with the minima occurring during the dry season. No explanation is offered for these observations. As Alheit (1982) has shown for Bermuda, the feeding activities of reef fish can reduce the benthic abundance in lagoon sediments very strongly (again, by a factor of about 5) within a few months.

COMMUNITIES OF LARGE MOBILE INVERTEBRATES

Crustacea

Three groups are especially important as fisheries resources and in the ecology of tropical seas: spiny or rock lobsters, swimming crabs, and penaeid shrimps or prawns.

Rock lobsters (*Panulirus* spp.) occur everywhere on hard substrates in the tropical oceans and are an important part of the reef fauna, with an ecology resembling their subtropical relatives (*Jasus* and *Palinurus*). Despite their wide distribution, commercial concentrations occur rather rarely, and the number of species occurring abundantly in each region is not large, so that only about 11 species account for almost all the catches of tropical rock lobsters (Postel, 1966; Mistakidis, 1973; Longhurst, 1971a). Female *Palinurus* carry their fertilized eggs for several months, and their leaf-shaped *Phyllosoma* larvae remain in the planktonic phase for as much as 1 year, during which they may be transported great distances. Some eventually settle and are recruited to the parent stock, by utilizing opposing current flow at different depths or by retention within a semipermanent gyre downstream from an oceanic island.

Panulirus longipes, P. ornatus, P. penicillatus, and *P. versicolor* occur on reefs from Mozambique to the Red Sea and Gulf of Aden, including the Mascarene Islands. In the western Pacific, two species (*Panulirus cygnus* and *P. versicolor*) occur around Australia and the archipelagic islands. The same species, together with *P. penicillatus,* occur over most of the central Pacific, though *P. marginatus* dominates at Hawaii. In the eastern Pacific, the tropical *Panulirus inflatus* occurs south of the Californian *P. interruptus,* and its range reaches to the Peruvian upwelling region. In the Atlantic, *Panulirus argus* dominates the Caribbean stocks, while *P. regius* is the only species off West Africa.

Resource partitioning by rock lobsters occurs within each region (Postel, 1965). On the northwest African coast in the Cape Verde area at the northern end of the Gulf of Guinea, *Palinurus charlestoni* occurs only on rocky reefs and especially those where the talus slope is very steep. *P. charlestoni* is restricted to the offshore Atlantic islands, while *Palinurus mauretanicus* occurs only on the deep *Dendrophyllia–Lophohelia* coral banks at the edge of the Mauretanian continental shelf. The tropical *Panulirus regius* occurs widely on hard sandy bottoms and shallow rocky reefs both in the Cape Verde region and right through the tropical Gulf of Guinea south to the Congo.

Though *P. regius* occurs alone on the West African coast, on the other side of the continent four Indian Ocean species coexist the whole length of the coast from Mozambique to Somalia, with some interesting local specializations. The west coast of Madagascar is inhabited by a *P. versicolor–ornatus–longipes* association, while *P. penicillatus* (also the dominant species of the Red Sea and Somalia) occurs alone on the east coast, except for a short section in the extreme southeast where a very abundant population of *P. hormarus* is dominant. Off Australia, there is an ecological distinction between the ecology of *P. longipes–versicolor–ornatus–penicillatus: P. longipes* occurs only on reefs, while the other three occupy more varied habitats.

Tropical swimming crabs are shorter-lived organisms than rock lobsters, and most species are probably annual. Though some genera, like *Euphylax* of the eastern Pacific, are pelagic and oceanic in their distribution, most are neritic and even estuarine. In these habitats they form an important component of the predatory biomass, preying on a great variety of smaller invertebrates and small fish. The most important Indo–West Pacific swimming crab is *Neptunus pelagicus* (Mistakidis, 1973) which occurs throughout the western archipelagos. In the Atlantic, blue swimming crabs of the genus *Callinectes* dominate the fauna on both sides of the ocean. *Callinectes sapidus* of the western Atlantic is the basis of a major blue crab fishery in the Caribbean and Gulf of Mexico; *C. latimanus* (estuaries) and *C. gladiator* (continental shelf) are the equivalent species in the Gulf of Guinea.

Penaeid shrimps (or prawns as they are called in Australia) are abundant wherever softer deposits occur on tropical continental shelves. Penaeids do not carry their fertilized eggs, but release them at once into the plankton after fertilization. Most species have long reproductive periods with one or more spawing peaks often associated with an offshore migration to deeper water. Large, commercial penaeids of the inner continental shelf are usually classified in two principal groups, white shrimp and brown shrimp, corresponding to subgenera of the genus *Penaeus*. White shrimp (e.g., *P. setiferus–vannamei–stylirostris–merguiensis–indicus*) occur close to the coast, are active by day, and feed on benthic microorganisms; brown shrimp (e.g., *P. duorarum–esculentus–brasiliensis–latisulcatus*) occur deeper, are more nocturnal, and are more predatory on benthic invertebrates.

Related to this classification, the most significant ecological distinction between species of penaeids is whether their planktonic larvae remain in the coastal plankton or whether they enter lagoons and estuaries to pass their juvenile instars. In those species with an estuarine juvenile stage, the nauplii and early mysid stages occur in the neritic plankton but later larvae concentrate around the mouths of estuaries and lagoons, which they enter on the flooding tide, in some cases using vertical migration between surface and bottom flow to move up the estuary (Rothlisberg, 1982; Orsi, 1986). The larvae very soon settle among vegetation in shallow water and move progressively deeper into the lagoon as they develop. Return to the sea occurs during periods of high freshwater discharge, usually within 3–6 months of entry.

There are, of course, variants from the classical tropical penaeid life cycle: some species occupy deep water near the shelf edge, some are coastal but do not enter estuaries as juveniles, and in others both adults and juveniles occur preferentially in brackish water. Coastal species whose larvae do not enter estuaries occur both as specialists alongside typical species, and also preferentially in regions where the estuarine habitat is not readily available.

As Kutkuhn (1966) was one of the first to show, there is in fact a complete gradation of habitat selection from those species whose larvae never enter brackish water to those in which both adults and juveniles occur preferentially in estuarine conditions. One end-member of this series comprises a few penaeids (e.g., *Hymenopenaeus, Plesiopenaeus,* the royal red shrimps of commerce) occurring together with some specialized carideans on the deep continental shelf and upper slope, not normally at less than 200 m, and whose larvae do not move inshore. Though Kutkuhn (1966) describes habitat differences between species of *Penaeus* itself, the species reviews in Mistakidis (1973) of Indian Ocean penaeids suggest that both *P. monodon* and *P. indicus,* which dominate the populations on the Indian coast, both have estuary-seeking larvae and estuarine juvenile stages. However, Garcia and LeReste (1981) indicate that *Penaeus semisulcatus* in the Persian Gulf and *P. esculentis* off Western Australia may pass their juvenile stages largely in the coastal regions. Both in the Gulf of Guinea and off western India, species of *Parapenaeopsis* (*P. atlantica* and *P. stylifera,* respectively) occur in coastal and neritic regions, but their larvae and juveniles do not enter estuaries (Longhurst, 1971a; Mistakidis, 1973); this is probably a generic characteristic. Adults of *Metapenaeus* occur more consistently in brackish water than *Penaeus,* though again there is variation within the genus in habitat selection. The species reviews in Mistakidis, for instance, show a clear distinction on the western Indian coast among *Metapenaeus dobsoni, M. monoceros,* and *M. affinis*; larvae of the last species do not enter estuaries nearly as routinely as the first two species.

As with most other marine biota, the number of species of large penaeids is greatest in the western Pacific and least in the Gulf of Guinea. In the Australian region seven species of *Penaeus* and four of *Metapenaeus* figure in the commercial landings, while in the Gulf Guinea there is only one important species of *Penaeus* and one *Parapenaeopsis.*

In the Indian Ocean, the most important species along the African coast is *Penaeus indicus,* followed by *P. japonicus* and *P. latisulcatus*; in the Red Sea and along the southern coast of Arabia, *P. semisulcatus* dominates as assemblage of about seven species. The same species dominates in the Persian Gulf, except in the Straits of Hormuz where *P. merguiensis* (= *P. indicus?*) is locally more important. There are immense resources of penaeids off western India and Pakistan, where *P. indicus* and *P. monodon* are most important, followed by *Metapenaeus dobsoni* (important as a juvenile, brackish water fishery), *M. affinis,* and *Parapenaeopsis stylifera.* In the Bay of Bengal, a wide range of species is associated with the highly variable salinity conditions. *Parapenaeopsis stylifera* occurs only at sea, *Penaeus indicus* and several species of *Metapenaeus* occur both on the shelf and in lagoons, and the small carideans *Acetes indicus, Palaemon tenuipes,* and *Hippolysmata ensirostris* occur in great quantities at river mouths; of these, *Acetes* dominates numerically.

The western Pacific probably holds the greatest diversity of penaeids, 20 species having been taken in a United Nations Development Programme (UNDP) fisheries survey off Vietnam, for example, though the dominant species remain *Penaeus indicus, P. merguiensis,* and *P. monodon,* together with *Metapenaeus sinulatus* (Mistakidis, 1973). The tropical Australian resources are dominated by three species pairs: *Penaeus merguiensis–indicus, P. esculentus–semisulcatus,* and *Metapenaeus ensis–endeavouri.* Similar species lists occur throughout the western archipelagos of the Pacific Ocean. There are also important local fisheries for *Acetes* and allied small shrimps in the Philippines and elsewhere in the western Pacific (Omori, 1975).

Penaeids are an important resource along the whole coast of Central America, but the species are different from those of the western Pacific. Catches are dominated by white shrimp (*P. vannamei, P. occidentalis,* and *P. stylirostris*), brown and pink shrimp (*P. californiensis* and *P. brevirostris*), and sea-bobs (*Trachypenaeus* spp., and *Xiphopenaeus riveti*). *P. occidentalis* dominates off northern South America, while northwards *P. vannamei* and others are dominant. Stocks are centered on regions of soft continental shelf deposits and are somewhat localized.

On the Atlantic coast of Central America, penaeids are also important and large fisheries exist. It was in the Gulf of Mexico that the tropical shrimp fisheries had their modern origin to satisfy the markets of North America. The port of Ciudad del Carmen, on the Gulf of Campeche, is the self-styled "Shrimp Capital of the World" and its main boulevard is dominated by a large monument to *Penaeus duorarum*, the pink shrimp. This species occurs in two major concentrations, one on Campeche Bank and one along the coast of southwest Florida. In the Caribbean–Gulf of Mexico region, the most important species are *P. duorarum* (pink shrimp), *P. aztecus* (brown shrimp), and *P. setiferus* (white shrimp). Populations of sea-bobs (*Xiphopenaeus*) and spotted pink shrimp (*P. brasiliensis*) occur also. In the tropical South American region, *P. setiferus* is replaced by *P. schmitti, P. brasiliensis* increases its importance southwards, and *P. aztecus* and *P. duorarum* occur as regional subspecies.

On the West African coast, as already mentioned, only two species (*Penaeus notialis* = *duorarum* and *Parapenaeopsis atlantica*) are important in the tropical region, though smaller populations of *P. kerathurus* also occur. General abundance of penaeids is very much constrained to areas adjacent to river mouths, where muddy deposits occur and access to lagoons is available for *P. duorarum*. Massive populations of *Acetes* occur in river mouths, and *Parapenaeus longirostris* is the regional deep-shelf specialist, occurring only a little shallower than the *Hymenopenaeus–Plesionika* fauna and having a range from the Mediterranean to South Africa.

The deep-water prawns, penaeids and carideans, are much less well-known, but may represent important resources that are already exploited in some places, as off the Hawaiian islands, where about 20 deep-water shrimp vessels were working in 1983–1984. Some species are cosmopolitan and diversity seems to be rather similar everywhere; off both north-eastern Australia (Potter and Dredge, 1985) and in the Gulf of Guinea there are five to six species, and both *Plesiopenaeus edwardsianus* and

Aristeomorpha foliacea occur in each region. Deep-living species of *Penaeus, Hymenopenaeus, Aristeus,* and *Plesiopenaeus* are characteristic of the upper slope, together with many species of caridean shrimps, which group extends to greater depths.

Throughout the western Pacific from Hawaii to the Marianas and New Caledonia, about 17 species of caridean shrimps occur at 400–800 m (King, 1986) wherever the slope is free of coral rubble. These include seven species of *Plesionika,* of which three (*P. ensis, P. longirostris* and *P. martia*) are the most important and eight of *Heterocarpus,* of which five have commercial potential. It is *H. laevigatus* (which FAO calls officially the "smooth nylon shrimp"!) which is the most likely candidate for exploitation. Species of *Plesionika* usually occur less deep (200–400 m) than *Heterocarpus* (400–900 m) and species are depth specialists usually being ranked, in order of increasing depth, as follows: *P. serratifrons* (150–300), *P. longirostris* (350–600), *H. gibbosus* (400–600), *H. sibogae* (450–700), *H. laevigatus* (500–800), and finally *H. dorsalis* (800–900). These depth strata are maintained over very large distances, from Hawaii to the Marianas; at eight isolated island groups, *H. laevigatus* occurs at each preferentially between 500–900 m (King, 1986).

The market demand for tropical shrimp exceeds supply, and pond production and sea ranching make up a larger proportion of the total supply each year; this phenomenon is also of significance ecologically, for massive alteration of coastal wetlands is likely to occur as pond production grows (Terchunian *et al.,* 1986). This is exemplified by the situation in Ecuador, where an injection of foreign capital into *P. vannamei* farming has destroyed great tracts of coastal mangal; production from these farms has now reached 30,000–35,000 tons annually compared with a total production in the western hemisphere of only 36,000–43,000 tons. *Penaeus* is the genus most frequently farmed, and larvae for stocking are produced either by capture from the neritic plankton, by spawning wild-caught fertilized females, and (increasingly) by induction of spawning in pond-reared stock through maturation induced by unilateral eye-stalk ablation. In the People's Republic of China and Japan, many tens of millions of pond-produced larvae are released annually for growth in coastal waters to augment natural stocks (New and Rabanal, 1985). There appear to be no statistics of the efficacy of this operation.

Cephalopods

These molluscs, by virtue of their convergent evolution with vertebrates that has given them so many fish-like attributes, merit special attention as components of tropical marine ecosystems. We shall return in

Chapter 10 to consider their special characteristics relevant to population biology, but here we note especially their aptness for rapid three-dimensional movement, their facultative binocular vision and efficient means of grasping prey, their schooling habit, and their reliance on cannibalism as attributes that place them directly in ecological competition with fish in the same size range. On the other hand, they possess other attributes that distinguish them to some extent from most fish: an ability to capture and consume prey almost the same size as themselves, and an often terminal reproductive pattern.

Tropical cephalopods fall into several ecological groups, for the discussion of which the review edited by Caddy (1983) is a good starting point. Five groups are usually distinguished: the nautiloids of the western Pacific, and four ubiquitous groups—cuttlefish, neritic squid, oceanic squid, and octopuses—each with their distinctive ecology and biology. Each group is fished in the tropics (though nautiloids only as decorative objects), and especially in the case of pelagic squid there is impressive potential for future expansion of catches.

Five or six species of *Nautilus* inhabit reef slopes at 300–500 m throughout the western Pacific, where they are taken in baited traps that must be left for several days to be successful. *Nautilus* therefore probably have a low population density, and assemble (like deep sea grenadiers and amphipods) at the bodies of dead fish sunk from above, attracted by olfaction. They are long-lived (15–20 years), mature late, but have a reproductive period extending over several years.

About six species of cuttlefish occur on the tropical continental shelves, each having a rather wide distribution: *Sepia officinalis* in the eastern Atlantic and *S. phaeroensis–lycidus–recurvirostris* in the Indo-West Pacific all occur as large populations. *Sepia officinalis* off Senegal undertakes both north-south and inshore–offshore migrations. In this, and in other species of cephalopods, the population comprises several overlapping cohorts, their origins separated by a few months.

Neritic squid of several genera are abundant over all tropical continental shelves. *Loligo duvauceli* and *L. chinensis* are important in the Gulf of Thailand, *L. edulis* throughout the western archipelagos and the northern Indian Ocean, and *L. brasiliensis* in the eastern Atlantic. *Sepioteuthis arctipinnis* is locally important in the Indian Ocean, as *Loliguncula brevis* is in the tropical western Atlantic.

Oceanic squid are difficult to study, and comprehensive data are absent for tropical species, except for *Dosidicus gigas* of the eastern Pacific; they undertake long migrations, of which the best mapped are those of the subtropical *Illex illecibrosus* between its spawning grounds off Florida and feeding grounds off Newfoundland. *I. coindetii* is a prominent Atlantic tropical species, and *Symplectoteuthis oualaniensis* occurs widely in the western Pacific.

Octopuses are a very homogeneous group, apart from a few aberrant pelagic and abyssal genera, and species tend to be ubiquitous. *Octopus vulgaris* occurs in all three oceans from temperate to tropical latitudes. Three species (*O. vulgaris–aegina–ocellatus*) occur throughout the Indo-West Pacific. *Octopus* and *Eledone* are thought to be terminal spawners, the female dying after her brooded egg mass is completely hatched.

CORAL REEF COMMUNITIES

Reefs are to be regarded both as a special form of benthic community, in fact as an end-member of the series of global enthic epifaunal communities, as well as a feature of the geography of tropical oceans. As benthic communities, their most puzzling attribute has been how they achieve the energy flow needed to maintain extremely high rates of production in an apparently oligotrophic oceanic environment.

The architects and builders of coral reefs are a very wide variety of organisms and chemical processes that combine to build carbonate structures that may be thousands of meters deep. The most important building organisms are the hermatypic corals belonging to many genera, the less massive encrusting coralline rhodophyte algae, as well as foraminifera, hydrocorals, and many molluscs that also contribute calcareous matter to the formation of a reef. This biological mosaic grows upon and continually extends the calcareous mass derived from their combined skeletal material. Solution channels in the coral rock come to be refilled by the precipitation of calcium carbonate from seawater penetrating the porous structure, and by the carrying-in of smaller fragments of coral material, so that solution and infilling are in balance. Reef growth in the vertical plane is controlled by rising sea levels, while wave action breaks off large masses from reefs fronts and the damage is repaired by horizontal reef growth.

The number of species and genera involved in reef formation is very large, and reefs are the classical marine example of high diversity of ecological niches in the tropics, parallel to the tropical rain forests on land (see Robinson, 1978; Connell, 1978). What is not so obvious, however, is the remarkable uniformity of specific composition of reef-forming communities over very large areas within ocean basins, compared with the very distinct differences between ocean basins. From the Red Sea to East Africa, and eastwards through the Indo-Pacific archipelagos to the Tuamoto Islands, the same species tend to occupy the same ecological niches in all reef formations. This fauna is attenuated in regions marginal for coral growth, but is everywhere quite distinct from the reef fauna of the Atlantic, which has its principal center of distribution in the Caribbean; there is generic substitution in similar ecological niches between the two oceans.

Many niches are not filled in the contemporary Atlantic reefs, though the Atlantic Tertiary fauna much more closely resembled that of the present-day Indo-Pacific; Wells (1957) lists 11 instances of an ecological niche in the Atlantic being filled by generic substitution for a taxon now confined to the Pacific. This attenuation by relatively recent geological events is also reflected in Atlantic weakening of important genera that occur in both oceans. Thus, *Acropora* and *Porites* (perhaps the most important reef-building genera) include only six species together in the Atlantic compared with about 180 in the Indo-Pacific. Significant endemism occurs in the disjunct coral reefs off Brazil, separated from the Caribbean reef province by a great distance of shelf unable to support hermatypic corals.

The nadir of hermatypic reef formation in the tropics is probably the eastern Atlantic where reefs occur in shallow water around only three offshore islands in the eastern Gulf of Guinea (Sao Thomé, Principé, and Annobon) where a very few genera, mainly *Porites, Sclerastrea, Favia, Montastrea,* and *Oculina* occur with some nonhermatypic *Astrangia, Phyllangea,* and *Tabastrea* (Thiel, 1928). Altogether, the Atlantic fauna comprises only 35 species in 26 genera of corals, compared with 700 Indo-Pacific species belonging to 80 genera; the center of coral diversity is today in the western Pacific (Fig. 4.3) related to the largest, deepest, least variable pool of warm tropical surface water existing in the present-day ocean. As well as a weakening of the coral fauna, other organisms are reduced or absent in the Atlantic reef ecosystem: calcareous algae, giant clams and anemones, coral-gall crabs, cypraeid and conid gastropods, alcyonarians, and soft corals. Gorgonians are one of the few exceptions to this rule, being far more important on Atlantic than Pacific reefs. Coralline sponges (Sclerospongiae) contribute significantly to limestone formation on the fore-reef slope of some Caribbean reefs, and appear to be a remnant of the stromatoporoid fauna of some Ordovician reefs (Hartman and Goreau, 1970).

Whether in the Atlantic or the Pacific reef ecosystems, the principal ecological characteristic to be noted is the extent to which organisms, both animals and plants, are mutually dependent. Commensalism and symbiosis have developed to a greater degree of complexity in these ecosystems than in any other animal communities; it is unfortunate that the general literature on commensalism in marine communities is so dominated by the much sparser examples to be found in cooler seas. As we shall see in Chapter 5, animal–plant symbiosis is the key by which coral reefs achieve the internal nutrient recycling to maintain their high biomass in an oligotrophic ocean. Commensalism between animals must be presumed to keep the wheels of internal recycling turning even more efficiently.

Antibiosis, or the chemical defense of space, is used by sessile organisms such as soft corals (Alcyonacea) to avoid predation and overgrowth by hard corals. These latter compete for space by using sweeper tentacles and mesenteric filaments, while soft corals exude toxins capable of causing necrosis in the tissues of hard corals even without actual contact. These two opposing tactics are at the base of the mosaic of sessile organisms that is the fabric of the coral reef community.

Living on the corals themselves, often in a species-specific relationship, is a great diversity of crustacea. One subfamily (Pontoniinae) of palaemonid shrimps are very frequently symbionts of sponges, actinians, molluscs, and the corals themselves on coral reefs: Bruce (1975) lists 14 Indo-Pacific genera of pontoniine shrimps having scleratinian coral hosts, five related to other coral genera, five to bivalve molluscs, four to sponges, and six to echinoderms. These shrimps have reduced armature, rounded exoskeletons, and live in a characteristic location on their hosts; little is known of their feeding habits, but many have specialized, scoop-like mouthparts. Other palaemonid shrimps (e.g., *Periclimenes, Lysmata*) are conspicuously colored and establish "cleaning stations" to which reef fish congregate to have their external parasites removed. Coral-gall crabs (*Hapalocarcinus*) exist within cavities, fashioned by their presence, in characteristic places on the branches of corals, often having only a tiny pore giving access to the outside. Their morphology is much reduced and they are thought to be filter feeders. Other crabs, such as *Trapezia,* occur on the surface of corals and feed by scraping mucus and algae from its surface; dromid crabs exist preferentially on dead corals, and majids on exposed rocky surfaces.

Invertebrate predators on coral reefs include the spectacular crown-of-thorns starfish (*Acanthaster plancii*) which encloses coral branches by everting its stomach and digesting their tissues *in situ*. Coral limestone and coral skeletons are relatively soft rocks and attract a great variety of boring invertebrates, whose activity adds to the spatial complexity of a reef. The most important of these organisms are sponges and algae, whose tunneling is done by chemical means, and molluscs and polychaetes, which bore mechanically.

A number of general morphological features can be recognized in mid-Pacific reefs that are reflected in all reefs, whether coastal or oceanic (Wells, 1957); these features result from the interaction of biological growth and physical forcing by wind, waves, and currents. The reef front, facing out to deep water, comprises a talus slope, often with a terrace about 10 m below low-tide level, and is sculptured into a series of surge channels that have evolved in such a way as to present the most effective possible breakwater (Munk and Sargent, 1948). Roofing of such channels by coral growth may occur, and within such overhung structures, and exposed to strong surges, are the most abundant assemblages of sponges on coral reefs. The reef front, especially on windward sides of atolls, is capped with an algal ridge, dominated by *Archaeolithothamnium*, *Lithophylum*, and *Porolithum*, encrusting red algae adapted to withstand the greatest wave action. On the back of the ridge, some corals (e.g., *Acropora*, *Pocillopora*, *Millepora*) occur, together with lighter colored algae. Zonation of the seaward slope below the algal ridge on the windward reef responds to illumination and wave action: an upper surge-channel zone, followed by a zone of abundant and often branching growth of corals (*Echinophyllia*, *Porites*, *Pocillorpora*, *Fungia*, *Acropora*, *Pavona*, and others), and finally a deeper zone where their growth is inhibited by low illumination, and delicate forms such as *Leptoseris* dominate. On the leeward reef, the seaward slope is somewhat different because of the much slighter wave action. The reefs are generally narrower, the talus slope is less well developed, and the reef front is much more vertical. Because of the protection from swell, large bracket-shaped formations of *Acropora* occur at quite shallow levels and extend down into an *Echinophyllia* zone where massive species of *Favia*, *Favites*, *Platygyra*, *Porites*, *Symphyllia*, and others occur. Other invertebrates (alcyonids, holothurians, crinoids, *Trochus*, etc.) are abundant on the ledges of this cliff-like habitat, along with unattached corals (e.g., *Fungia*).

Behind the algal ridge, a reef flat extends lagoonwards through a number of zones, and it is here that coral reefs are most easily observed because of their near-exposure at low tide. The windward reef flat usually bears the largest development of reef islands, in the lee of which, sloping down to the lagoon, the richest development of coral fauna occurs. This coral-algal zone is, as Wells (1957) points out, usually the situation where the richest growth of corals occurs with as much as half of the reef-flat surface actually covered with coral colonies. Branching species of *Acropora* are prominent, and *Favia*, *Pocillopora*, *Platygyra*, *Lobophyllia*, and many other genera also occur. This zone is also particularly rich in other invertebrates, and as the reef-flat deepens toward the lagoon the dominant fauna of the flats changes, with forms like the giant clams (*Tridacna*) beginning to appear. Within the lagoon, coral knolls and micro-atolls may rise toward the surface, and the floor is mostly of white coral sand and may bear sea-grass meadows. *Halimeda* and other macroalgae may (as at

Funafuti, according to Davis, 1928) sometimes form a soft green turf on the floor of the lagoon, below which there may be up to 25 m depth of decaying, peat-like algal material. Lagoons within atolls commonly have depths of slightly less than 50 m, and lagoon depth in groups of atolls tends to be similar.

Halimeda and other calcareous green algae (*Penicillus* of the Caribbean, *Tydemania* of the Indo-Pacific) deposit large quantities of interlocking crystals of calcium carbonate in the form of aragonite within their thallae. In so doing, they contribute very significantly to the formation of sand, especially reef lagoons. *Halimeda* meadows appear to be sited in locations where tidal-jet pumping through reef entrances is able to entrain small but regular amount of subthermocline (nutrient-rich) water into the lagoon (Wolanski *et al.*, 1986).

Where water movement across reef flats is moderately active, free-moving spherical growths up to <5 cm in diameter of corals (coraliths) and red algae (rhodoliths) may be formed and become so numerous as to carpet the reef-flat more or less completely (Scoffin *et al.*, 1985). Because of active flushing through the upper parts of the coarse calcareous sediments on which they lie, such spheres are able to maintain a complete envelope of living tissue even when shifted only infrequently by water movements. As they grow, and move less in the surging currents, they tend to assume a discoid shape.

COMMUNITIES OF PLANKTON ORGANISMS

A review of tropical plankton communities, or species assemblages, is much more easily made than for the tropical benthos: We do not have to contend with the east–west diversity gradient that appears to exist for benthos and fish, and the plankton literature is far more extensive.

Assemblages of single-celled phytoplankton have not been so extensively studied as zooplankton communities, perhaps mostly because of the ease with which a single measurement (of chlorophylls or other pigments) has come to be accepted to represent the amount and kind of phytoplankton present. Nevertheless, we know that tropical phytoplankton communities are relatively rich in small cells; in the open, oligotrophic tropical ocean, diatoms are comparatively unimportant, while dinoflagellates, coccolithophores, and small naked flagellates dominate the eukaryotic plant plankton. In fact, as has been shown very recently (Li *et al.*, 1983; Platt *et al.*, 1983) extremely small (<1 μm) prokaryotic plant cells (e.g., *Synechococcus*) are relatively more important in low than in high latitudes and may contribute very large percentages to total plant production in tropical seas.

Only in upwelling situations and tropical estuaries are diatoms as relatively important as in high-latitude seasonal blooms and it is only in such
environments that pulsed, seasonal phytoplankton production is an important feature of tropical oceans. That large, often centric diatoms dominate the phytoplankton of tropical estuaries has been reported many
times: for the Ponggol estuary in Singapore (Chua, 1973), the Sierra
Leone estuary in West Africa (Bainbridge, 1960a), the Gulf of Nicoya,
Costa Rica (Hargraves and Viquez, 1985), and Trinidad (Bacon, 1971).
The genera most frequently cited include *Coscinodiscus, Chaetoceros,
Rhizosolenia,* and *Skeletonema* and the presence of abundant large plant
cells is an important factor in tropical estuarine ecology.

Also characteristic of tropical oceanic surface waters are much larger
chlorophyll-bearing protists: large Radiolaria and Acantharia, sarcodinid
protists with symbiotic plant cells in their protoplasm. Large Acantharia
(>200 μm) may contain up to a thousand dinoflagellate and prymne-
siophyte cells, and it has recently been reported that they also contain
very small cyanophyte-like autofluorescing cells. Like the symbiotic plant
cells in hermatypic coral tissues, these algal cells function by very closely
coupling the physiological demands of plant and animal cells.

Especially in the northwest Atlantic subtropical gyre (the so-called Sargasso Sea) pelagic specimens of the littoral macroalga *Sargassum* may be
common; these occur at the surface of the ocean as loosely growing algal
spheres from about 10 cm to about 1 m in diameter, and may become
aggregated in windrows. Associated within the macroalgal spheres is a
small, specialized fauna of fish and crustacea. Also occurring in the same
habitat are filamentous cyanophytes (e.g., *Oscillatoria*) which occur as
small golden-brown tufts, sometimes in sufficient quantity to color the
water. These cells may contribute significantly to total production (Carpenter and Price, 1977; Carpenter and McCarthy, 1975) in some parts of
the tropical ocean, especially in the Sargasso and Arabian Seas.

These plants are the principal food of, and therefore determine the
nature of, the zooplankton communities that form the main subject of this
section, which is based largely on a recent study of the global composition
of zooplankton by analysis of 38 regional plankton surveys, totalling 4178
stations from all latitudes and all oceans (Longhurst, 1985a). This study
shows that there are significant and predictable differences in the composition of tropical and high-latitude plankton communities, and also that
significant systematic differences exist between oceans. These variations
are of ecological, rather than taxonomic, significance, and help us understand tropical fisheries ecology.

Perhaps fortunately for the progress of plankton ecology (unlike benthic ecology), the discipline never lost its way in circular discussions of

the community concept, and how to classify species assemblages; in this section we use the word "community" deliberately loosely, and with no significance other than as a descriptor of recurring assemblages of species having certain ecological characteristics. In fact, it is doubly fortunate that plankton ecologists have not indulged in the formal classification of such assemblages seriously, for we now know that the principal characteristic of the spatial distribution of plankton in the ocean—on scales at which we are likely to sample them—is that species, or groups of species, lie in a vertical series of horizontal layers. To sample the complete range of species or ecological types in the upper part of the water column (above those depths known to be occupied by bathypelagic forms), it has been customary until recently to sample by means of an open, towed net hauled vertically or obliquely from some depth, like 200–300 m. We now realize that such a sample, besides suggesting an erroneous view of average distributions, also includes many organisms that never encounter each other in nature, and integrates many species groups that would be considered different "Petersen communities" if they were from the benthos.

Table 4.2 shows how zooplankton communities are composed globally in relation to latitude (polar, temperate, tropical), habitat (oceans, continental shelf, bays, and reefs), taxonomic group (Medusae, Polychaeta, Copepoda, Doliolida, etc), and feeding group (gelatinous and raptorial predators, herbivores, etc). This tabulation confirms the generally accepted view of the global dominance of copepods, though it makes it clear that the greater diversity of tropical plankton (diversity, that is, in the range of plankton forms represented) means that copepods are much less dominant in tropical plankton than in high latitudes. Expressed as organic carbon, copepods form only one-third of the mean plankton biomass in tropical oceans, and only slightly more on tropical continental shelves. Higher relative percentage biomass of siphonophores, polychaetes, ostracods, penaeids, and thaliaceans in tropical plankton accounts for the difference in relative copepods dominance. Diversity, or at least the number of species present in the community, is certainly reduced in estuarine and coastal lagoon plankton (e.g., Bainbridge, 1960a) and may be lower in enclosed seas, such as the Red Sea and Persian Gulf, than in the open ocean (Kimor, 1973).

Tropical neritic and estuarine regions may have relatively high numbers of cladocera (mostly *Evadne*); these are ephemeral summer forms in temperate plankton but occur in all seasons in the tropics. Along the coast of the Gulf of Guinea, they range from 25 to 40% of the plankton numerically, with peak abundance during and just after the rainy season. Though they occur everywhere, cladocera have not been reported in such relatively high abundances in the tropical Pacific. On the other hand, there are

TABLE 4.2. Global Taxonomic Composition of Plankton Communities, as Relative Numerical Abundance of Taxa[a]

	MEDU	SIPH	CHAE	POLY	CLAD	OSTR	COPE	AMPH	MYSI	EUPH	LUCI	PTER	APPE	THAL	DOLI
Oceans															
Polar	0.06	0.15	3.88	0.23	0.00	2.45	86.32	0.56	0.00	0.52	0.02	2.44	3.32	0.02	0.00
Temperate	1.39	0.45	3.99	0.54	3.12	0.47	80.45	0.81	0.02	2.77	0.33	1.76	2.79	0.52	0.12
Tropical	0.85	2.61	6.38	0.85	0.14	3.39	72.31	0.54	0.15	3.02	1.30	1.70	3.85	2.10	0.66
Conshelves															
Temperate	0.22	0.59	3.32	0.39	7.07	0.31	80.10	0.78	0.07	1.02	0.27	1.50	3.51	0.46	0.05
Tropical	0.67	2.11	5.83	0.43	6.46	5.72	62.95	0.82	0.02	1.17	3.15	0.60	5.45	5.15	0.10
Bays															
Temperate	0.00	0.00	0.84	0.84	4.60	0.00	86.69	0.84	1.71	0.66	0.02	0.55	0.05	0.01	0.00
Tropical	0.00	0.00	7.74	0.00	2.21	0.00	84.38	0.88	0.73	0.00	2.54	0.00	1.89	0.00	0.00
Reefs															
Tropical	0.22	1.48	5.66	0.58	0.10	2.63	71.41	0.71	0.04	1.10	13.35	0.73	0.74	1.07	0.00
Global %	0.54	1.47	4.96	0.59	2.33	2.37	76.55	0.67	0.17	1.98	1.25	1.28	4.07	1.29	0.41

[a] From Longhurst (1985a).

special abundances of ostracods (mostly *Cypridina*) in many data sets from the Indian Ocean and especially the northern Arabian Sea, where they have been recorded as far exceeding even the abundance of copepods (Longhurst, 1985a). *Sergestes* and *Lucifer* are warm-water specialists, the latter being very important in tropical neritic plankton everywhere, perhaps especially where salinity is low, as off Sierra Leone, the Amazon mouth, and Hong Kong, for example. Because they are large organisms, they form more than half the biomass in some plankton samples and range from 10 to 25% in the mean values for neritic plankton for all three oceans. Salps are also characteristic of tropical plankton, where they occur about as frequently as appendicularians, and, once again, because they are large organisms, and sometimes occur in swarms, they reach very high relative biomass values in some data sets: the 31.2% of biomass off the coast of Bengal shown in the tabulation is a mean of 47 stations taken between July and August in this region.

The most significant result of this particular analysis is the final translation of the data into relative importance of feeding groups in tropical plankton. The simplest set of categories for such groups is the following: gelatinous predators—medusae, ctenophores and siphonophores, and large scyphomedusae often deliberately excluded from plankton samples; raptorial predators—all chaetognaths, some cladocerans, some copepods, some euphausiids, all amphipods, all annelids, and all gymnosomatous molluscs; macro-filtering herbivores—some cladocerans, some copepods, and a few thecosomatous molluscs; gelatinous herbivores, tunicates (appendicularians, thaliaceans, doliolids) which are mostly mucus-net filterers capable of utilizing very small particles; omnivores—some copepods, some euphausiids, and all mysids and penaeids; detritivores—most ostracods, to recognize their special affinity for large aggregated organic particles.

Figure 4.5 shows in schematic form the relationship of these groups to the categories of phytoplankton and seston which form their food in the tropical ocean. It is at once obvious from Table 4.3 that this figure represents a very different ecosystem from that of high-latitude plankton: in particular, the numbers and biomass of predators of all kinds is relatively much higher than in colder oceans: There is about twice as much biomass of predators relative to herbivores and detritivores in tropical as in polar oceans, and on tropical continental shelves as on temperate continental shelves.

These relative percentages of predators, herbivores, and detritivores in the tropical plankton could be translated into an interpretation of tropical species lists compared with temperate plankton. Tropical lists include raptorial euphausiids as well as filter-feeding species, many genera and

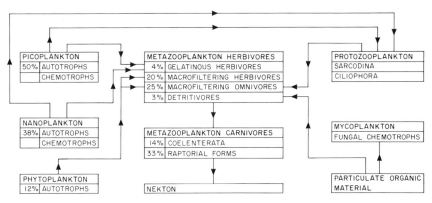

Fig. 4.5. Trophic group analysis for idealized warm-water oceanic plankton ecosystem. (From Longhurst, 1985a. Reprinted with permission from *Progress in Oceanography* **15**, Copyright 1985 Pergamon Journals Ltd.)

species of raptorial copepods, a great variety of siphonophores, and so on. We shall return later to the consequences of the community composition of tropical plankton for the production of organic material in tropical oceans, and the consequence of a great diversity of links in the food web for the stability of tropical ecosystems.

Associated with these organisms, which comprise the mesoplankton, are communities of microzooplankton (organisms of <200 μm): larval

TABLE 4.3. Global Latitudinal Variation in Relative Importance of Plankton Trophic Groups, as Relative Biomass[a]

	Gelatinous predators	Raptorial predators	Gelatinous herbivores	Grazing herbivores	Grazing omnivores	Detritivores
Oceans						
Polar	0.88	21.72	0.21	56.13	18.08	2.98
Temperate	4.08	33.65	0.68	19.80	39.80	1.98
Tropical	14.30	33.05	4.12	20.48	24.64	3.44
Conshelves						
Temperate	3.62	14.01	1.07	43.08	37.90	0.31
Tropical	9.84	28.14	6.28	7.84	42.58	5.31
Bays						
Temperate	0.00	21.61	0.29	26.24	39.18	0.4
Tropical	1.14	53.50	0.56	18.91	25.42	0.8
Reefs						
Tropical	4.30	21.54	1.00	6.94	38.53	1.27
Global %	4.77	28.40	1.78	24.93	33.27	2.06

[a] From Longhurst (1985a).

stages of metazoan plankton (>100 μm) and many species of protistan plankton–Ciliata, Tintinnida, Radiolaria, Acantharia, and Foraminifera. Few data yet exist on this important component, but Beers and Stewart (1971) found that microzooplankton biomass in the eastern tropical Pacific (10°N–12°S) was about 24% of that retained by a 300 μm plankton net. Microzooplankton biomass was positively correlated with phytoplankton standing stock, and was about 34% of it; perhaps as much as 70% of the net phytoplankton production is consumed in the open ocean by micro-zooplankton.

We must now note in passing the fact that all of this discussion has concerned the community of plankton organisms living in the upper 250 m (or the whole water column in shallower water) integrated within single plankton net tows at each station; we have already commented that this would now be expected to produce an artificial assemblage of organisms, in which several different species assemblages should be recognized. Figure 4.6 shows this for a plankton station in the eastern tropical Pacific Ocean, chosen for the relative simplicity of its ecological regime. Analysis of the co-occurrence of species in 42 individual depth horizons over the upper 250 m by the Bray-Curtis method shows that there are several independent species groups, arranged in a vertical series across the thermocline. Figure 4.7 shows how herbivorous, omnivorous, and predatory copepods are differently distributed in the vertical plane in relation to the thermocline and how detritivorous ostracods occur below the thermocline where large organic particulates are always especially abundant. Table 4.4 and Fig. 4.8 show how the upper 250 m of the water column in the open ocean may be regarded as a series of six depth zones each having different ecological characteristics, and how each zone tends to be the special realm of plankton species having similar feeding requirements.

Vertical migration between these depth zones in the tropics has a simpler and more predictable pattern than in high latitudes, where the most abundant copepods descend in a seasonal and ontogenetic migration to great depths (500–800 m) to overwinter and exhibit diel migration only during a part of the summer season, and over a smaller depth range. In the tropical ocean, two forms of diel migration usually occur throughout the year, and rearrange the plankton species associations by day and night in a rather predictable fashion. The first type is the interzonal migration of a population of mesopelagic plankton and nekton which resides during the daytime in species-specific layers at 250–500 m; such layers frequently comprise euphausiids, sergestid shrimps, small bathypelagic fish, and co-pepods, often of the genera *Metridia* and *Pleuromamma*. These layers, easily identified as the deep-scattering layer in echo-sounder traces, rises to the surface layers at dusk and descends again at dawn. These inter-

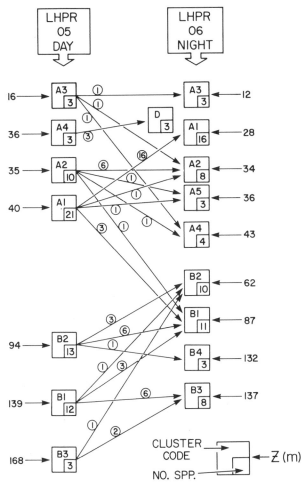

Fig. 4.6. Rearrangement of zooplankton recurrent species assemblages by diel vertical migration. Eastern tropical Pacific; numbers within circles show number of species moving between assemblages by diel migration. Z, Depth (m). (From Longhurst, 1985a. Reprinted with permission from *Progress in Oceanography* **15**, Copyright 1985 Pergamon Journals Ltd.)

zonal migrants, in the terminology of Vinogradov, form layers of organisms that can be traced continuously for thousands of miles across the open ocean in low latitudes and that, as we shall see in the next section, are a most important element in the food chain of tropical oceans. By their diel migration, interzonal migrants may translocate as much as 20% of the total biomass above 500 m into the epiplankton zone at night. The second

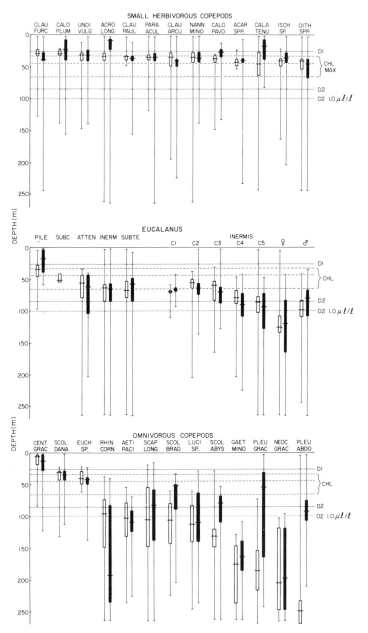

Fig. 4.7. Vertical distribution of exemplary species of small and large herbivorous cope-pods and of ominivorous copepods. Each species is illustrated by day (open) and night (closed) icons, showing modal depth, depth range of central 50% of population, and extreme depth range. For the water column, we show top (D1) and bottom (D2) of the thermocline, the top, peak, and bottom of the chlorophyll maximum layer, and the oxygen isopleth for 1.0 μl l^{-1}. Data from eastern tropical Pacific station. (From Longhurst, 1985a. Reprinted with permission from *Progress in Oceanography* **15,** Copyright 1985 Pergamon Journals Ltd.)

TABLE 4.4. Principal Ecological Characteristics of Six Depth Zones in Open Ocean Habitat in Eastern Tropical Pacific[a]

Zone	Depth (m)	Salin. (S ‰)	Temp (°C)	Brunt–Väisälä (cm h^{-1})	NO$_3$ (µg at l^{-1})	CHL-A (µg at l^{-1})	PP h-1 (mg C m^{-3})	POC (µg C l^{-1})	Zooplankton species/meter Day	Night
Neuston layer	0–10	34.13	>28.0	<0.01	—	<0.02	<0.20	0.34	0.00	15.90
Mixed layer	10–35	34.29	>28.0	5.15	—	0.18	0.29	0.46	9.20	15.30
Upper thermocline	35–55	34.84	27.20	15.60	9.85	0.48	1.10	0.44	14.80	16.70
Lower thermocline	55–85	34.73	22–15	7.40	25.55	<0.40	<0.36	<0.30	13.90	18.80
Upper mesopelagic	85–110	34.80	15–13	0.80	31.40	—	—	<0.03	14.10	12.90
Lower mesopelagic	110–150	34.89	13–11	0.30	>32.00	—	—	—	5.50	4.90

[a] From Longhurst (1985a).

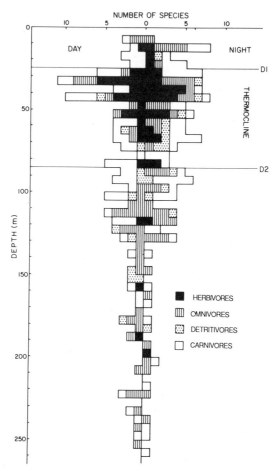

Fig. 4.8. Preferred or modal depths of 135 individual zooplankton species, for the same eastern tropical Pacific oceanic station as Figs. 4.8 and 5.4. (From Longhurst, 1985a. Reprinted with permission from *Progress in Oceanography* **15**, Copyright 1985 Pergamon Journals Ltd.)

type may be termed zonal migration for, within each depth zone, some individual species perform vertical diel migrations of lesser distance than is involved in interzonal migration, and which do not carry them into a different depth zone. This does not involve all species within each zone, and there is little commonality between species in their depths of daytime and nightime residence within each depth zone.

We have better seasonal data for biomass changes in plankton communities than for benthos in the tropics. In the eastern tropical Pacific,

across a large grid of open ocean stations sampled during the EASTRO-
PAC expeditions for more than 1 year, rates of primary production had
two maxima, varying by a factor of about 4 (Fig. 4.9), but without such a
large difference in chlorophyll standing stock; zooplankton varied apparently synchronously with chlorophyll (within the precision of the two-
monthly sampling intervals) and micronekton was suitably lagged by
about 2 months. Elsewhere in the tropical ocean, as off Hawaii (Shomura
and Nakamura, 1969), the temporal signal is small, and apparently not
regularly seasonal. In the tropical ocean (10–15°S) to the south of Java,
Tranter (1973) found seasonal variation in standing stocks of about the
same amplitude as those of Blackburn in the eastern tropical Pacific,
though more logically related to the monsoon wind regime than to high-
latitude seasonal rhythms. From an analysis of very many plankton samples, Longhurst (1985a) suggests that there is no measurable seasonal
variability in open tropical ocean zooplankton composition between major taxonomic groups. A number of multiyear, seasonal studies have been
performed in the vicinity of tropical islands (Bermuda, Barbados, Jamaica), but these are very difficult to interpret; apparently rapid changes
in biomass may not represent seasonal succession, but rather a changing
pattern of mesoscale eddies around the island, even if the broad outline of

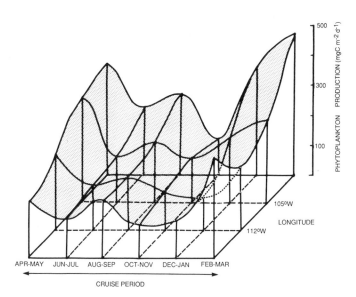

Fig. 4.9. Seasonal and spatial variability in primary production of phytoplankton in the
eastern tropical Pacific Ocean. (From Owen and Zeitszchel, 1970.)

TABLE 4.5. Seasonal Changes in Tropical Plankton Community Composition[a]

	MEDU	SIPH	CHAE	POLY	CLAD	OSTR	COPE	AMPH	MYSI	EUPH	LUCI	PTER	APPE	THAL	DOLI
Open oceans															
Atlantic															
Wet	0.49	10.64	7.43	0.28	0.00	0.48	24.28	5.15	0.00	37.85	10.50	1.47	0.04	1.38	0.00
Dry	0.99	8.32	6.27	0.45	0.00	0.64	35.85	3.89	0.00	19.27	11.86	4.84	0.04	7.58	0.00
Pacific															
Wet	0.34	12.12	8.12	0.84	0.01	2.25	28.06	1.23	0.00	43.71	0.80	1.17	0.20	1.10	0.04
Dry	1.22	10.18	11.73	0.80	0.00	2.26	30.48	1.69	0.00	38.70	0.66	1.17	0.19	0.86	0.02
Indian															
Wet	3.60	5.47	11.03	0.90	0.00	4.44	43.92	1.99	0.00	18.54	3.58	4.51	0.07	1.95	0.00
Dry	4.09	6.09	8.55	3.72	0.02	5.88	28.43	2.25	0.00	35.80	1.83	1.86	0.07	1.42	0.00
Continental shelves															
Atlantic															
Wet	1.85	10.67	8.49	1.39	0.01	3.98	38.56	2.54	0.00	14.21	13.12	1.70	0.29	3.18	0.00
Dry	2.08	12.98	10.37	1.63	0.06	2.17	34.73	3.91	0.00	9.47	16.43	1.40	0.36	4.41	0.00
Pacific															
Wet	1.10	1.34	11.40	0.07	0.61	1.95	52.97	2.94	0.00	0.49	24.72	1.10	0.08	1.23	0.00
Dry	0.00	0.71	12.37	0.00	2.40	2.70	51.22	3.85	0.00	0.00	25.69	0.91	0.09	0.08	0.00
Indian															
Wet	0.00	8.56	10.06	0.49	0.00	12.49	48.71	4.00	0.00	0.10	13.28	1.32	0.22	0.74	0.00
Dry	2.01	7.09	7.67	0.23	0.01	2.15	22.74	6.93	0.00	14.54	15.04	0.28	0.10	21.22	0.00
Bays and estuaries															
Atlantic															
Wet	0.00	0.00	21.07	0.00	0.32	0.00	62.26	10.12	0.00	0.00	6.15	0.00	0.08	0.00	0.00
Dry	2.28	0.00	9.14	0.00	0.21	1.61	35.51	4.04	0.00	0.00	45.44	0.75	0.09	0.91	0.02

[a] From Longhurst (1985a).

the seasonal variability is repeated from year to year. Translocation of areas of high or low productivity is a very different process from seasonal succession, resulting in variation of standing stock, within an ecosystem.

Such variability, of small magnitude compared with those in temporate latitudes (primary production rate varies by a factor of about 16 in the North Atlantic at midlatitudes) may be due partly to nonseasonal variability of wind stress at the sea surface. As McGowan and Hayward (1978) showed for the central gyre of the North Pacific, a between-year difference in wind stress is the best explanation for a between-year, same-season variation of a factor of 2 observed in standing stocks of plankton in this region. Under such circumstances, one could expect to observe a version of the temporal-spatial sequence of community development described and modelled by Vinogradov *et al.* (1973); from a plume or front of upwelled water, a succession of phytoplankton dominance, herbivore dominance, predatory plankton dominance, and finally micronekton development may be expected. This, in fact, appears to have been observed *in situ* over the EASTROPAC monitoring pattern, and in the development of a plume of upwelled water in a low latitude eastern boundary current (Longhurst, 1967b).

Coastal and estuarine plankton in the tropics perhaps has more predictable seasonal variability than oceanic plankton, but the amplitude of the seasonal signal may not be significantly greater. Table 4.5 describes the seasonal phyletic changes to be anticipated in coastal plankton; examination of the individual data sets shows that there is less seasonal change in composition in eastern boundary current upwelling regions than in river-dominated coastal stations. As we shall discuss later, species shifts may occur at seasonal upwellings, but apparently not relative abundances of higher taxa.

In estuarine-neritic regions, the amplitude of the chlorophyll signal is usually less than a factor of about 3 (e.g., for the Cochin estuary, the continental shelf of the Laccadive Sea, and the Gulf of Guayaquil), though as much as a factor of 10 has been recorded in a West African estuary dominated by massive seasonal river discharge; even here the zooplankton response is damped to the more usual factor of less than 5 (Longhurst, 1983). The phyletic changes seem to be driven principally by seasonal blooms of cladocera and by variations in the abundance of the larvae of benthic invertebrates.

Chapter 5

Biological Production: Benthos and Plankton

In this chapter we review what is known of the production of organic material by tropical benthos and plankton: Material produced in this way is the source of energy for the growth of fish. Not only in the tropics, but everywhere, we have much more information about the kinds, numbers, and biomass of animals and plants that comprise marine ecosystems (and terrestrial, too, for that matter) than we do about their rates of growth, reproduction, and mortality (e.g., Fasham, 1984). This situation can be redressed in two ways. We can hope that studies of the physiological basis of growth, scaled by body size, temperature, and other fundamental variables, will allow extrapolation from generalities to the growth of organisms in actual ecosystems, and we can accumulate enough case histories by direct study of actual ecosystems so that we can hope to predict the rates of growth, reproduction, and death of at least the dominant organisms in other similar ecosystems.

Generalizing is not easy for ecosystems of temperate seas that have been studied extensively, and is very difficult indeed for tropical ecosystems. Such information as is available to us also contains a paradox: We know a great deal more about rate processes for the two ends of the food chain—microorganisms and fish—than we do for the intermediate links, the plankton and benthos. This is because the need to quantify population parameters of commercial fish has drawn much more biomathematical research effort in that direction than for the intermediate trophic levels, and because the availability of direct methods of measuring the parameters of growth of marine plants and bacteria has made this an especially attractive field of study for biological oceanographers, in addition to its intrinsic importance as the base of marine food webs. Some of the methods used in analyzing the population parameters of tropical fish are also

available for invertebrates, as we shall discuss when we introduce population biology formally in Chapters 9 and 10.

MEASURING PRODUCTION
AND CONSUMPTION RATES

This book is not the place to discuss in detail the methods by which the
growth and reproduction rates of microorganisms, principally bacteria,
cyanobacteria, and algae, can be measured. A variety of radioactive
tracer techniques are now available for labeling the uptake of inorganic
nutrients or organic substrate to measure the new material produced au-
trophically, or chemotrophically, by bacteria; such techniques are also
available to partition new organic production between lipid, protein, or
carbohydrate fractions. The basic method for measuring phytoplankton
carbon assimilation, by quantifying their rate of uptake of a radioactive
carbon isotope, is well known, though still being debated 30 years after its
introduction by E. Steeman Nielsen. Both incubator and *in situ* methods
are still in use at sea. Alternatively, carbon fixation may be quantified if
the rate of release of oxygen during photosynthesis is known; recent
refinements of the Winkler titration have now made this measurement
sufficiently precise as to be useful for this purpose. Mass budgeting of
oxygen, as deduced from oxygen sections for ocean basins and assumptions concerning its residence time, have also recently been used to infer
plant production rates (Jenkins, 1982; Jenkins and Goldman, 1985). Finally, particle counting and sizing techniques based on optical or conduc-
tivity sensors, some originally developed for medical purposes, enable us
to measure the change in cell numbers during an incubation period, or
physically to sort cells for subsequent experimental incubations.

   In recent years much emphasis has been placed on understanding phytoplankton production in the oligotrophic tropical ocean as well as in the
principal upwelling ecosystems of low latitudes; the tropical open ocean
ecosystem has been a particularly fruitful environment for studies of nutrient requirement and uptake by autotrophic microorganisms, of the relations between production rate, light, and temperature, and of the equiva-
lence between measurements of plant biomass by chlorophyll and its
content of organic carbon and nitrogen. It is also in the tropical ocean that
much of the recent work on size-fractionation of production between algal
and cyanobacterial cells of different dimensions has been done. There is
no longer a high latitude bias in the general understanding of how plant
material is produced in the ocean.
   Primary production of coral reefs (by encrusting and symbiotic algae)

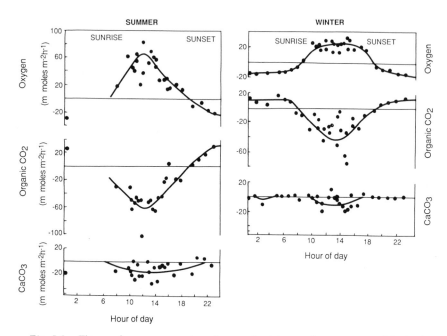

Fig. 5.1. Fluxes of oxygen, organic carbon, and calcium carbonate on a reef flat in the Hawaiian islands in summer. Fluxes are either from the water (negative values) or benthos (positive values). Each data point indicates one transect across the flat. (From Atkinson and Grigg, 1984.)

has usually been measured by treating the reef as if it were a plant community and measuring the rates of change of oxygen, CO_2, and $CaCO_3$ levels over a diel cycle in the water passing across it; various flow respirometry techniques of increasing sophistication have been used since the early experiments by Sargent and Austin (1949) in the Pacific. Figure 5.1 shows a typical output from a set of measurements of this kind. From such a cycle of oxygen production, with a maximal rate at midday, and oxygen consumption at a stable rate during the hours of darkness, it is possible to calculate gross and net community production, turnover rates, and efficiency of production (e.g., Lewis, 1977). Carbon fixation can then be converted to equivalent production of proteins and lipids. Lewis tabulates data from 17 coral reef sites, showing mean production rates of 3588 gC m^{-2} y^{-1} for reef flats, and 2070 gC m^{-2} y^{-1} for sea-grass meadows.

For zooplankton and benthos, the technical situation is not at all so clear; a number of methods exist, some applicable to both groups of biota, but none have been used so widely as the standard plant production measurements, and for none of them can we have the same confidence as for phytoplankton. Because particle sizing and counting is now routine, most zooplankton physiology has concentrated on the mechanics and efficiency of herbivore filtration systems. Conceptually, at least, the simplest technique for measuring growth is by the detailed analysis of size frequency distributions arranged sequentially in time, as we discuss in Chapter 9. For plankton, it is very difficult to be sure that the same discrete population is sampled during the study: analysis of size (or stage) frequency distributions has therefore been used most frequently in coastal bays and inlets where a self-contained population occurs. By using very short sampling intervals the technique recently has been used successfully on two oceanic copepods: *Calanoides carinatus* (Petit, 1982) and *Eucalanus pileatus* on the Congolese continental shelf. In the latter study, with almost daily sampling, Dessier (1985) found that the mean generation time over a 7-month period was 17 days. Using an algebraic form of the Allen curve (Nees and Dugdale, 1959), Dessier calculated production on an area (22 mg C m^{-2} d^{-1}) and volume (1.3 mg C m^{-2} d^{-1}) basis.

For benthos, the size-based analysis methods are basic and successful, since population stability is much greater and cohorts can be identified and followed with relative ease (e.g., Pauly and Calumpong, 1984, for tropical sea hares). Adequate criteria for dimensions can be obtained for both hard- and soft-bodied benthic organisms, and for temperate benthic communities, it is possible to perform cohort analysis on all significant species, individually, for whole benthic communities (e.g., Warwick *et al.*, 1978; Warwick and Geroge, 1980): 15 species were required for a *Venus,* 9 for an *Abra,* and 7 for a *Macoma* community. Related to these techniques, and producing conceptually similar information, is the measurement of growth of individual organisms by tagging, or by reading the natural growth checks on suitable hard parts; accumulation of sufficient data of this kind to overcome statistical variability in individual growth curves can produce data comparable to that from size-frequency analysis. Indeed, methods now exist for the simultaneous analysis of size-frequency and size-at-age data as well as tag-return data which optimize the use of available information (Brey and Pauly, 1986).

Tagging has been generally unproductive in tropical seas for most fish, except tuna, because, among other reasons, cleaner fish tend to nibble and destroy the tags (Watts, 1959); nor have natural growth marks in skeletal structures been exploited widely. Such marks were first recorded in the carbonate skeletons of corals, where daily growth checks can be identified both in living and fossil forms. However, even though this technique is capable of producing data on the overall growth in colony size (from about 1.0–25.0 mm, or 1000–35,000 g m^{-2} of carbonate per year) these rates do not, as Lewis (1977) points out, necessarily contain any direct information on the energy requirements or organic carbon budget of a coral reef. For a very few ($N = <10$) species of tropical molluscs of interest to local fisheries (e.g., Okera, 1976), growth marks on shells have been used to derive growth rates, but these studies are insufficient to draw any general conclusions concerning community growth rates for benthos in the tropics.

We should note here, parenthetically, the remarkable potential afforded by growth rings, visualized by fluorescence, which occur in the skeletons of the massive neritic coral *Porites*. As Isdale (1984) and Boto and Isdale (1985) have shown in *Porites* colonies on the Queensland coast, we can deduce data on ambient sea temperature, salinity, and the occurrence of fulvics and humics from river run-off from these rings. These proxy environmental data for the neritic zone can be derived over periods of many hundreds of years, and with a precision of about 1 month. A marine analog of dendrochronology in the tropics can now be predicted to be a growth industry!

For zooplankton, a technique is available based on differences in the nitrogen : phosphorus ratio of food and of excretory products for groups of individuals of a species or community. Based on an original suggestion by Ketchum (1962), the technique is now being used extensively in the tropics by LeBorgne (1978, 1982) and others, and is adding to our understanding of tropical production : biomass ratios more rapidly than any other method. The technique requires bottle incubation to measure excretion rates and the $C:N:P$ ratio of food and excreta; this can be done at sea, or the $N:P$ ratios of herbivores and predators can be determined on frozen samples. From these ratios can be calculated both net (K_1) and gross (K_2) growth efficiencies.

Another set of approaches to the measurement of growth and growth efficiency of zooplankton and benthos is based on the empirical relations between body weight, respiration rate, and environmental temperature. In a paper probably destined to become a classic reference in the subject, Ikeda (1985) has reported metabolic rates (oxygen uptake, ammonia, and phosphate excretion) for a comprehensive selection of epiplankton communities, each comprising about 50–150 species ranging over six orders of magnitude in size of individuals, and occurring in various latitudes (and hence environmental temperatures from -1 to $+30°C$). These measurements reveal that 85–95% of variation in metabolic rate can be attributed to body size and environmental temperature. Phyletic differences appear to be based principally on the different chemical composition of the phyla: Euphausiids and salps have oxygen uptake rates different by a factor of 17 when measured as wet weight, or 4 on a dry weight, and only 0.8 on a

carbon or nitrogen basis. Ikeda has two principal findings: a strong negative correlation between adult body mass and water temperature (that
tropical plankton is smaller, group for group, than cold water plankton
has long been known in general terms), and a strong positive correlation
between body size and metabolism, clearest for oxygen metabolism which
is itself the best integrator of total metabolism. Ikeda has extrapolated
from his results to a geographical expression (Fig. 5.2). He places global
plankton biomass maps (which show relatively very high values in high
latitudes) in their correct perspective. Relative production rates, per unit
area, are much more similar in high and low latitudes. This illustration is
not intended by Ikeda to be more than an indication of relative community
productivity, yet it does serve to place tropical plankton production in a
proper global perspective for almost the first time.

   Ikeda's study of zooplankton, based on his own original and standard-
ized observations, also gives us greater confidence in more holistic approaches, such as the studies of Banse and others, on the general question
of the relationship between biomass (measured easily and frequently) and
production (measured, as we have seen, seldom and with greater difficulty). The allometric basis of physiological parameters of growth in marine ecosystem has been reviewed very recently by Platt (1985), who

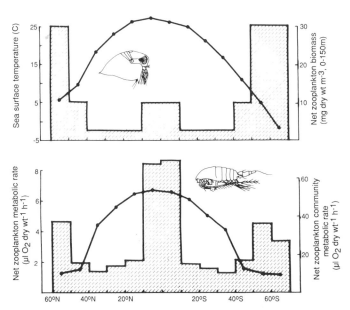

Fig. 5.2.   Zooplankton community biomass (top figure) and production (lower figure) as
a function of water temperature and latitude. (From Ikeda, 1985.)

relates the Elton–Odum ecological pyramids of biomass to community particle size spectra in a useful manner, and argues persuasively for the utility of particle-size analysis; Platt writes

A description in terms of biomass goes a long way towards providing a taxonomy of more practical utility . . . once a size spectrum has been constructed . . . we already have considerable potential information about the physiology of the community . . . therefore it is worthwhile to consider basing pelagic ecosystem studies on size dependent principles.

The production : biomass (or P/B) ratio is both a useful general expression, and a parameter capable of direct measurement as Kimmerer (1983) has done recently for tropical copepods. As proposed by Boysen–Jensen (1919), a minimum estimate of production may be obtained if change in biomass can be measured during a full year for any population ($P = B_e + B_2 - B_1$ where B_e is biomass eliminated, and B_1 and B_2 are biomass at the beginning and end of the year). Reviewing P/B for many species of temperate organisms living at 5–20°C, from copepods to molluscs, ranging in size by a factor of 10^5, Banse and Mosher (1980) showed that body mass at maturity (M_s) expressed in kilocalories is an efficient scaling factor for P/B, which itself varies by 10^2. Banse and Mosher comment further that ". . . P/B of invertebrates living at annual mean temperatures >25°C may be elevated over those of temperate species of the same M_s while those of polar forms are depressed . . . ." Later, Banse (1982) suggested that very small invertebrates (protozoa, nematodes, harpacticoids) of the meiobenthos and microplankton have P/B ratios that are lower in relation to their M_s than would be expected by extrapolation from the earlier study. Robertson (1979), independently of Banse's approach, showed that there is a simple relationship between P/B and the life span of many marine benthic organisms from temperate regions: the longer-lived, the lower the value of P/B. This relationship applies even within a single species according to the average age of the individuals in the local or regional population being studied; that this is a significant factor in P/B variability is shown by populations of a small estuarine bivalve, *Macoma baltica*. A Canadian population with an individual longevity of 2 years had a P/B of 1.53, while a British population with a longevity of 7+ years had a P/B of only 0.91.

P/B is a such a convenient parameter in production studies that it is useful to review what values for tropical benthic and planktonic ecosystems can be assumed in the absence of direct measurements, as is frequently the case. Assembling P/B ratios for tropical plankton, the study of Atlantic plankton by LeBorgne (1982) is the best available starting point, since it was done with a method we can trust, and covered both micro- and mesoplankton along a transequatorial transect of 42 stations from 5°N

TABLE 5.1. Daily *P/B* Ratios for Zooplankton at 19 Stations in
the Gulf of Guinea[a]

Stn.	Lat.	Temp. (°C) range (0–100 m)	Daily *P/B* ratios Microzoo. (%)	Mesozoo. (%)
1	03°50′ N	14–24	130.9	14.5
2	02°00′ N	15–29	65.3	35.3
3	00°30′ N	—	131.2	—
4	00°00′	16–29	119.8	62.2
5	00°20′ S	15–20	69.2	26.7
6	00°30′ S	16–28	—	27
7	00°30′ S	14–23	141.1	21.8
8	02°00′ S	14–22	94.7	—
9	02°00′ S	15–25	141.7	27.6
10	02°30′ S	14–22	166.6	49.6
11	03°00′ S	14–28	230.2	32.8
12	03°35′ S	14–22	50.9	21.9
13	05°00′ S	14–22	90.5	27.9
14	06°00′ S	14–22	108.3	37.8
15	09°20′ S	16–27	—	28.2
16	10°15′ S	16–25	34.2	40.8
17	11°10′ S	18–25	—	18.7
18	12°00′ S	18–23	—	56.3
19	15°00′ S	18–21	—	38.4
Mean daily *P/B* ratio (%)			112.47	33.38

[a] From LeBorgne (1982).

to 15°S. *P/B* ratios (Table 5.1) for whole communities over the depth range 0–100 m (copepods ranging from 73 to 89%), for which LeBorgne provides feeding-group analysis, give daily *P/B* ratios ranging from 0.43 to 2.30 for microzooplankton and of 0.53 to 0.95 for mesozooplankton. These correspond with doubling times of 0.4–2.9 and 1.6–6.9 days, respectively. As can be seen from Table 5.2, and from other studies (e.g., Miller *et al.*, 1977), doubling times for tropical plankton of 2–7 days are now well supported, with daily *P/B* ratios converging on values between 0.25 and 0.50. In addition to community measurements, LeBorgne also offers two species-specific estimates: *Salpa fusiformis,* with a doubling time of 0.7 days, corresponding closely with previous estimates by Heron (1972), and its nudibranch predator *Glaucus atlantica,* with a doubling time of 6.3 days.

For tropical benthos, things are not so good. We have been unable to locate any comparable studies of benthos communities for tropical regions. The best that can be hoped for, therefore, is to scale benthic com-

TABLE 5.2. Review of Data on Daily P/B Ratios for Warm Water Zooplankton[a]

Tropical region	Kind of plankton	Daily P/B ratio (%)	References
Coral Sea	Microzooplankton—small copepods	22–28	Greze (1978)
	Microzooplankton—small chaetognaths	15–20	
	Mesozooplankton—herbivore copepods	6–10	
	Mesozooplankton—carnivore copepods	5–8	
Gulf of Guinea	Large copepods	2–8	Malovitskaya (1971)
Florida	Chaetognaths	20–41	Reeve and Baker (1975)
	Siphonophores	6–20	
Equatorial Pacific	Copepod larvae	40–85	Shushkina and Kisliakov (1975)
	Chaetognaths	5–17	
	Euphausiid larvae	30	
Pacific central gyres	Copepod larvae	50–100	
	Chaetognaths	40–60	
	Euphausiid larvae	30	
Equatorial Pacific	Herbivore mesoplankton	15–99	Vinogradov et al. (1976)
	Carnivore mesoplankton	23–33	
	Protozoan microplankton	84–150	
Peru upwelling	Omnivore mesoplankton	3–86	Shushkina et al. (1978)
	Carnivore mesoplankton	14–65	
Pacific atoll lagoon	Large copepod	9	Gerber and Gerber (1979)
	Small copepods	30	
	Heteropod mollusc	3	
Coastal lagoon	Mixed mesoplankton (*Acartia* dominant)	21–86	LeBorgne and Dufour (1979)
Tropical Atlantic	Mixed microzooplankton	34–230	LeBorgne (1982)
	Mixed mesozooplankton	15–62	
Atlantic coast USA	Mesoplankton (*Acartia* spp.)	37–143	Durbin and Durbin (1980)

[a] From LeBorgne (1982).

munity P/B from comparable temperate community studies, such as those of Warwick *et al.* (1978) for all the communities in the Bristol Channel and Celtic Sea. From these, annual P/B values can be derived for the communities occupying the principal grades of deposit: continental shelf mud, 1.28; continental shelf sand, 0.56; estuarine mud, 1.1. These values represent the differences between *Venus* communities on sand and shell sand, and *Abra* and *Macoma* communities with many fast-growing polychaetes and small bivalves.

By what factor should these values be scaled to represent tropical benthic communities? Longhurst (1983) suggested, by reference to Q_{10} changes that might be expected over the temperature range involved, and to the ratio between P/B for temperate and tropical demersal fish that feed on these benthic communities, that a factor of about $\times 2$ was appropriate; studies such as Tranter (1973) and Robertson (1979) confirm that a factor of between $\times 2$–$\times 3$ would probably be about right (see also Chapter 9). Resolution of this important hole in our understanding of the biological energetics of tropical oceans awaits a benthic Ikeda or LeBorgne, or a tropical Warwick. Until then, perhaps we can work with values derived from the above discussion, annual P/B = offshore sand 1.2–1.7, offshore mud 2.6–3.8, estuarine mud 2.2–3.3.

ECOSYSTEM TROPHODYNAMICS: OPEN OCEANS

Our discussion of the plankton communities of the tropical ocean was restricted to the larger plant cells and macroscopic zooplankton. Classically, these organisms have been regarded as the base of all pelagic food chains. In fact, it is now known that this is not really a good description of the origin of new organic material, and its transformation, in the pelagic ecosystem. Though there is overlap between them, it is now known that three principal pathways exist by which organic material is produced and transformed by consumption by larger organisms. The first process originates in the utilization of dissolved organic material (DOM) by bacterial and fungal cells of very small size (the bacteroplankton, of order 0.2–2.0 μm) and their consumption (e.g., Azam *et al.,* 1983; Sorokin, 1981; Williams, 1981) by a great variety of heterotrophic protists, mainly choanoflagellates and colorless chrysomonads. These protists are in turn consumed directly by larger metazoal zooplankton, or at least by their larval stages. Because the DOM that fuels this process originates chiefly in the metabolism and exudates of plant cells, it is often referred to as the microbial loop.

The second process is based on the growth of very small prokaryotic and eukaryotic cells capable of photosynthesis; these are categorized as the picoplankton, of order 0.5–2.0 μm, and are dominated by prokaryotic cyanobacteria and algal eukaryotic prasinophytes. Picoplankton are consumed, like bacteroplankton, principally by protists (e.g., Iturriaga and Mitchell, 1986); though they may appear in the guts of metazoan zooplankton, this is probably mostly a secondary effect of the gut contents of the prey of the larger organisms. Some metazoa, especially salps and tunicates, may be able effectively to utilize these very small plant cells. The third process is the classical food chain, based on the consumption by metazoal zooplankton of phytoplankton cells, both in the nanoplankton (pigmented flagellates) and larger (coccolithorphores, dinoflagellates, diatoms) size classes. Allocation of classes of organisms to each of these three pathways is not simple; dinoflagellates, for instance include both important components of the oceanic autotrophic phytoplankton and some large genera (*Noctiluca, Gyrodinium*) that are phagotrophic and participate in the consumption of bactero- and picoplankton (e.g., Kimor, 1981).

These new concepts render the question of size-fractionation of phytoplankton production in the open ocean even more cogent. It is now known that extremely small (<1.0 μm) coccoid cyanobacteria are important in tropical pelagic plant production. As Li *et al.* (1983) and Platt *et al.* (1983) showed, such cells are light-saturated at depths of the deep chlorophyll maximum, and at a variety of sites in the warm-water sphere (eastern tropical Pacific, eastern Atlantic, Mediterranean) they comprise a large fraction of total chlorophyll (25–90%) and of total inorganic carbon fixation (20–80%). As Wyman *et al.* (1985) have observed, the light-harvesting pigments of cyanobacteria are dominated by phycoerythrin and this pigment functions not only in collecting quanta for photosynthesis but also as a nitrogen reserve. These recent findings about cyanobacteria extend the lower limit of the range of particle sizes involved in plant production; as recently as 1980, Malone described phytoplankton size fractionation for biomass in the open ocean as about one-third netplankton (>20 μm) and two -thirds nanoplankton (2–20 μm); the new revision downward of the percentage contribution of large cells capable of being fed upon by copepods requires a total revision of our analysis of pelagic ecosystem functioning. We already know that copepods cannot grow on a culture of cyanobacteria (Johnson *et al.*, 1983) and we must presume that these cells are grazed initially only by heterotrophic nanoplankton (e.g., Sieburth and Davis, 1983). Though it is beyond the scope of this text to analyze further the ecology of the microcosm of very small planktonic organisms, we should just emphasize that it is realized increasingly that very large percentages of the total energy exchange in the planktonic ecosystem occurs in this microcosm: Bacteroplankton, zooflagelates, and ciliates comprise the largest fraction of all planktonic respiration. For further discussion and description of the bacteroplankton, its nutrient sources and its consumption by very small protists, the reader is directed toward the work of Sorokin *et al.* (1985), Sieburth (1984), and Azam *et al.* (1983).

Despite the new understanding of the importance of the microbial loop in recycling organic material back into the pelagic food chain, our review of ecosystem trophodynamics of the oligotrophic tropical ocean shall take as its starting point a model driven by phytoplankton photosynthesis. We assume that the best generalization for the tropical ocean is a simple two-layered ecosystem—an upper nutrient-limited and a lower light-limited layer, as described by Eppley (1981). We also assume that bacterial processes will be found to track phytoplankton processes closely, and that plant growth can serve as a indicator of microbial ecosystem function.

To understand such a system, typical (as we shall see) of much of the open tropical ocean, it is first necessary to introduce the simple concept of new and regenerated primary production by phototrophic cells. This concept, introduced by Dugdale and Goering (1967) and more recently reviewed by Smetacek (1985), goes far to explain the real significance of global maps of chlorophyll, or plant biomass in the oceans. New production is driven by nutrient input to a pelagic ecosystem from below the euphotic zone by upwelling and turbulent diffusion, or from river discharge of terrestrial nutrients, while regenerated production results from the biological recycling of nutrients within the pelagic ecosystem. Nitrogen is generally regarded as the nutrient most likely to be limiting. New production may therefore alternatively be defined as that driven by inorganic nitrate, while regenerated production is driven by organically produced ammonium, urea, and other forms of organic nitrogen that are available to plant cells. In a steady-state, open-ocean ecosystem in production balance (as in the open tropical ocean), the flux of organic detritus downward to the deep ocean must balance the upward nutrient flux at the base of the pycnocline, together with such nitrogen as is fixed within the mixed layer by cyanobacteria.

To put these two forms of plant production in the open ocean in context, Eppley (1981) suggested three indices: new/total production, which can be measured by sinking flux and growth of consumers as a percentage of total production, and which ranges from 0.05 to 0.8 depending on the nature of the sedimenting particles. Regenerated/total production, which represents the amount of recycling, and hence zooplankton grazing activity values ranging from 0.20 (upwelling) to 0.95 (open tropical ocean). Regenerated/new production represents the number of times that a nitro-

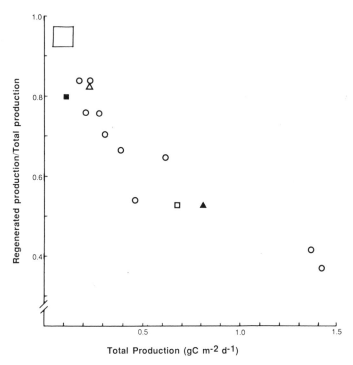

Fig. 5.3.   Ratio of regenerated as a function of total production for phytoplankton. Open circles, Califonia Current; large square, Caribbean; small open square, Monterey Bay; solid triangle, eastern tropical Pacific; open triangle, Mediterranean. (From Eppley, 1981.)

gen atom is likely to be recycled before it sinks out to deep water, incorporated into an organic particle; Eppley suggests values as low as <1 for upwelling, and as high as <20 in the open tropical ocean. Figure 5.3 shows how the first of these indices is related to total production over the range 0.2–1.5 gC m^{-2} d^{-1}, representing oligotrophic ocean to eutrophic coastal waters—the lower the total production rate, the more regeneration occurs. In areas of high regeneration rate, ambient nitrogen levels may be so low as to be unmeasurable, reflecting the almost instantaneous manner in which available nitrogen atoms are taken up by plant cells, and suggesting that the distributions of plant cells, heterotrophic bacteroplankton, and zooplankton must be very intimately related.

The simplest form of oceanic planktonic ecosystem occurs in those parts of the tropical ocean where the mixed layer is isothermal and lies above a single thermocline toward the base of which the depth of 1% light penetration occurs. The profiles for oxygen, nutrients, bacteria, plant cells, and zooplankton in such situations forms an ordered microcosm, with predictable depth relations between each profile. Such an ecosystem has been termed the Typical Tropical Structure (TTS) by Herbland and Voituriez (1979) from their Gulf of Guinea studies; its main features are illustrated in Fig. 5.4 for a site in the eastern tropical Pacific. Perhaps only in the Red Sea, where a thermally layered ocean typical of

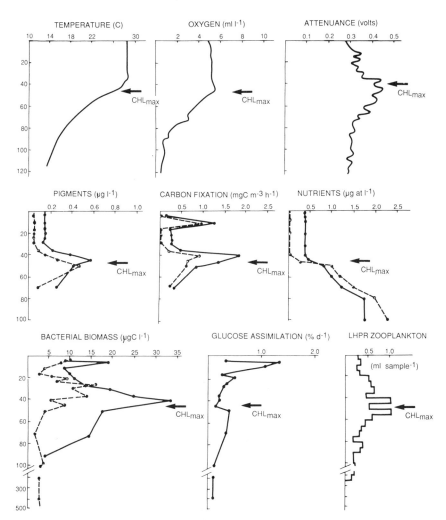

Fig. 5.4. Example of ecosystem profiles where the temperature structure is a Typical Tropical Profile: BIOSTAT station, eastern tropical Pacific, a stable situation northwest of the Costa Rica Dome. (From Longhurst, 1985b.)

the tropics is replaced by a relatively isothermal water column, is this generalization unlikely to be valid. It is, however, most remarkable that plankton profiles in the Red Sea are not very different from those generally typical of the tropical ocean, seeming to be ordered in the vertical plane principally in relation to light rather than temperature or density (e.g., Weikert, 1980).

A deep chlorophyll maximum is an almost universal feature in open ocean profiles, but varies spatially and temporally with consequent changes in the depth of maximum primary production rate. In the TTS, the depth of highest production is on the upper edge of the chlorophyll maximum. Herbland and Voituriez define the TTS in the following terms—the depths of the nitracline, oxycline, and chlorophyll maximum are statistically identical; this depth is determined by the maximal thermal gradient. Pigment concentration in the chlorophyll maximum, and integrated chlorophyll and primary production below unit surface area are negatively correlated with this depth. LeBorgne (1981) confirms that zooplankton carbon is also positively correlated with chlorophyll over wide areas of the open tropical Atlantic. So, in the tropical ocean, the deeper the thermocline, the deeper the chlorophyll maximum, the less chlorophyll and zooplankton, and the smaller the rates of primary and (probably) secondary production per unit area.

There is some evidence (Longhurst and Herman, 1981) that the depth of highest concentration of zooplankton is actually coincident with the depth of highest production rates, either because zooplankton follow a feeding clue to this depth, or because phytoplankton growth there responds to maximum rates of nitrogen excretion. Figure 5.5 shows regional profiles of zooplankton biomass in the eastern tropical Pacific Ocean, and how these respond to changes in the depth and thickness of the thermocline. Zooplankton vertical migration in the TTS, as we have already noted has two components: Interzonal migrants move from the thermostad into the thermocline and mixed layer at night, and there is some upwards zonal movement of epiplankton species. The deep chlorophyll maximum must be carefully analyzed in any ecological budgeting of the TTS both for chemical composition and particle size. In some situations it represents also the depth of maximum plant biomass, in units of carbon or nitrogen, but in others it may simply be a reflection of changed cell carbon/chlorophyll ratios.

It is not, of course, only the layers of high concentrations of plant cells (or of depths at which plant growth is especially active) that have significance in feeding relations for herbivores in the pelagic realm. The layers of high zooplankton and micronekton abundance are also the feeding sites for predators; a few studies have been done, mostly in colder seas, which

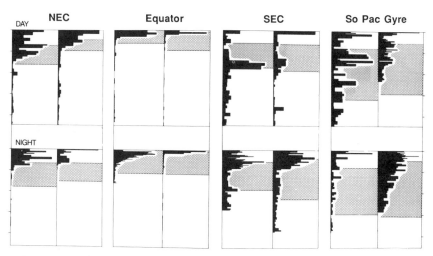

Fig. 5.5. Regional differences in zooplankton profiles as function of the thermocline topography induced by the zonal current system at about 110°W. NEC, North Equatorial Current; SEC, South Equatorial Current; So Pac Gyre, South Pacific Gyre. Profiles are of numbers of organisms, relative scale within each region. The thermocline region is shaded. (From Longhurst, 1976. Reprinted with permission from *Deep-Sea Research* **23.** Copyright 1976 Pergamon Journals Ltd.)

show that it is only in such layers of micro- and macroplankton that food particle abundance is higher than a critical lower threshold, below which the gathering of small particulate food is not energy effective. There is reason to suppose that it will be found that it is at such layers, clearly visible in echograms, that large pelagic tuna aggregate for feeding; large filter-feeding organisms like whale sharks, baleen whales, megamouth sharks, and other micronekton feeders probably utilize layers of plankton and micronekton often lying at thermocline depth in the tropics.

Larger organisms associated with a TTS form a clearly defined two-layered system based, respectively, on the epiplankton and the interzonal plankton and micronekton. Associated with the epiplankton, permanently occupying the depths above the planktocline, are tunas (as top predators) feeding on smaller fish (Gempylidae, Bramidae) and small euphausiids (especially *Stylocheiron*) themselves feeding on the various components of the smaller epiplankton described in the previous chapter. Lying deeper, in the daytime below 250 m in the eastern parts of tropical oceans, and below 450 m in the western, are the interzonal fish (Myctophidae and Gonostomatidae) and larger euphausiids (*Euphausia*). This population rises at dusk to feed nocturnally on the smaller species of the epiplankton,

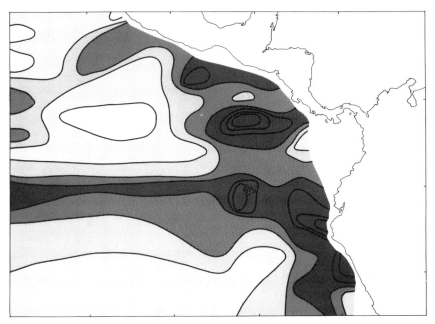

Fig. 5.6. Distribution of surface values of primary production of organic carbon by phytoplankton as mgC m^{-2} d^{-1}, isopleths at 100, 200, 300, 500, 700, 1000, and 1500 units. (Redrawn from EASTROPAC Atlas, Love, 1971.)

but are not themselves fed upon by tunas and fish of the surface layer, which cease feeding at night. As Roger (1977) and Legand *et al.* (1972) emphasize, this two-layered ecosystem is organized so as to transport energy downward below the euphotic zone, where it is utilized by the larger bathypelagic, nonmigrating predators and by the omnivorous and carnivorous deep zooplankton. Only in special cases, where some surface species (bigeye tuna, sperm whales) are involved, is there active transport of energy vertically upward by the feeding excursions of larger biota.

Figure 5.6 illustrates the variability in primary production in open-ocean ecosystems of the kind we describe here for the eastern tropical Pacific: Most of the area produces <200–300 mgC m^{-2} d^{-1}, with regions of equatorial and coastal divergence (where new production is important) rising to >500 mgC m^{-2} d^{-1}. Zooplankton grazing is in approximate balance with organic production in the open ocean, for grazing pressure not only co-occurs in the vertical plane with plant production, but over large areas of the tropical ocean, as we have noted above, there is a linear relationship between zooplankton and plant biomass per unit area (Le-Borgne *et al.*, 1983).

Given the range of mechanisms by which new production may be fuelled in the open ocean, it is not surprising that integrations of autotrophic production and heterotrophic consumption demonstrate a wide range of differently balanced budgets. However, in tropical oceans it is not usual to find production greatly in excess of consumption as frequently occurs in high latitude spring bloom situations, or in pulsed coastal upwelling ecosystem; indeed, some models, such as some of those proposed by Sorokin *et al.* (1985) for the Indian Ocean, appear to confirm that relatively few plant cells are lost to the epipelagic ecosystem of the oligotrophic ocean by sinking before they are consumed by herbivores.

Where dynamic processes disturb this equilibrium, and cause nitrate and other inorganic nutrients to be transported up into the euphotic zone, new production is intensified and the area supports larger standing stocks of pyto- and zooplankton than surrounding regions. Such is the case in the thermocline domes caused by anticyclonic gyres that we have already discussed, where the TTS is much modified and all profiles are highly variable in time and space. In the centers of midlatitude, subtropical cyclonic oceanic gyres, where the thermocline is deeper, sometimes much deeper, the TTS situation is stretched downward and new production is very low. The equatorial divergence has been much studied, especially by Soviet oceanographers, whose results make it clear that this ecosystem greatly resembles that of a coastal upwelling in low latitudes. There is considerable regional variability due to the decreasing intensity westwards of the physical divergence; from about 95°W to about 155°W biomass and production rates of both phyto- and zooplankton decrease, the percentage of nanoplankton increases, and the percentage of predatory zooplankton increases. Comparable changes occur poleward from the linear center of the divergence as the plankton ecosystem ages (Vinogradov, 1981).

The presence of oceanic islands has for long been known to enhance the biological productivity nearby; that is, their presence modifies the nature of the TTS. The shoreward gradient in productivity and zooplankton standing stock was termed the "island mass effect" by Doty and Oguri (1956). Since then, the effect has been noted in relation to a number of islands, especially in the Pacific: the Revila Gigedos, the Marquesas, Hawaii, and (farther west) Tahiti. It has also been noted for the Caribbean region. By far the best statistical evidence for the existence of this general phenomenon comes from a recent study by Dandonneau and Charpy (1985) who used the values of nearly 5000 randomly distributed surface chlorophyll measurements taken by merchant ships in the southwest Pacific, for each data point of which they calculated a distance to the nearest island. Near the equatorial divergence, the upwelling process masks the island mass effect, as does winter overturn south of the study area; the island mass effect is characterized by a negative slope for the chlorophyll concentration/distance to nearest island regression. Such a condition is most clearly seen in relation to high islands (Vanuatu, Fiji, Samoa), and Dandonneau and Charpy ascribe the effect principally to terrestrial run-off, since the southwest Pacific is a region of weak and variable currents.

LeBorgne and others (1985) made similar observations on the south coast of New Caledonia and at the isolated Mare Island in the Loyalty Islands. An island effect of increased chlorophyll biomass south of New Caledonia was ascribed to the effects of a wide lagoon and terrestrial run-off. At Mare, only mesoscale variability in chlorophyll could be detected rather than a sustained gradient in any direction from the island.

The principal mechanism by which an isolated, small island can increase biological activity in the surrounding waters, where the current regime is stronger and perhaps more predictable, is not far to seek, though there must be many individual variations on the theme. As Barkley (1972) showed for Johnston Atoll in the western Pacific, a simple vortex street can be created by a point-source obstruction to the linear flow of an ocean current. The simplest example of a vortex street is, of course, the flapping of a flag in a stiff breeze, as it sheds eddies alternately to left and right. The existence of a series of alternate cyclonic–anticyclonic eddies behind islands lying in ocean currents has now been confirmed abundantly by satellite imagery of sea-surface temperature, and an analysis of how such a system produces itense near-shore mixing zones near just on the rise of isolated islands has been demonstrated by Hogg et al. (1978) for Bermuda and by Osborn (1978) for Santa Maria, in the Azores. One of the striking consequences of the 1982–1983 ENSO was that the eddies normally lying westward of the Galapagos, and seen clearly as both a thermal and chloro-phyll signal in satellite imagery, were changed to a northeasterly orienta-tion. This simple explanation of a dimly perceived effect provides a satis-factory explanation for the biological basis for many island fisheries (Bakun, 1986).

ECOSYSTEM TROPHODYNAMICS: CONTINENTAL SHELVES AND COASTS

To understand the productivity of continental shelves and coastal eco-systems, we need to understand the differences (as for pelagic ecosys-tems) between new production, fueled by physical inputs of nutrients from sources external to the ecosystem, and regenerated production, fu-eled biologically by recycling of nutrients already accumulated in the ecosystem. In the open ocean, the sea floor is so far below the photic zone that we can usually ignore its influence when budgeting energy flow in the pelagic ecosystem but in coastal areas we must consider the influence of the sea bottom which may interact with biological production in the overlying water column.

Continental Shelf Nutrient Cycle

Where the water is sufficiently shallow for tidal mixing from the roughness of the sea floor to reach up into the photic zone, or completely to destroy the layered density structure of the water column, then benthic biota (from bacteria to starfish) play a role in the regeneration of nutrients. Benthic regeneration, comparable to the recycling of nutrients in the plankton ecosystem, is one factor in the high productivity of continental shelves (Rowe and Smith, 1977) and nitrogen sections across banks and onto continental shelves illustrate the production of compounds such as ammonia and urea by the benthic ecosystem and their utilization by phytoplankton and littoral algae. In fact, this benthic production can be measured directly by means of bell-jars, or experimental benthic chambers, placed on the sea floor (Zeitzschel, 1981). The extent to which nutrients (nitrogen is most usually limiting) are regenerated by the benthic community by decomposition of organic material produced in the photic zone above will depend entirely on the density and turbulence structure of the water column; where a two-layered structure exists, with a chlorophyll maximum within the discontinuity layer, benthic regeneration is constrained by the amount of turbulent mixing and eddy diffusion that occurs at the discontinuity layer. Where a well-mixed, single-layered water column occurs, the nutrients regenerated by the benthos are available with little delay to the plants of the photic zone. It is the dynamics between these two regimes that determines to what extent continental shelf productivity is enhanced by the continuous recycling of nitrogen between benthic and planktonic ecosystems; these dynamics also lie behind the fundamental importance of the fronts between the tidally mixed and thermally stratified regions of continental shelves that we have already discussed. Figure 5.7 shows how chlorophyll values are highest where the mixed and stratified waters meet, and how nutrient concentrations are low (because taken up by plants) in the upper layer on the stratified side of the front. Though we have no tropical examples of this enrichment process, it must occur everywhere on continental shelves wherever a distinction can be made between inshore mixed and offshore stratified water so that shelf fronts (Chapter 3) are formed; it is of great importance that this process should be looked for in tropical seas.

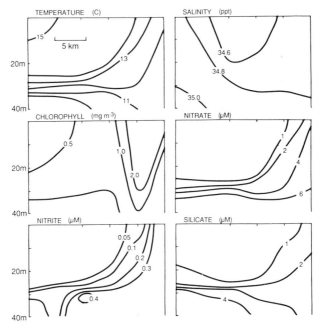

Fig 5.7. Sections through a tidally induced continental shelf front, to illustrate the surfacing of isopleths for nutrients within it and the chlorophyll maximum that occurs there. (From Holligan, 1981.)

Regenerated or biologically recycled production can function to slow the passage of nutrients through an ecosystem, and hence allow the build-up of biomass and diversity, but it must ultimately be fueled by the import of fresh nutrients to replace what are lost. Import occurs by mechanisms we have already described in physical terms: (1) from dynamic events at the shelf edge resulting from the interaction between tides and topography, (2) from river discharges, and (3) from coastal upwelling processes driven by local wind stress, long waves of distant origin, and interaction between currents and topography. More than one mechanism may be involved in transport of nutrients over a continental shelf, and then up into the photic zone.

First, dynamic processes at the shelf edge may, as we have discussed, cause nutrients to be injected onto the continental shelf, initially below the thermocline. This process may represent an important flux capable of accounting for the bulk of the nutrients demanded by the apparent level of new primary production that occurs on the shelf (e.g., Sandstrom and Elliott, 1984); this calculation is based on the nutrient content and volume of slope water pumped onto the shelf by the large amplitude internal waves and solitons, and on the nutrient content of the shelf water that it displaces.

Second, water discharged from rivers and lagoons tends to have relatively high turbidity, so that phytoplankton in the estuarine regions may be light limited. Consequently, levels of inorganic nutrients in estuaries may remain higher than ambient levels on the shelf, as we shall see when we discuss some examples of shelf ecosystems. In the northern Gulf of Mexico, shelf break and river discharge inputs of nutrients dominate the budget, but their effects can be distinguished (Walsh, 1983). The most extreme example of the consequence of changing river discharge by building dams and barrages is, of course, the now-classical collapse of the *Sardinella* fishery of the eastern Mediterranean when closure of the Aswan High Dam stopped the annual Nile floods, stopped the consequent annual bloom of phytoplankton to the east of the delta, and eventually caused the decline of the fish stocks of the region. Thus, the new Egyptian freshwater fishery in Lake Nasser was developed at the expense of the *Sardinella* fishery in the Mediterranean!

There is a general relationship between the amount of water discharged onto continental shelves and the production there of fish that has been studied particularly well in the Gulf of Mexico. Soberón-Chávez and Yáñez-Arancibia (1985) review this on a global basis and in greater detail for the Mexican states on the western coast of the Gulf of Mexico. Here, for several regions and over a period of 8 years, a positive correlation existed (Fig. 5.8) between both mean and annual river discharges and catch-rate of neritic fish on the adjacent shelf. Since these fish stocks are dominated by species completing much of their growth and reaching maturity in about 1 year, it is reasonable to suppose a causal relationship. This study was expanded by Deegan *et al.* (1986) for 64 estuaries on all coasts of the Gulf of Mexico, confirming the relationship between river discharge budgets, estuary area, intertidal area, and fishery harvest.

However, though rivers bring inorganic nutrients and suspended particulate organic carbon (POC) to assist in fueling the coastal ecosystems, the relationship is not simple. On some tropical coasts, as along the Gulf of Guinea or on the Sunda Shelf outside the Gulf of Thailand, the organic inputs at each small river mouth undoubtedly cause local enrichment of the shelf ecosystem. Where the discharge of organic material is very large, parts of the shelf ecosystem may be overloaded with reduced organic muds, as we have described for the Niger, Amazon, and Ganges. Off the Ganges, in the northern Bay of Bengal, this appears to result in a region of very low benthic productivity to the east of the mouths of the Ganges, but a very high standing stock of benthic organisms fueled by

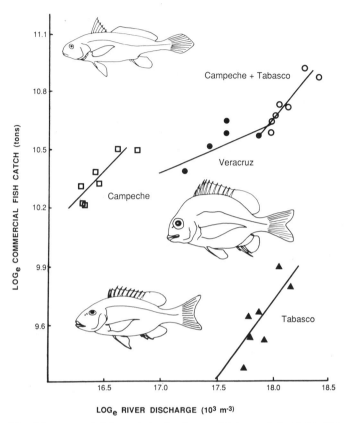

Fig. 5.8.   Mean annual fish catch in the demersal fishery of the Gulf of Mexico as a function of annual river discharge of freshwater in four subregions; the relationship is positive in each (Redrawn from Soberón-Chávez and Yáñez-Arancibia, 1985.)

riverine eutrophication to the west. As Parekular *et al.* (1982) show, the areas of highest benthic production in the Indian Ocean are fueled by the Indus and Ganges discharges and by coastal upwelling along the southwest coast of India, though Kurian (1971) showed that even here the largest aggregations of benthic biomass occurred near the mouths of creeks and rivers. This recalls the similar finding, on a much smaller geographical scale, of higher benthic biomass off the Gambia River in West Africa than on the shelf to the south, where only small rivers discharge onto the shelf (Longhurst, 1959).

The distribution of benthic biomass downslope on tropical continental shelves appears to reflect the importance of inshore, shallow water primary production, or the discharge of organic material from rivers, or both

TABLE 5.3.   Abundance Indices for Benthic Communities in Western Gulf of Guinea. Biomass as Wet Weight in Grams[a]

	Guinea–Sierra Leone			Liberia		
	Spp/sample	ind./m	g/m	Spp/sample	ind./m	g/m
Shallow shelf communities						
Estuaries						
Shelf	4.06	122.00	137.10	13.17	84.52	11.22
0–20 m	17.32	110.30	15.15	15.14	83.85	15.16
20–40 m	22.79	140.00	11.96	24.45	86.85	15.15
40–60 m	17.25	43.50	8.50	22.00	74.00	16.48
>60 m	5.00	20.00	4.76	—	—	—
Deep shelf communities						
50–100 m	5.50	16.00	0.85	19.00	72.00	12.76
100–150 m	7.00	45.10	3.02	23.60	41.50	3.02
150–200 m	8.00	72.00	7.04	5.50	31.00	1.51
200–250 m	3.00	15.00	3.50	8.00	22.00	2.90
250–350 m	2.00	23.00	3.03	—	—	—

[a] From Longhurst (1959).

possibilities. Off Sierra Leone, there is a striking break in the biomass values (Table 5.3) where the shallow shelf communities, which resemble Petersen communities, change to the communities of the deep shelf. This depth corresponds approximately to the bottom of the thermocline and hence the seawards limit of the inshore mixed water column where benthic regeneration of nutrients can be directly utilized by phytoplankton production. Kurian (1971) noted the same phenomenon along the southwest coast of India, where the benthic biomass of the deeper parts of the shelf is very low in the 50–100 m zone.

The third source of nutrients is, of course, coastal divergence and upwelling, and it is this mechanism that has been most studied. There is a large literature on coastal upwelling studies and these have been dominated by studies in low latitudes. As we have already discussed, coastal upwelling is a complex process and Fig. 5.9 shows how different the possibilities are, in several geographical regions, for bringing nutrients into the coastal photic zone from below the offshore thermocline. Much of the current understanding of phytoplankton physiology has come from studies of the behavior of algal cells in upwelling water. A recent study of how nutrients are upwelled and utilized in the Peruvian upwelling (MacIsaac et al., 1985) encapsulates much of this knowledge. It shows how four stages in utilization can be recognized in any mass of upwelled water. In stage I, as the water first reaches the photic zone, phytoplankton cells also reaching the surface are shifted down, that is to say they are physio-

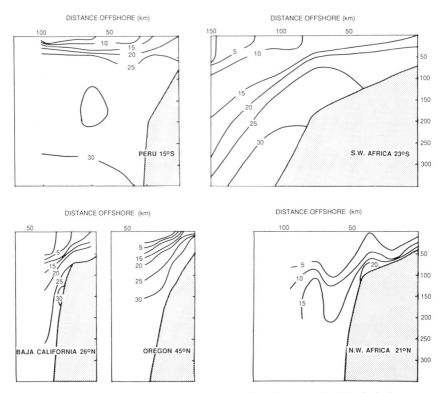

Fig. 5.9. Typical temperature sections normal to the coast in all principal eastern boundary current upwellings to illustrate regional differences. (From Walsh, 1977, "The Sea: Ideas and Observations on Progress in the Study of the Sea" (E. D. Goldberg, ed.), Copyright © 1977 John Wiley & Sons, Inc.)

logically adapted to low nutrient demand, and growth is slow in this stage; in stage II, they undergo light-induced shift up to a physiological state of high nutrient demand; in stage III, growth rates are maximal, organic macromolecules are rapidly synthesized, and nutrient levels in the water are as rapidly depleted; finally, in stage IV, nutrient levels are low, plant cells shift down again to lower rates of nutrient uptake and growth, and a phytoplankton population based on nutrient recycling develops. The whole of this sequence normally occupies about 10 days; the duration of the initial stage of light-induced shifting up is about 3–4 days after the cells have reached ambient light levels of about 50% of surface irradiance, though this may be reduced depending on the physiological state of the upwelled cells—this itself depending on the light regime they have recently experienced while being transported toward the coast below the upwelled and offshore-directed surface water. There is also some sugges-tion that the maximum rate of nutrient uptake achieved in stage III is determined by the initial nutrient concentration; this appears to determine the difference in production rates achieved, for example, between North-west Africa and Peru.

The planktonic herbivores associated with upwelling areas must de-velop a life history strategy that enables them to inject a viable population sufficiently early into the upwelling plume as to utilize the production of plant cells before they sink out of the photic zone and are lost to the benthic ecosystem that occurs below the upwelling region. This is espe-cially critical where the upwelling process is not continuous, which is most often the case. The herbivorous calanoid copepods that dominate the zooplankton in its herbivore stage utilize diatom blooms in recently upwelled water by maintaining a population of resting late copepodite larvae in the water below the photic zone during the nonupwelling season. This population migrates to the surface waters, feeds, and reproduces as soon as upwelling occurs. Mensah (1974) has described this process for *Calanoides carinatus* off Ghana, expanding the earlier descriptions for the same species off Nigeria and Ghana by Bainbridge (1972); C5s rise to the surface in the upwelling period, feed heavily on diatoms and dinofla-gellates, moult into adults, of which the females feed heavily on the same diet and reproduce almost at once. Young copepodites (C1–C4) feed in the surface waters, and C5s descend to deep water at the shelf edge, there to maintain themselves (we have to suppose) on bacteroplankton until the next upwelling event. Smith (1984) has observed essentially the same process, for the same species, in the Somali current, where it is a good indicator of upwelling centers. Here, *Calanoides carinatus* lives deep during the northerly monsoon, but in the southerly monsoon occurs in great numbers off Ras Haroun (10°N) and at the 5°N upwelling centre. In the Baja California upwelling system, at about 25°N, the same process has been found for *Calanus pacificus*, which occurs as dense layers (up to 370 m^{-3}) of C5 copepodites in the oxygen minimum layer below the upwelling region during the nonupwelling season (Longhurst, 1967b). Undoubtedly, herbivores with life history strategies of this kind must exist in all tropical, intermittent upwelling systems, and should be looked for there.

Coastal Ecosystems of Level Sea Bottoms

The relative roles of benthic and pelagic components of coastal ecosys-tems are clearly different from those in the open ocean, and the two components are much more intimately coupled over continental shelves; classically, it had been assumed without any direct evidence that production of plant material and its consumption by herbivores was essentially in balance over continental shelves and that it was the fate of every plant cell to be consumed by a herbivore, this being thought to constitute the fundamental difference between marine and terrestrial ecosystems. On land, it is the fate of most plant material to be decomposed by bacteria and fungi. However, in fact, recent studies have shown that this difference exists between many coastal ecosystems (where production and consumption are unbalanced) and oceanic ecosystems (where they are more closely in balance). Heinrich (1970, 1977) made it clear that, in his view, tropical neritic ecosystems are essentially unbalanced as between plants and herbivores, phytoplankton being abundant and underutilized by herbivores. Though comprehensive studies of all components of ecosystems are still lacking for the tropics, much of the available evidence points in this direction; Smayda (1966) for the Gulf of Panama, Qasim (1973) and others for the Cochin lagoons and estuaries, Flint and Rabelais (1981) for the Gulf of Mexico, Longhurst (1983) for the Sierra Leone estuary and continental shelf, and Walsh (1983) for the coasts of Texas and Peru. All these studies concluded that pelagic herbivory could account for no more than a small part of phytoplankton production, often no more than 10–20%, and that the fate of the remainder was to be consumed by the invertebrates, protists, or bacteria of the benthic ecosystem or to be slumped off the shelf to the deep-sea deposits, as also suggested by Joiris et al. (1982) with reference to differences between the southern and northern North Sea.

What is most difficult to quantify, as Walsh (1983) attempted to do for the eastern coast of North America, is the extent to which the phytoplankton cells which sink to the continental shelf are utilized by bacterial respiration, or are exported by turbidity flow to the continental slope and the deep ocean floor. What is known of the formation of slope sediments during geological time, and of hydrocarbon deposits within them, suggests that export of organic material must generally occur from continental shelves. On the other hand, a recent attempt (Rowe et al., 1986) to quantify this process on the continental shelf of New England failed to demonstrate the passage of exported organic material across the shelf-break front. Instead, though the study confirmed that the spring phytoplankton bloom exceeds the immediate demands of all herbivores by a factor of at least 2, the unconsumed plant cells can just balance heterotrophic demand by benthos, plankton, and bacteria during the remainder of the year. This process is thus clearly not yet well resolved and will undoubtedly be much studied in the coming years.

We have not yet discussed the mechanisms by which material is lost from continental shelf ecosystems, mechanisms whose existence would cause the system eventually to run down if new material were not imported to balance losses. Such export can be by downstream advection of biological material along the shelf or out over the deep sea, or it may be by the sinking of organic material onto the sea bottom, there to accumulate (as we have described for some tropical shelves as banks of thixotropic reduced mud) or to slump over the continental edge to the deep sea floor.

We shall now examine two specific examples from tropical coasts to show the complex and important coupling between benthos and plankton, without an understanding of which there can be no quantification of the relations between autotrophic production, and the production of fish and other top predators in tropical ecosystems of continental shelves. Our two examples will be the ecosystem of the continental shelf and estuary of Sierra Leone, West Africa, at about 8°N, and the Peruvian coastal upwelling at 5–15°S.

For the Sierra Leone coast, an analysis is available (Longhurst, 1983) of benthic–pelagic coupling and balance between producers and consumers to which we shall refer briefly and return to in Chapter 8. The study models three components of the coastal ecosystem: the estuary of the Sierra Leone river, the inshore sands, and the inshore muddy regions of the continental shelf.

Table 5.4 shows the principal characteristics of the three ecosystems. The standing stocks of all consumers and predators are weighted by the actual contribution to biomass of individual species of plankton, benthos, and fish. From this information can be calculated rates of growth in common units for all the components of the three ecosystems: for all herbivores, a demand factor for the phytoplankton that is required to support their growth was calculated from K_1, gross growth efficiency, and from this it was possible to sum the total demand by herbivores (plankton and fish) against the productivity of phytoplankton for each ecosystem and season. As Table 5.5 shows, the production of phytoplankton greatly exceeds the demand by all herbivores; it is only in the wet season estuary, when plant growth is greatly reduced by low solar radiation and nonphytoplankton turbidity, that production and consumption are approximately in balance and that otherwise between 72 and 90% of production must be exported to a sink on the continental shelf or in the deep ocean.

This conclusion is very similar to that of Walsh (1983) who has modeled the Gulf of Mexico shelf ecosystem within the influence of the Mississippi river (Fig. 5.10); here, a high rate of primary production of 100 gC m^{-2} y^{-1} drives an ecosystem also containing an abundant phytophagous clupeid fish (*Brevoortia*) comparable with the population of *Ethmalosa* off Sierra Leone. On an annual basis, 77% of the primary production in the northern

TABLE 5.4. State Variable and Forcing Functions for Sierra Leone Coastal Ecosystem Model[a]

| | | Estuary | | Continental shelf | | | |
| | | | | Mud | | Sand | |
		Wet	Dry	Wet	Dry	Wet	Dry
Forcing functions							
Mixed layer temperature	(°C)	27.00	28.60	26–26	28–29	26–27	28–29
Mixed layer salinity	(S ‰)	19.70	31.30	31–34	35.00	31–34	35.00
Surface radiation	g-cal cm^{-2} d^{-1}	113.00	437.00	113.00	437.00	113.00	437.00
Transparency	(k)	0.79	1.00	0.20	0.20	0.20	0.20
Phosphate	g-at P l^{-1}	0.43	0.54	0.38	0.38	0.20	0.20
State variables of biological systems							
Phytoplankton	mgC m^{-2}	612.00	3281.00	540.00	819.00	540.00	819.00
Zooplankton herbivores	mgC m^{-2}	9.00	21.00	11.00	66.00	11.00	66.00
Zooplankton omnivores	mgC m^{-2}	28.00	21.00	27.00	33.00	27.00	33.00
Zooplankton predators	mgC m^{-2}	13.00	20.00	19.00	42.00	19.00	42.00
Benthos	mgC m^{-2}	460.00	460.00	1240.00	1240.00	1400.00	1400.00
Demersal fish, mobile prey	mgC m^{-2}	104.00	104.00	126.00	126.00	30.00	30.00
Demersal fish, benthic prey	mgC m^{-2}	156.00	156.00	189.00	189.00	40.00	40.00
Clupeid fish, filter feeders	mgC m^{-2}	714.00	714.00	—	—	—	—
Clupeids, zooplanktivores	mgC m^{-2}	—	—	501.00	501.00	—	—
Predatory fish	mgC m^{-2}	—	52.00	63.00	63.00	63.00	63.00
Fish, benthic algal browsers	mgC m^{-2}	20.00	20.00	—	—	—	—

[a] From Longhurst (1983).

TABLE 5.5. Biological Production (mgC m^{-2} d^{-1}) and Relative Utilization of Phytoplankton by Herbivores, Sierra Leone Coastal Ecosystems[a]

| | | | Continental shelf | | | |
| | Estuary | | Mud | | Sand | |
	Wet	Dry	Wet	Dry	Wet	Dry
Biological production by each trophic group						
Phytoplankton	73.00	835.00	277.00	450.00	277.00	450.00
Zooplankton herbivores	6.00	14.00	5.00	26.00	5.00	26.00
Zooplankton omnivores	19.00	14.00	11.00	14.00	11.00	14.00
Zooplankton predators	9.00	14.00	7.00	17.00	7.00	17.00
Benthos	3.00	3.00	9.00	9.00	4.00	4.00
Demersal fish, mobile prey	1.10	1.10	1.30	1.30	0.30	0.30
Demersal fish, benthic prey	1.60	1.60	2.00	2.00	0.40	0.40
Clupeoid fish, filter feeders	6.50	6.50	0.00	0.00	0.00	0.00
Clupeoids, zooplanktivores	0.00	0.00	4.60	4.60	0.00	0.00
Predatory fish	0.00	0.50	0.70	0.70	0.70	0.70
Fish, benthic algal browsers	0.10	0.10	0.00	0.00	0.00	0.00
Relative utilization of phytoplankton production						
By pelagic ecosystem	5	92.4	24	15	24	15
By benthic ecosystem	1.4	16.4	8	3.2	3	5
To organic carbon sink	89.6	−8.8	68	81.2	72.5	78.8

[a] Output from simulation model outlined in Fig. 5.4, data stratified by region, substrate and season. (From Longhurst 1983.)

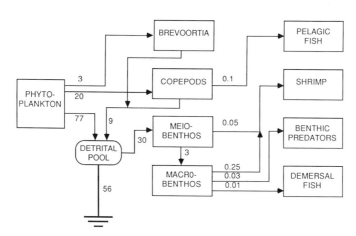

Fig. 5.10. Carbon-based Gulf of Mexico ecosystem model showing flux between trophic groups, and emphasizing the detrital food chain based on excess of phytoplankton production over herbivore consumption; units are gC m^2 y^{-1} Pelagic fish box includes all except *Brevoortia*. (From Walsh, 1983. Reprinted with permission from *Progress in Oceanography* **12**, Copyright 1983, Pergamon Journals Ltd.)

Gulf of Mexico sinks directly to the bottom, and 56% of the total (or 73% of what sinks) is not utilized by the benthic community and is available for burial or export.

The Sierra Leone data may also be used to analyze the relative role of the benthic and pelagic food chains, to the extent that they can be considered separate systems, in utilization of primary production. Benthic and demersal fish biomass is approximately equal to plankton and pelagic fish biomass in the estuary, but significantly higher on the inner continental shelf; however, benthic biomass turns over much more slowly than pelagic biomass, and the tabulated data show how pelagic exceeds benthic production, as we shall discuss in Chapter 8. Thus, despite the overproduction of phytoplankton material for planktonic production, and its direct translocation into the demersal food chain, the relative productivity of the benthic food chain is constrained by other factors, presumably inherent to the kinds of biota of which it is composed or caused by predation pressure by demersal fish.

The balance between production and the consumption by herbivores in many tropical coastal ecosystems often appears to resemble the Sierra Leone example and the Gulf of Mexico model. For the Gulf of Panama, Smayda (1966) estimated that zooplankton grazing could consume only a small fraction of the daily production of phytoplankton, less than 10% of the phytoplankton standing stock being grazed daily; the major seasonal phytoplankton production peak that occurs in January–March during the upwelling season is not reflected in a peak in zooplankton biomass. The Gulf of Guayaquil, 800 miles to the south of Panama, also appears to resemble the Sierra Leone estuary in its pelagic production system; though the data obtained by Stevenson (1981) will not permit an exact comparison, it is reasonable to interpret his results as another example of phytoplankton overproduction. In the Cochin estuaries, Qasim (1970) calculates that more than 90% of phytoplankton production sinks to the sediments as cells and fecal pellets, and that only 24% of it is consumed by planktonic herbivores. Qasim comments on the very high organic content of the surficial sediments, and describes the rapid growth of *Metapenaus* fed solely on sedimented organic material; he therefore concludes that a very high proportion of phytoplankton production is utilized by the benthos and alternate food chains. The opinion is reinforced by Madhupratap *et al.* (1977), who emphasize the importance of the benthic–demersal food chains in the utilization of phytoplankton production on the Cochin coast. Flint and Rabelais (1981) reach the same conclusion for the coast of the Gulf of Mexico and, independently of Walsh, suggest that penaeid production there is based on the detrital pool, itself related to "excessive primary production," as it may also be in West African lagoons.

The Peruvian upwelling ecosystem is one of the most intensively studied regions of the oceans; not only Peruvian biologists from the Instituto
del Mar del Perú, in Callao, but also oceanographers from many foreign
institutes have worked there, sometimes with their own ships. The incentive, of course, has been to try to understand how to manage the fishery
for *Engraulis ringens,* which was at one time by far the largest single-
species fishery in the world. Walsh (1983) has reviewed and modeled the
entire ecosystem, emphasizing the coupling of its benthic and pelagic
components, and the paths by which phytoplankton energy is dissipated.
Though there is a great deal of spatial variability over the 50 × 2000-km^2
area of upwelling, Walsh assumes an overall production of phytoplankton
biomass of 1000 gC m^{-2} y^{-1}; using assumptions for the food requirements
of the each of the four components of the herbivorous plankton (nauplii,
small and large copepods, euphausiids) and based on observations of their
abundance and distribution, he calculates an "ingestion flux" for this
component. Similarly, he calculates a demand rate for phytoplankton by
the stocks of partly phytophagous anchovy existing before and after the
collapse of the population, and for the ecosystem farther to the south
where anchovy were always relatively less abundant, and large stocks of
Sardinops occur.

Figure 5.11 shows that in the northern region, when *Engraulis* was
abundant, Walsh believes that most primary production was grazed down
by anchovies and plankton herbivores, reaching the continental shelf only
as fecal pellets; he believes that only 32% of the plant material reached the
sediments in this manner, the rest being accounted for by respiration and
growth of the pelagic component of the ecosystem. Further, he calculates
that only a very modest 8% was unutilized by the benthos and in bacterial
respiration, so being available for burial or export.

However, as Fig. 5.11 also indicates, off the southern coast of Peru,
where *Engraulis* was not abundant, and herbivorous plankton was the
principal pelagic herbivorous component, the balance was quite different.
Now, almost half the total production passes directly by sedimentation to
the detrital pool, so that when added to what sinks as fecal pellets, about
60% of all carbon fixed by phytoplankton was exported by turbidity currents to the deep sea or was buried at the edge of the shelf. To support his
model, Walsh gives direct evidence, from experiments at sea, that in the
absence of anchovies 40–81% of the primary production was lost or unac-
counted for each day.

Ecosystem Trophodynamics of Coral Reefs

Coral reefs have seemed paradoxical since the early naturalists first
wondered how such an exuberant and diverse mass of life could be sup-

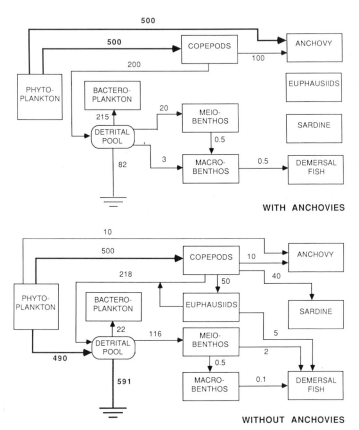

Fig. 5.11. Trophic group analyses based on carbon flux models of Peruvian Current pelagic ecosystems with and without a major *Engraulis ringens* population; it is suggested that in the former case the flux of material through sardines and euphausiids is very small. Units are gC m^2 y^{-1}. (From Walsh, 1983. Reprinted with permission from *Progress in Oceanography* **12**, Copyright 1983, Pergamon Journals Ltd.)

ported by the clear tropical ocean waters surrounding them. As we shall now discuss, the answer to this paradox lies partly in an exquisitely complex recycling of material within the reef, so that organic material is gradually accumulated by the community as a whole, partly by the capture of new nitrogen from the surrounding ocean and partly by the fixation of atmospheric nitrogen by the primitive marine plants that are a major component of the ecosystem.

Coral reefs, as we have already discussed, are a complex network of plants and animals with intimate relationships between so many species that the community really defies detailed quantitative analysis. The relationships include many plant–animal, and animal–animal symbiotic relationships, and it is only in recent years (see, for example, Lewis, 1977, 1981; Muscatine and Porter, 1977) that even broad agreement has been reached as to the principles guiding the balance between production and consumption in reef ecosystems.

As Lewis has shown in his reviews, there is now reasonable agreement that primary production levels on reefs may be very high: a mean value of about 7 gC m^2 d^{-1} (Kinsey, 1979) is to be expected and a range of about 1.5–14.0 gCm^2 d^{-1} encompasses most estimates, which is 1–2 orders of magnitude higher than we expect from the phytoplankton in the waters surrounding the reef, and can be compared with the highest levels of carbon fixation in terrestrial ecosystems. The very high values (>17 gC m^2 d^{-1}) that Smith (1981) records for the Houtman Albrouhos Islands can be attributed to the extent of three-dimensionality in the thickets of branching *Acropora* which cover these reefs. Production on coral reefs is partitioned between many forms of plant life: macroalgae (*Sargassum*) and calcareous green algae (*Halimeda*), filamentous cyanophytes forming a superficial layer especially in dead coral rock, mats of filamentous cycnophytes on consolidated sediments, coralline red algae in a variety of situations, sea-grass meadows in sheltered parts of lagoons, having an association with cyanophytes in their root-masses, and, finally, the wide occurrence of symbiotic dinoflagellate (*Gymnodinium microadriaticum*) cells within the tissues of scleractinian corals, giant clams (*Tridacna, Hippopus*), as well as alcyonarian, anthozoan, and scyphozoan coelenterates. A prokaryotic green symbiont (*Prochloron*) occurs within the tissues of several species of large ascidians and in many sponges on coral reefs. The general nature of these symbiotic relationships is, of course, that the host organisms receive carbohydrates from the algal cells, which themselves receive nitrogen and phosphorous compounds, and other molecules essential for photosynthesis from the soft tissues of the hosts.

Dependence on the algal cells in their tissues, of course, restricts the growth of the hosts to the shallow, sunlit parts of the reef and induces morphological adaptations in corals that we shall discuss below. The giant clam (*Tridacna*) is a highly evolved cockle (Cardiacea) in which the umbones and shell hinges have migrated 180°, so as to expose to sunlight the special tissues that carry the dense masses of algal cells. Light control in *Tridacna* is highly sophisticated: groups of hyaline organs, evolved from the cardiacean eyes, focus light on the algal-bearing tissues while shading is accomplished when required by special pigmented folds of siphonal tissue. In addition to utilizing the products of algal photosythesis that appear often to be available in amounts in excess of the hosts requirements and that lead to very fast growth, *Tridacna* also filter-feeds in the normal manner. It is a spectacularly successful mollusc, individuals weighing as much as 150 kg. Sadly, *Tridacna* is now rare over much of its range through overexploitation for its shells (Munro and Gwyther, 1981). Recent studies by Chalker and others (1983) have shown that hermatypic corals can be regarded for some purposes as plant communitites having light- and shade-adapted forms, whose biochemistry is finely tuned to the light regime of their habitat. Light saturation (P-I) curves for *Acropora* colonies on the Great Barrier Reef, collected over a range of depths representing 3–83% of incident illumination, show quite clearly that the compensation intensity, the intensity at which photosynthesis is 95% light saturated, and the respiration rate all decreased with decreasing light intensity. Chalker's studies suggested that the absolute values for saturating light levels and maximum photosynthesis rates show that endosymbiotic algae within the tissues of hermatypic corals are, in fact, photosynthetically intermediate between sun and shade floras of other plants, while adjusting to the light regime from the surface down to about 50 m.

Reef-flat species of coral are restricted to those able to resist the effects of UV-A and UV-B (280–400 nm) light. For those whose tissues must be transparent to accommodate symbiotic algal cells, resistance to UV light is achieved by incorporation of compounds in their tissues that selectively absorb light in the 310- to 340-nm range.

However, not only are there floristic/faunistic and physiological adaptations to varying light levels, but also morphological. As Vareschi and Fricke (1986) describe for the Red Sea coral *Plerogyra sinuosa,* some species have special expandable tentacles which form baloon-shaped vacuoles with high concentrations of zooxanthellae that increase the effective photosynthetic area of the coral by a factor of about 5. Such vacuoles are deployed in daylight hours and retracted at dusk. Pigment concentration is similar, but vacuoles are larger with increasing depth (or decreasing light intensity), and adapt themselves rapidly to changing levels of illumination. Vareschi and Fricke interpret this in terms of a simple adaptational mechanism to optimize light exposure of the zooxanthellae.

A large percentage of the organic material complexed by endosymbiotic algae in corals during photosynthesis is translocated, principally as glycerol, glucose, alanine, and leucine and is available to be taken up by the tissues of the host coral; up to 40–60% of the complexed organic material may be translocated in this way (Taylor, 1969; Muscatine and Porter, 1977) and the level of translocation itself appears to be controlled by, or is stimulated by, the coral tissues themselves. Muscatine and Porter (1977) calculated, from a set of assumptions concerning algal and coral respiration, carbon assimilation, and translocation, that translocation might provide for all coral respiratory requirements for basal and reproductive metabolism. A study of *Pavona praetorta*, an Indo-Pacific coral, by Wethey and Porter (1976) shows that only 30% of the carbon fixed by its zooxanthellae would satisfy the carbon needs of the coral organisms. On the other hand, zooplankton feeding, some of which comes from the benthic zooplankton community rather than plankton from the water passing over the reef, appears not to supply more than 10–20% (and sometimes a great deal less) of the daily energy requirements of coral animals (Johannes and Tepley, 1974; Porter, 1974).

An additional complexity makes it even harder to quantify the role of algal metabolites in the energy balance of corals; zooxanthellae have an essential role in facilitating skeletal formation by carbonate and other crystal growth in the tissues of the coral animals. Calcification occurs up to 10 times faster in the light than in the dark (hence the daily growth rings in coral skeletons) and very slowly in corals without symbiotic algae. Various authors (see Lewis, 1981) have attributed this process either to algal metabolites stimulating calcification or to the action of algae in reducing CO_2 and coral metabolites in calcification centers.

Though, as we shall see in Chapter 8, holistic models of coral reef carbon budgets are not fully satisfactory, there is a growing body of evidence to suggest that carbon fixation by plants on coral reefs is not nitrogen limited, even in an oligotrophic ocean atoll. On some reefs, primary production is dominated by the algal turf in which cyanobacteria (or blue-green algae) are an important component, and harvests of 30 g m^{-2} d^{-1} (dry weight) have been recorded at ambient nitrogen ($NO_2 + NO_3$) levels of <0.2 $\mu g.l^{-1}$ (Adey and Steneck, 1985). Such production must depend on fixation of molecular atmospheric nitrogen by cyanobacteria within the algal mats; consumption of algal turf by herbivores and regeneration of nitrates then supplies nitrogen to algae unable to fix it from the atmosphere. That such algae are not nitrogen limited is indicated by the existence of a nitrate gradient across reefs and nitrate export from the back-reef (Wiebe *et al.*, 1975; Wilkinson and Sammarco, 1983; Adey and Steneck, 1985). It is also possible that nitrates are supplied from within the reef by flow of interstitial water from land-based aquifers in the case of some fringing reefs (D'Elia *et al.*, 1981) or, in the case of atolls based on submerged volcanic peaks, as at Mururoa, by the process of endo-upwelling described by Rougerie and Wauthy (1986). Heat flow from the volcanic base drives interstitial water, originating as subthermocline, nutrient-rich seawater, upwards to the living structure of the atoll covering the fractured mass of limestone below.

Even if nitrogen is not apparently limiting, at least on some reefs, an atmospheric source for phosphorus is not available. As Adey and Steneck (1985) calculate, however, a sufficient input exists for at least some reefs just in the constant flow of oligotrophic ocean water across the reef; for this element at least, the model of Ryther (1959) for lack of nutrient limitation on reefs may be correct.

As Grigg *et al.* (1984) discuss, it is probably not ambient nutrient levels that limit the level of primary production on coral reefs, since there is a nonlinear response of production rate to the addition of artificial inorganic nutrients (Kinsey and Domm, 1974). Rather, they believe it is the extent of the photosynthetic surface that is the major factor: the greater the surface area provided by tree-like branching of corals, the greater the photosynthetic rate and the more fish can the reef support.

Algae, cyanophytes, and coelenterates together dominate both the community structure and the production of new organic material within it; coelenterates, mostly corals, are joined by a great variety of other invertebrates and fish in the transformation of this plant material into new animal material, and in the utilization of external sources of organic material, POC and zooplankton, in the ocean water passing across the reef. The complexity and diversity of this community of secondary producers is legendary; one living coral head, broken up and sorted through, may yield 1000–2000 specimens of 50–100 species of invertebrates. A total of 1441 specimens of 103 species of polychaetes (Grassle, 1973) is a remarkable, but perhaps not unusual, example. Porifera, Sipunculoida, Polychaeta, Mollusca, and Crustacea form the principal inhabitants of the interstices in coral colonies and these, together with the coral species themselves, include many filter and suspension feeders on particulate organic material (POM) and plankton. The intricacy of the internal energy transfer mechanisms is extraordinary, and includes the utilization of organic exudates between large organisms. As Hammond and Wilkinson (1985) have discovered, an epibiont holothurian (*Synaptula lamperti*) occurs at high densities (<200 m^{-2}) on the surface of sponges on the Great Barrier Reef, especially the large horny sponge *Ianthella basta*. The holothurians directly ingest organic exudates from the surface of the sponge. It is remarkable, considering the efficiency of corals as plankton traps, that exceedingly dense ($<1.5 - 10^6$ m^{-3}) swarms of specialized copepods apparently existing largely on POM in the form of coral mucus can occur in the water over coral reefs (Hammer and Carleton, 1979), as we have already noted. The direct consumption of plant material is performed by other benthic invertebrates, mainly echinoderms (especially *Diadema*), crustaceans, and gastropods. Herbivorous fish, especially Scaridae and Acanthuridae, browse on plant material, especially algal mats and mats of cyanophytes.

Another remarkable example of mutualism in the coral ecosystem is the synchronic reproduction of many coral genera and species. Babcock *et al.*

(1986) observed mass spawning of 105 coral species on the Great Barrier Reef on only a few nights in the year so that spawning products (comprising spherical egg–sperm bundles) were released at times of low-amplitude tides, and between sunset and moonrise. This does not include planulaters, or coral species which brood their larvae, and also appears not to occur on Red Sea reefs, but has been reported to a lesser degree in the Caribbean. Not only will such synchrony maximize the chances of fertilization, but planktonic predators over and around the reef will be saturated, and a high proportion of eggs will survive. Such observations recall the well-known reproductive behavior of the palolo worm (*Eunice viridis*), a polychaete whose reproduction is timed to occur precisely on two nights of each year, at the October full moon, over Samoa–Fiji region reefs. The reproductive portion of the worm appears in immense numbers at the sea surface to release eggs and sperm, swamping predation and ensuring fertilization.

A wide spectrum of predators comprise the third major component of the coral food web. The dominant group is, of course, the corals themselves, within which a wide range of plankton-catching polyps occurs; it is the coral species with the largest polyps that are most capable as plankton-catchers, and they mostly catch small crustacea (copepods, zoeas, other decapod larvae, etc). Coral polyps are also able to ingest bacteria and POM from the water passing over them, and can obtain POM from the surface of the sediments around them by means of extruded mesenterial filaments, and finally are capable of taking up dissolved organic materials from sea water directly. However, it is significant that their stable carbon isotope composition rather closely resembles that of their endosymbiotic algal cells, whose translocated metabolic products appear to dominate the growth of coral polyps.

As Odum and Odum (1955) calculated in a classical paper on reef energetics, the producers, herbivores, and predators on a reef form an Eltonian pyramid of the anticipated proportions—herbivores were about 19% of the biomass of primary producers and predators were about 8% of the biomass of herbivores. At the base of this Eltonian pyramid of numbers, the meiofauna of nematodes, harpacticoids, turbellarians, gastropods, and other minute interstitial fauna provides a significant source of recycled nitrogen within the reef. Gray (1985) found from $2-12$ mg N m^{-2} h^{-1} flowing from this source at several Great Barrier Reef sites.

The balance between gross production and community respiration suggest a P/R ratio of 0.6–2.6 ($\bar{x} = 1.43$) for 14 sites listed by Lewis (1981). Such a value is supportive of the concept of an open, steady-state ecosystem in which the number of mechanisms, including many cases of mutualistic endosymbiosis which retain nutrients within the system, is so great that a spectacular standing stock of biomass can be built. It is the very complexity and beauty of a reef that enables the engimatic situation of such biological richness in an apparently oligotrophic ocean to be maintained.

Where the reef ecosystem is larger than an isolated atoll, coral bank, or fringing reef around a small high island, another dimension is added; as we have already noted, upwelling events occur at the outer edge of the Great Barrier Reef off Queensland, and these have consequences quite different from the simple passage of ocean water across an isolated atoll. This upwelling (Andrews and Gentien, 1982) delivers nutrient-rich water from the continental shelf thermocline directly to the entire width of the reef so that it is immediately available to the living coral polyps and their symbiotic algae. The nitrogen flux established in this way is of the order 20 μg at 1^{-1} and is capable of supporting about 175 gCm^2 y^{-1} which, as Andrews and Gentien point out, is an order larger than expected for phytoplankton.

So far we have considered only inputs and recycling; the open ecosystem also loses material downstream that cannot again be recycled internally. Apart from the removal of biomass by such top predators as are migratory and visit the vicinity of reefs to prey on the multitude of small fish around the coral, yet which cannot be considered as residents within the ecosystem, the main export is as POC, which passes out of the lee drainage channels mostly, it seems, in the form of flocs of mucus. There is some uncertainty in the literature as to the extent to which this material can be utilized by small fish and plankton but it appears to comprise the principal export currency of reefs.

Chapter 6

Species Assemblages in Tropical Demersal Fisheries

In this chapter we shall discuss the communities of fish on which tropical fisheries are based, and the extent to which fish communities follow the same pattern as the communities of benthic and pelagic invertebrates from which they obtain their food. We are, however, at a double disadvantage in discussing fish communities because the number of regional surveys of fish communities in which fish have been treated as individual species, rather than as marketable components, is very small compared with regional surveys of invertebrates. Another disadvantage is that a large proportion of the numerous demersal surveys conducted in various tropical areas have remained undocumented or reported only in the form of summary reports not permitting community analysis. This applies especially to the earlier surveys conducted before the start of large-scale trawling. What follows, then, describes the fish communities of the tropics in what was probably their natural state, before heavy fishing pressure caused the modifications to which we shall be able to allude, though not describe in detail; if there is any merit in this situation, it is that we can assume that given sensitive management, or the relaxing of fishing pressure for some other reason, tropical fish communities would tend to revert to the natural state that we describe. That is, we assume that there is no natural stable state that resembles the present fishery-induced structure of many demersal fish communities. The principal agents for ecological change in tropical fish communities have undoubtedly been the rapid development of small inshore trawling fleets that supply urban centers with fresh fish, and the rapid growth of trawl fishing for tropical shrimp (mostly *Penaeus*) with the use of small mesh trawls and the capture and rejection at sea of very large numbers of demersal fish—the by-catch problem.

DEMERSAL FISH OF THE CONTINENTAL SHELVES

As we shall see, the nature of demersal fish communities of continental shelves reflects some of the same principles that we discussed for benthic communities: an equatorward gradient of increasing diversity in species composition, an east-west gradient in zoogeographical diversity with far fewer taxa in the eastern Atlantic than the western, and with many families being represented in the western Indo-Pacific that are absent in the Atlantic. Similar zoogeographic forces appear to have established the nature of demersal fish communities and the benthic communities of the tropical continental shelf.

The fish fauna of tropical continental shelves is remarkably consistent, and many of the same families are represented over similar deposits, and in similar water masses, throughout the tropics. The environmental factors chiefly determining what fish occur in an area seem to be the amount of organic mud in the bottom deposits, the occurrence of isolated patches of rocky or biogenic reefs, the occurrence of brackish, estuarine conditions associated with coastal lagoons and river mouths, and the nature of the oceanic water mass lying over the continental shelf. The fish fauna of coral reefs is the acme of fish community diversity, an extension of the fish fauna associated with scattered rocky reefs on open continental shelves.

Elasmobranchs are almost everywhere quite diverse and often abundant; apart from the large, mostly pelagic sharks that are not usually taken in demersal fishing, several small species of Squalidae occur in all tropical oceans. Especially in brackish conditions, large saw-sharks (Pristiophoridae) occur in the Indo-Pacific, and saw-fish (Pristidae), related to rays, occur in all three oceans; these great fish are regularly taken in trawls and other demersal gear in estuarine fisheries and some even occur in freshwater in some regions. Several families of bottom-dwelling Rajidae occur throughout the warm oceans: 45 species of 9 genera of guitarfishes (Rhinobatidae), 35 species of 9 genera of electric rays, some with very reduced vision, (Torpedinae), >100 species of 6 genera of rays (Rajinae), 35 species of 5 genera of stingrays (Dasyatinae) especially in brackish water, and finally 20 species of 3 genera of round rays (Urolophinae). Though some of the genera of Rajidae also occur in subtropical seas, most are restricted to tropical seas. Because of their low fecundity, elasmobranch populations are especially sensitive to fishing pressure.

There is a great diversity and range of genera among tropical Clupeiformes, not only in the pelagic ecosystem, but also in the benthic–demersal ecosystem of continental shelves. It is not really satisfactory to attempt a formal separation between pelagic and bentho-pelagic fish, because clupeids and engraulids very frequently occur in demersal trawl catches in tropical seas: several species of *Ilisha, Pellona, Pellonula, Opisthonema, Opisthopterus* and *Sardinella,* and *Stolephorus* dominate this component of the tropical fish resources. Associated with these clupeids are two groups of eel-like bentho-pelagic predators: the pike eels (Muraenesocidae) with 7 species of 3 genera mostly in the Indo-Pacific, and the unrelated hairtails or cutlass-fish (Trichiuridae) with 17 species of 6 genera of very elongated, silvery fish in all tropical oceans.

One family of the mostly freshwater siluroid catfish occurs on tropical continental shelves, where individuals are often abundant; numerous genera and species of sea catfish (Ariidae, Tachysuridae) occur inshore, and many species are estuarine. *Arius, Bagre,* and *Tachysurus* are the dominant genera of this group; many, or most, species are reddish or dark brown in color, and have dangerous barbed spines as the first rays of their pectoral and dorsal fins. Deckhands on tropical trawlers have to know how to deal with the wounds they inflict on bare, uncautious feet. Small catfish of the genus *Plotosus* may also be very abundant in and around coralline areas.

Two groups of Scorpaeniformes occur abundantly in the tropics, usually on sandy bottoms: Triglidae include 14 genera of benthic-feeding fish, often with sensory rays anteriorly on the pectoral fins, and some of which (e.g., *Trigla, Lepidotrigla, Prionotus,* etc.) occur in all three tropical oceans; Platycephalidae include 22 genera which occur today only in the Indo-Pacific, where some (e.g., *Platycephalus*) may be very abundant.

Three families of Myctophiformes include abundant continental shelf species in the tropics that are significant fisheries resources, though the greatest importance of the family in the tropics is in the bathypelagic fauna of the open oceans: Synodontidae (= Synodidae) include 4 genera with 34 species of which some (e.g., *Saurida, Trachinocephalus*) are abundant in Indian, Pacific, and Atlantic Oceans; Harpodontidae comprises one genus (*Harpodon*) with three species occurring abundantly in rather brackish water in the Indo-Pacific where they are known as Bombay Duck; finally, one of the three genera of Chlorophthalmidae, or greeneyes, may be abundant on deep continental shelf areas in all three tropical oceans.

Of approximately 150 families of Perciformes (the largest group of fish, and the largest vertebrate order), about a dozen families dominate the demersal fish of the continental shelf in the tropics. In some regions, where the tropical surface water is an especially shallow layer, other families more typical of the subtropics occur deep on the continental shelf right through the tropical latitudes. The tropical Perciformes of continental shelves may conveniently be grouped into three ecological types, though the range of morphology is relatively small—species associated with inshore muddy deposits, those of sandy bottoms of the continental shelf, and those associated with more rocky ground.

On muddy, inshore grounds, where the water tends to be brackish and turbid, the fish fauna is often dominated by Sciaenidae (28 genera, 160 species) throughout the tropics. Drums and croakers (e.g., *Sciaena, Pseudotolithus, Johnius, Umbrina*) may be viewed as the gadoids of tropical seas, and some attain large size; there is usually some specialization between fast bentho-pelagic fish-eating species, and benthic-feeding forms with inferior mouths and sometimes with gular barbels. Occurring with these croakers are golden-brown threadfins (Polynemidae, 7 genera, 35 species) with inferior mouths and many elongated, sensitive rays anteriorly on the pectoral fins, after the manner of triglids; as with sciaenids, in each region one species usually reaches very large size. Laterally flattened spadefishes (Ephippidae, 15 species of 7 genera), silvery or grey in color, though the young almost always bear vertical dark bars, also occur everywhere in this fauna.

On sandy grounds, a wider range of perciform families occurs, generally tending to be pink or silvery in color; these are active, large-eyed species adapted to clear water and lighted conditions, feeding mostly on benthic epifauna and the vagile benthos, especially decapod crustacea. Breams or dorades (Sparidae, about 30 genera, about 100 species) in all oceans, threadfin breams (Nemipteridae, 3 genera and many species) only in the Indo-Pacific, grunts (Pomadasyidae, 21 genera, 175 species) in all three oceans, and a range of genera of mostly smaller fish—Priacanthidae, Mullidae, Gerreidae, and (in the Indo-Pacific) the slipmouths, slimies, or ponyfish (Leiognathidae, 3 genera, 20–30 species).

The resources of rocky grounds, both in estuaries and offshore in all three tropical oceans, are dominated by three families of large bass-like fish, known usually as groupers and snappers, many species of which may attain a very large size indeed and some may be highly colored. Typically, the mouth is large and the prey is mostly other species of fish. Serranidae has many genera (e.g., *Liopoma, Epinephelus, Mycteroperca, Paralabrax, Plectropomus, Serranus*) and more than 300 species; Lutjanidae has 23 genera (e.g. *Caesio, Lutjanus, Ocyurus*) and more than 200 species, some of which are responsible for carrying ciguatera toxins (p. 247); Lethrinidae has 2 genera (*Lethrinus, Neolethrinus*) and about 20 species.

Pleuronectiform flatfish occur throughout tropical seas, ranging from predatory halibut-like species to worm-eating tongue-soles. The predatory, left-eyed flounders Psettodidae (*Psettodes*, with 2 species in the eastern Atlantic and Indo-Pacific), Citharidae (2 genera, 3 species) and Bothidae (36 genera with 212 species) occur in most tropical seas. The right-eyed flounders (Pleuronectidae) have many species in all latitudes, and representatives occur on all tropical continental shelves. Thirty-one genera and 120 species of soles of the family Soleidae have many representatives in tropical seas, though the 5 genera and 90 species of tongue soles of the family Cynoglossidae are the most important tropical species that have a sole-like form.

Finally, a number of highly evolved genera are included in tropical continental shelf stocks. Balistidae (trigger fishes), Tetraodontidae (puffers), Ostraciontidae (boxfish), and Zeidae (John Dories), may all occur at times abundantly in trawl catches on the open continental shelf; these are outliers of the fish fauna of coral reefs where, as we shall see, most of the morphological evolution of teleosts has occurred and where an enormous diversity of form, function, and behavior exists.

In the following account of tropical fish communities, we shall identify four basic kinds of assemblages: the fish of inshore and estuarine muddy habitats and turbid water characterized by sciaenids; those of sandy deposits and clearer water characterized by sparids; those of rocky reefs characterized by lutjanids; and, finally, those of coral reefs with no single characterizing family. For Blaber and Blaber (1981) the fundamental distinction is rather between turbid water and clear water associations; there is merit in their emphasis for the interpretation of how estuaries are used as nursery areas, as we shall discuss below.

From this general description of the species common in tropical trawl-catches, we can now proceed to a more detailed analysis of the natural fish communities in each region of the tropical oceans. If management of a fishery is to be successful it must be include a recognition of the place of each target species in a multispecies fish community, so as to take account (so far it is possible in a practical management scheme) of interspecific relations. In confronting this problem, we must consider if tropical diversity raises a difference in kind, or only quantity, compared with temperate, low-diversity multispecies fisheries management, and, further, if the east–west gradient in diversity of landings from the eastern Atlantic to the western Pacific has implications for management principles. At its simplest, can we manage the Guianian fishery of the western Atlantic for 20 species of co-occurring sciaenids by the same principles as the fishery for five or six species off Nigeria in the eastern Atlantic in otherwise similar habitats?

TROPICAL ATLANTIC DEMERSAL FISH

Eastern Atlantic

For the same reasons as for benthic communities, we shall begin our
review with the Atlantic, and especially the eastern Atlantic: The fauna
has less diversity than anywhere else in the tropics, the region has been
better surveyed than any other tropical continental shelf, and what has
emerged reflects rather clearly some important general principles that
may be harder to discern in more complex situations in the Pacific or
Indian Oceans.

The continental shelf of the Gulf of Guinea was surveyed both piecemeal and comprehensively from 1950 to about 1966. There exist accounts
of formal resource suveys, with detailed biological data on fish communities, covering Guinea (Postel, 1955), Gambia and Sierra Leone
(Longhurst, 1963), Ghana and the Ivory Coast (Salzen, 1957), Togo and
Dahomey, now Benin (Crosnier and Berrit, 1966), Nigeria (Longhurst,
1964a), Cameroon (Crosnier, 1963), and the Congo, now Zaire (Poll,
1951). Finally, during 1964 and 1965, the detailed and comprehensive GTS
or Guinean Trawling Survey (organized by the Commission for Technical
Cooperation in Africa, of France, Britain, and Portugal) covered the
whole coast from Senegal to the Congo; two matched French trawlers
with international scientific crews made 1-h, depth-controlled trawl hauls
at eight standard depths covering the whole extent of the shelf on each of
63 transects arranged to give a 40-mile line interval along the 2500-mile
coastline of tropical West Africa; the 480-haul survey was duplicated for
the wet and dry seasons. Associated with each trawl haul were oceanographic measurements of surface and bottom temperatures and salinities,
water color and clarity, and oxygen. The total fish and invertebrate catch
from each haul was identified to species, counted, and measured; the
published data (Williams, 1968) undoubtedly represent the most comprehensive survey yet performed on a tropical fisheries ecosystem, and de-
serve to be much better known; in fact, this survey is far more comprehensive than is available for most temperate continental shelves, since it
is completely unbiased by relative fishing success, a bias which, as we
shall see, renders many other tropical fishery surveys difficult to interpret.

From the individual resource surveys, and even before the GTS was
undertaken, a concensus had emerged that the West African tropical fish
fauna could be represented as a set of communities, variously named by
different authors: depth (as an indicator of the water mass present over
the bottom) and the nature of the substrate appears to determine which

community occurs. In shallow water, on relatively soft bottoms, the fish fauna is dominated by Sciaenidae (especially *Pseudotolithus senegalensis*), and including Ariidae (*Arius heudelotii, A. mercatoris, A. latiscutatus*), Drepanidae (*Drepane africana*), Polynemidae (*Galeoides decadactylus*), Cynoglossidae (*C. senegalensis, C. monodi,* and *C. brownii*), and Pomadasyidae (*Pomadasys jubelini*), all generally brown, grey, or silver in coloration in keeping with the relatively turbid, muddy water of their habitat.

Somewhat deeper on the continental shelf, on sandy bottoms and where water clarity is much greater, a fauna of red or pinkish fish occurs, dominated by Sparidae (*Pagrus ehrenbergi, Dentex congoensis*), Triglidae (*Lepidotrigla cadmani*), Scorpaenidae, Platycephalidae (*Platycephalus gruveli*), etc. Deeper yet, at 200–300 m a deep shelf fauna of *Chlorophthalmus, Scomber* (yes, *Scomber*!), *Peristedion, Bembrops,* and *Antigonia* lies above the slope fauna that from 400 m downwards includes *Merluccius, Trigla, Halosaurus,* and others. It was also recognized that the fauna of rocky, deeply submerged reefs and banks differs from this pattern, and includes large red and brown snappers (*Lutjanus goreensis, L. agennes, Lethrinus atlanticus*).

It seemed possible, subjectively, that the inshore sciaenid fauna, which extends in modified form into the estuaries, is restricted to muds and muddy sands above the thermocline, and so within the tropical surface water; that the sparid fauna dominates on sandy, shelly deposits below the tropical surface water, and on some particularly shelly places in the warm water above the thermocline; and that the deep-shelf fauna is an outlier of the slope fauna below. Some of the later regional resource surveys, such as that off Nigeria, were planned deliberately to investigate the communities of demersal fish and their relationship to deposit and depth, and the general consensus was further formalized.

The detailed haul-by-haul distribution of the 200 most abundant species in the data from the GTS was analyzed using objective recurrent group analysis (Fager and Longhurst, 1968) for all depths, all transects, and both seasons. This analysis confirmed very closely the previous subjective analysis of fish communities, though estuarine regions were not sampled. The affinities between species groups within what had previously subjectively been identified as major communities were much stronger than between species groups in different subjective communities. With the addition of the estuarine community, which was not sampled by the GTS, we now know that six communities occurred off West Africa during the 1960s: an estuarine and a coastal sciaenid community having a muddy, turbid water habitat, and a lutjanid community in the warm water of the mixed layer and upper thermocline; a sparid community below the ther-

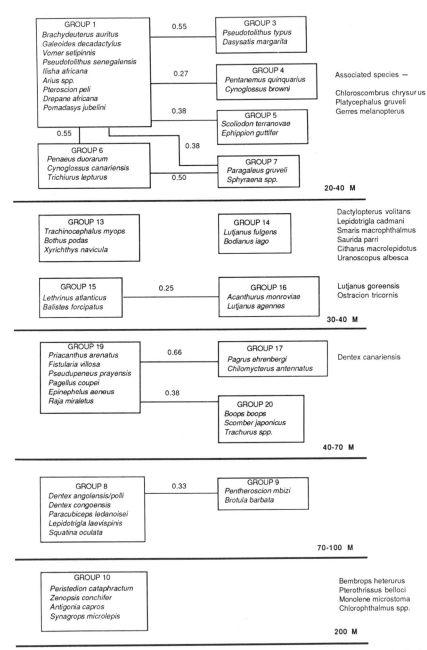

Fig. 6.1. Recurrent group analysis of demersal fish assemblages in the western Gulf of Guinea, stratified by depth, and showing relative affinity between species groups. (From Fager and Longhurst, 1968.)

mocline with deep shelf and slope communities lying still deeper (Fig. 6.1). The estuarine community is distinguished by a few important species substitutions: *Pseudotolithus typus* for *P. senegalensis* and *Pentanemus quinquarius* for *Galeoides decadactylus,* for example; the relative abundance of *Pteroscion peli, Arius* spp., and *Dasyatis margarita* are also greater than on the open shelf.

The recurrent species grouping showed minor associations of small clusters of species having ecological affinities related especially to depth and deposit, but also making sense in terms of interspecific interactions: it shows also a few species having no group affinities, often also having a eurybathic distribution. Statistical analysis of the depth distribution of each subgroup of two to three species showed that each had a favored, restricted depth range, above and below which it occurred randomly within the total depth range and over which it was viable. It was, as Fager and Longhurst (1968) put it, "the same sort of depth distribution one would expect from an underwater vehicle trying to maintain a certain depth, but subject to a random error in . . . depth measurement."

Figure 6.2 shows the practical consequences of the shallow layer of tropical surface water that occurs along the eastern margins of tropical oceans for the distribution of fishery resources. The warm surface water covers only the inner parts of the shelf, widening somewhat off Sierra Leone because of shelf topography and the presence of the Shoals of St.

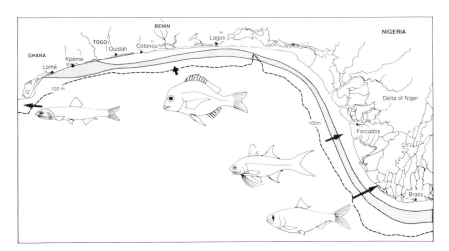

Fig. 6.2. Showing how the thermocline (shaded), the warm mixed layer, and the cold subthermocline water lie on the continental shelf of the eastern Gulf of Guinea. The fish symbols are illustrative of the recurrent species assemblages occurring in each indicated area. (Data from Williams, 1968.)

Anne and Sherbro Island, but very restricted along the whole coast east of Cape Palmas as far as the Bight of Benin. Only off the upwelling coast of Ghana is the pattern disrupted, and it is only here that the GTS (and earlier surveys) did not show a regular replacement with increasing depth of fish communities such as we have described. Off Ghana, the distribution of species is much less easy to analyze, probably because of the existence of a variable oceanographic situation and species distributions that shift in response to it.

There is now evidence that the heavy exploitation of the West African coastal demersal fisheries by fleets of small trawlers during the 1960s has caused a major change in the composition of the demersal stocks, though the evidence comes mainly from studies of a single species that has apparently become anomalously abundant. This is the trigger-fish, *Balistes carolinensis* (Fig. 6.3), which at the time of the GTS nowhere formed more than 5% of catches, and usually a much smaller proportion than that. As Cavarivière (1982) reports, this species began to proliferate off Ghana after 1971, as revealed by greatly increased landings in that country compared with landings of other species. By 1978, it had become an important element in the catches as far north as Senegal. It is now reported that from Senegal to Nigeria, this species forms as much as half of the demersal fish biomass in what used to be the biotope of the sciaenid community.

Balistes is a curious fish. Though basically demersal (or as we shall see, a reef-fish genus) it is reported to move into midwater at night to feed on plankton. It is a nest-builder, and there is some parental care of the reproductive products. Finally, it is extremely resistant to the rigors of being trawled and rejected at sea, its leathery skin slowing dessication, and its dangerous dorsal spine causing deckhands to search for it among

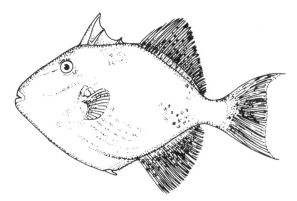

Fig. 6.3. *Balistes carolinensis*, Gulf of Guinea.

the catch and throw it over the side (alive!) before serious sorting begins. All this may predispose it to profit from the destruction of its competitors and predators, and to proliferate under conditions of heavy trawling pressure. The data, however, do not make it entirely clear if the supposed spread of the proliferation occurred as suggested, outwards from a center in Ghana. It would be very interesting indeed to have modern resource assessment data from the GTS region to confirm what has so far only been learned from catch statistics of commercial trawlers.

Analysis of the geographical distribution of the principal species in each community (Table 6.1) showed conclusively that there was a general correspondence between their bathymetric distribution within the tropical region, and their north–south distribution along the African coast. The species of the sciaenid community were, of course, confined to the tropical region between Cape Verde in Senegal and Cape Lopez south of the Congo mouth, though other species of sciaenid do occur inshore in the cooler water north and south, reflecting the general tropical to subtropical distribution of the family.

The distribution of the species of the subthermocline sparid fauna is strikingly different: There appears to be a clear case for equatorial submergence through the Gulf of Guinea of the sparid-dominated fauna from the cooler regions to the north and south. Sparids (*Dentex, Pagrus, Pagellus*, etc.) are important in the fish fauna from the Mediterranean southwards, and there is a remarkable similarity between the sparid fauna below the thermocline in the tropical region, and the fauna of the continental shelf off Mauretania–southern Morocco and Angola–Namibia. *Dentex canariensis, D. filosus, Pagrus ehrenbergi*, and species of *Upenaeus, Lepidotrigla, Smaris*, and *Brotula* all extend from cool north (Mauretania) to cool south (Angola) through the tropical region. Also

TABLE 6.1. Zoogeographic Relationship of Principal Species Comprising the Gulf of Guinea Demersal Fish Communities[a]

Tropical fish community	Tropical	Tropical/ temperate	Deep	Total
Sciaenid	15[b]	8	—	23
Lutjanid	5	1	—	6
Shallow sparid	1	4	—	5
Deep sparid	8	19	—	27
Deep shelf	—	2	5	7
Upper slope	—	2	14	16

[a] From Longhurst (1966).
[b] Number of species.

occurring in the subthermocline water in the Gulf of Guinea are some other typical eastern boundary current pelagic species: *Scomber japonicus, Trachurus trachurus,* and *Boops boops.* This submerged tropical community is strengthened by a few purely tropical congeners (*Dentex polli, D. angolensis,* and *Pagellus coupei*) while a few subtropical species do not penetrate equatorward into the Gulf of Guinea (*D. macrophthalmus, Pagellus mormyrus, Diplodus* spp.). The analysis of recurrent species groups also shows how a very few species of the tropical subthermocline sparid community have penetrated up through the thermocline and occupy hard sandy and shelly deposits unsuitable for the sciaenid community; these are *Pagrus ehrenbergi, Pagellus, Priacanthus,* and *Dactylopterus.*

Western Atlantic

Though there are more than twice as many species of tropical shore fishes in the western as in the eastern Atlantic (900 versus 380 species), and presumably about the same ratio in continental shelf species, the pattern of fish communites we have described for the Gulf of Guinea is very easily recognized in the trawling surveys published for the western Atlantic where the latitudinal extent of warm water, and its depth, is much greater. The demersal fisheries of the western Atlantic have been well surveyed, and the surveys have been well reported. The central tropical region between the Amazon and the Orinoco was explored in 1957–1959 (Mitchell and McConnell, 1959) and the results interpreted in an ecological framwork by Lowe-McConnell (1962), following earlier surveys from Trinidad to Guiana (Whiteleather and Brown, 1945). To this central study can be added other surveys off the Orinoco (Cervignon, 1965), Brazil (Lopez, 1964), the southeastern Caribbean (Richards, 1955), Venezuela (Simpson and Griffiths, 1967; Penchaszadeh and Salaya, 1984), and the Gulf of Mexico (Struhsacker, 1969; and contributions in Yáñez-Arancibia, 1985). Extensive, later surveys by the U.S. research vessel *Oregon* along the whole north coast of South America are also available as technical reports (see references in Struhsaker, 1969). Soviet-Cuban trawl surveys of Campeche Bank were reported by Sauskan and Ryzhov, 1977).

The detailed surveys off Guiana make it quite clear that two subsets of the sciaenid community exist along that coast that resemble the estuarine and coastal sciaenid communities of West Africa. Presumably because of the greater depth of the tropical surface water off South America, the inshore sciaenid community extended somewhat deeper than the Gulf of Guinea estuarine community, to about 60 m off northwest Guiana. In shallow water (<20 m) of high turbidity, on muddy deposits along the coast, the community is dominated by sea catfish (*Bagre, Arius, Selenaspis*), stingrays (*Dasyatis* spp.), flatfish (*Paralichthys, Achirus*), spadefish (*Chaetodipterus*), a threadfin (*Polynemus virginicus*), and 11 species of sciaenids (*Cynoscion, Macrodon, Nebris, Stellifer*, as well as other genera).

Further offshore, on firmer muddy deposits but still within the influence of the inshore turbid water, the fish are lighter in color and the community is dominated by three different species of sciaenids: *Micropogon furnieri, Macrodon ancylodon*, and *Cynoscion virescens*. Although *Arius* and *Dasyatis* still occur, they are less important and secondary to such forms as *Larimus breviceps, Nebris microps, Umbrina gracilicirrhus* (all sciaenids) and the carangid *Vomer*, the grunts *Genyatremus* and *Hemulon* spp., the threadfin *Polynemus*, and the spadefish *Chaetodipterus*. It is specially to be noted that of the 24 species of sciaenids that occur off Guiana, 14 are common or very common in trawl catches. A very similar set of species with only some differences (for example, the addition of *Cynoscion petranus* and *Umbrina coroides*) forms the sciaenid community as far south as Cape Frio; northwards, along the north coast of South America other important species are added to the Guiana list: two croakers (*Cynoscion jamaicensis* and *Ophioscion naso*), two drums (*Larimus breviceps* and *Stellifer stellifer*), and several catfish (*Arius* and *Notarius* spp.)

These tropical communities lying between the Orinoco and Amazon are separated by the reefs and sandy ground of the Caribbean region from the subtropical fauna of the Gulf of Mexico which is clearly a form of sciaenid community, in the sense defined for the West African coast, but has only a few species in common with the tropical South American community. Four species of croakers, and especially *Micropogon undulatus*, together with *Cysnoscion arenarius, C. nothus*, and *Leiostomus xanthurus*, dominate the subtropical sciaenid community. As in the tropics these are accompanied by a range of associated genera, recalling those of West Africa and South America—*Scoliodon, Chloroscombrus, Vomer, Selar, Ilisha, Dasyatis, Chaetodipterus, Polynemus, Trichiurus*, and *Bagre*. This community is best developed in the influence of the Mississippi discharge, within about 150 miles of the delta.

Farther south, Yáñez-Arancibia *et al.* (1985) have described a coastal fish community on level sea bottoms in the Gulf of Campeche that has the general characteristics of a tropical sciaenid community, though the occurrence of *Cynoscion* betrays that it is on the edge of the tropical region. In addition to small bentho-pelagic clupeids (*Harengula, Opisthonema, Cetengraulis*) and carangids (*Chloroscombrus*), the demersal fish are dominated by several species of sciaenids (*Cynoscion nothus, Stellifer colonensis*), polynemids (*Polydactylus octonemus*), ariids (*Arius felis*), flatfish (*Syacium guntheri, Bothus* spp., *Etropus crossotus*), cutlassfish (*Trichiurus lepturus*), as well as drepanids and others typical of this kind of fish community. The Gulf of Campeche fish fauna also seems to be something of a mosaic comprising a typical sciaenid community as described above, and some genera more typical of lutjanid and sparid communities: *Eucinostomus gula* and *E. argenteus; Haemulon aurolineatus* and *Stenotomus caprinus;* and *Upeneus parvus.*

The fish communities of the deeper parts of the shelf between eastern Venezuela and Cabo Frio are strikingly different from West Africa and there is very little evidence of submergence of the sparid fauna through the tropical region from cooler northern and southern regions. In fact, only about five species of sparids occur anywhere in the western Atlantic, and only a single species (*Calamus proridens*) was found in the trawl surveys off Guiana.

The muddy deposits and turbid coastal waters off Guiana are, as we have already noted, restricted to a coastal zone about 40 miles wide out to water variously 25–60 m deep; beyond this the deposits are principally sandy until at about 100 m a zone of nonliving coral is encountered, resembling that reported off Nigeria. Thus, the environment is not unlike that off the West African coast, but the fish communities deeper than the sciaenid fauna are quite different. Over the sandy deposits, beyond the influence of coastal mud and in clearer, green water, a fauna of pelagic species dominates: nine species of carangids, barracuda (*Sphyraena*), several clupeids, and some outlying lutjanids, from the typical lutjanid community occupying—as occurs on rocky reefs off West Africa—a Holocene coral reef along the shelf edge. There is a hand-line fishery for lutjanids near the shelf off Guiana and northern Brazil in 60–70 m of water: the red snapper, *Lutjanus aya,* together with smaller numbers of the blackfin snapper, *L. buccanella,* is a valuable resource throughout this region.

Does a sparid community, so prominent along the African coast, occur at all in the western Atlantic? In the subtropical communities of the continental shelf that occur at much higher latitudes than along the African coast, there is a clear analog of the African subtropical sparid community. Northwards from Florida, three sparids and some large sciaenids dominate the demersal catches, and replace the sciaenid community of the Gulf of Mexico, which extends somewhat onto the eastern coast of the Florida peninsula wherever suitable deposits occur. In this northern community, *Stenotomus capriscus* and *Archosargus rhomboidales* are the most abundant sparids, though *Pagrus pagrus* also occurs; the large *Cy-*

noscion regalis and *C. nebulosus,* together with several species of *Menti-cirrhus,* complete the analog. Exactly as on the African coast, this community extends poleward until it is replaced by a gadoid community, which occurs at about the latitude of New Jersey. West of the Mississippi influence, to Yucatan in the western Gulf of Mexico, a relatively small area of sandy and sandy mud deposits is occupied by a community in which sparids, especially *Stenotomus capriscus,* are dominant.

Although the evidence is less clear, the same situation seems to occur south of the tropical region, and south of Cape Frio. Along the coasts of Uruguay and Argentina, sparids again become important in the catches (Lopez, 1964) and we can assume that the importance of *Pagrus pagrus* and the closely related and ecologically similar *Cheilodactylus bergi,* together with a smaller abundance of large sciaenids similar to the northern subtropical species, is sufficient evidence of the occurrence there of a sparid community as we have defined it off western Africa. Thus, while it is possible to demonstrate the existence of northern and southern sparid communities in the western Atlantic Ocean, it is less certain to what extent this community occurs through the tropical region, apart from its apparent absence off northeast South America. Perhaps sparids and their associates, having a basically subtropical distribution, are unable to penetrate importantly through the tropics unless there is a combination of sufficiently cool, sufficiently lighted water below the thermocline. Thus, the greater depth of the Caribbean thermocline may explain why here the sandy deposits are occupied by pomadasyids, which are a tropical rather than a subtropical group. Pomadasyids on sandy bottoms, near reefs occupied by lutjanids and serranids, are an extremely important and characteristic component of the Caribbean demersal resources.

As we might expect from the greater relative area of reef, rock, and sand habitat in the western than the eastern Atlantic, the lutjanid community is much more relatively important than in the Gulf of Guinea, though it is rather more uniform than the corresponding sciaenid community, which is more regionally diverse than in the Gulf of Guinea. On the Campeche Bank, Sauskan and Ryzhov (1977) describe three statistical species associations in each of which lutjanids, sparids, and pomadasyids co-occur; the ''core species'' in these three groups are (1) *Haemulon aurolineatum, Lagodon rhomboides,* and *Calamus bajonado,* (2) *Ocyurus chrysurus, Haemulon plumieri, Lutjanus synagris, Anisotremus virginicus, Lachinolamus maximus, Priacanthus arquatus,* and *Aluter shoepferi,* and (3) *Priacanthus arenatus, Sphaeroides spengleri, Diplectum formosum, Chaetodon ocellatus,* and *Acanthostracion quadricornis.* The first two associations occur in shallower water than the third; though it is not easy to determine from the very attenuated description given of this survey, there does appear to be a similarity with the eastern Atlantic subtropical regions off Mauretania and Senegal where hard, sandy grounds are dominated by sparids, lutjanids, and associated genera.

Such associations of reef fishes, and of the sandy areas around reefs, occur very widely in the western Atlantic: on the Florida coast and keys, the Bahamas, all the Antilles and Cuba, and the coast of America from Yucatan to eastern Venezuela. As in the case of hermatypic corals, there has seemed to be a major gap in their distribution on the shelf influenced by the Amazon discharge so that the populations associated with the reefs on the Brazil coast south of the Amazon were effectively disjunct, though as we have noted, lutjanids are known from the deep shelf of Guiana. However, trawl surveys on the deep shelf of Amazonia at 40–80 m depth have shown the existence of many species of reef fish (or of the lutjanid community) on bottoms dominated by epibenthic sponges. Species of *Gymnothorax, Apogon, Epinephelus, Diplectrum, Serranus, Haemulon,* and other genera were found there by Collette and Rützler (1978) and this suggests most strongly that there is in fact no barrier to the distribution of reef fishes between Brazil and the Caribbean, as seems to be the case for hermatypic corals.

Thus, in the western tropical and subtropical Atlantic, we can distinguish four species assemblages of fish that comprise the demersal resources:

1. *Lutjanid community.* Fauna of rock, coral, and coral sand from Florida to Brazil, which is dominant on the Bahamas, the Antilles and other Caribbean islands, and on the coast from Yucatan to Panama. (Balistidae; Lagocephalidae; Lutjanidae, 14 spp.; Pomadasyidae, 3 spp.; Serranidae, 11 spp.; Synodidae.)

2. *Subtropical sciaenid community.* Fauna of soft deposits from the northern Gulf of Mexico to at least Cape Hatteras, and especially well developed near river mouths, including the Mississippi. (Branchiostegidae; Clupeidae; Gerreidae; Polynemidae; Serranidae; Sciaenidae, 9 spp.)

3. *Tropical sciaenid communities.* Fauna of the soft inshore muddy and soft sand deposits from the southern coast of the Caribbean to Cape Frio in Brazil. (Dasyatidae; Ariidae, 6 spp.; Clupeidae; Gerridae; Heterosomata, 3 spp.; Ephippidae; Pomadasyidae; Sciaenidae, 19 spp.)

4. *Sparid communities.* Fauna of the sandy and muddy sands of the subtropical regions north of Cape Hatteras and south of Cape Frio with very attenuated representation through the tropical region. (Ariidae; Carangidae, 8 spp.; Clupeidae; Mullidae; Sciaenidae; Sparidae; Synodidae, also known as Synodontidae.)

INDO-PACIFIC DEMERSAL FISH

Turning now to the vastness of the Pacific Ocean, from Panama westward to Mozambique, we are at a loss for data as comprehensive as that from the GTS. No trawling survey has covered a major part of the Indo-Pacific comprehensively and with full biological analysis as the GTS did for the eastern Atlantic. However, in the northern Indian Ocean and the archipelagic region to the east, there are sufficient accounts of resource surveys (some in the "grey" literature) to derive, piecemeal, a model of the distribution of demersal fish species in the Indo-Pacific.

There are several region-wide reviews of demersal fish resources, many published in the reports of the Food and Agriculture Organization of the U.N., and of its Regional Programs (i.e., the Indian Ocean Program, the Bay of Bengal Program, and the South China Seas Fishery Development and Coordination Program). For the whole Indian Ocean and Indonesian region, Shomura *et al.* (1967) and Cushing (1971b) reviewed the fishery resources, as did Rao (1969), and Hida and Pereyra (1966) for the Indian region; but it is from many individual works that we must try to construct a model of the fish species associations on which the fisheries are based. Because much of this literature is scattered, but is critical to the description of the demersal fish communities of the Indo-Pacific, it may be useful first to survey our sources.

For the Indian Ocean Banks, we have consulted early fisheries survey reports: Wheeler and Ommaney (1953) for Mauritius and Seychelles, Morgans (1964) for the North Kenya Banks, and Sivalingham and Medcof (1955) and Fernando (1974) for the Wadge Bank. For continental shelf fisheries, we have used the following sources among others: for Mozambique, Saetre and de Paula e Silva (1979); for the Gulf of Oman, Ali and Thomas (1979); and for the coast of Pakistan eastward to the Gulf of Kutch, Hussain *et al.* (1974); Kesteven (1971); Qureshi (1955). For the Indian coastline, there are many individual reports of trawl surveys, among which we have principally used Rao, Dorairaj, and Kagwade (1974) for the Bombay region north to the Kutch; Bapat, Radakrishnan, and Kartha (1974) for the Mysore coast; Narasimham *et al.* (1979) for Andhra Pradesh in the western Bay of Bengal; Bain (1965) for the north end of the Bay of Bengal; and Druzhinin and Hlaing (1974) and Pauly *et al.* (1984a) for the coast of Burma. Farther to the south, there is information on the demersal landings of the west coast of Thailand and Malaysia (Isarankura, 1978; Anonymous, 1974b), and the major studies of the Malacca Straits and South China Sea by Green and Birtwhistle (1927) and Ommaney (1961), and of the northern Australian region by Okera (1982) and Masuda *et al.* (1964). For the remainder of the South China Sea, and northwards, we have used Chang (1974) for the coast of Borneo, while for the Gulf of Thailand we have been able to refer to several studies: Ritrasaga (1976), Tiews (1962), Ruamrasaga and Isarankura (1965), and Menasveta and Isarankura (1968). Trawl surveys in the Philippines have been described by Warfel and Manacop (1950), and Manacop (1955). We have also used the numerous reports on demersal fisheries cited by Simpson (1982) and Menasveta (1970).

Taken together, the results of these studies do seem to reflect a series of demersal fish communities that resemble in general outline those that we have described for the Atlantic Ocean: unfortunately, it may now be too late for this to be done in detail, for the demersal fish stocks, except along the coast of Burma, are probably by now heavily modified by several decades of largely unregulated commercial exploitation by diesel trawlers. One factor that should be noted in the modification of tropical continental shelf fish stocks by trawling is the 8 : 1 ratio in landed value of penaeid shrimps (for export) and demersal fish (for local consumption). This has led to extensive trawling, directed at penaeids, throughout the tropics with rather fine-meshed trawl nets and on near-shore grounds, important as nursery areas for fish. Though in such areas the ratio of shrimps/fish in the catch is relatively high, compared with offshore, deeper grounds, it is precisely here where one might expect that shrimp trawlers might have their maximum impact on fish stocks by directly affecting juvenile survival and recruitment to the adult population. This, the "by-catch" problem in shrimp fisheries, is largely unsolved especially in regions, such as Southeast Asia, where the fish by-catch does have market value and there is often little incentive to develop savings gear to reduce the fish catch while maximizing the catch of shrimps.

There is evidence right through the tropical Indo-Pacific that on muddy deposits on continental shelves and sometimes restricted to inshore and brackish regions, a fish fauna occurs in which sciaenids are an important (or even dominant) component, and in which they are accompanied by rajids, ariids, polynemids, drepanids, pleuronectids, and a typical range of bentho-pelagic clupeids and carangids. The range of genera associated with this fauna is greater than in the Atlantic, and includes fish such as *Rastrelliger* and *Harpodon* (especially in the Bay of Bengal), and *Saurida, Lactarius,* and *Leiognathus* (both in the western Pacific and in East African estuaries and bays). On harder, sandy deposits right across the Indo-Pacific there is a fauna of bream-like fish (*Nemipterus, Pomadasys,* and others), together with snappers (*Lutjanus, Epinephelus,* and others), which occurs in its greatest development on rocky banks with coral heads; it shares many genera with the full development of this fauna on coral reefs, which is dealt with separately below. On the eastern and western margins of the Pacific Ocean, the relations between tropical and subtropical fish faunas closely resembles the situation in the Atlantic, with the sciaenid and lutjanid-dominated communities being replaced by subtropical assemblages of sparids and their associated genera.

Indo-Pacific Soft-Deposit Communities

The communities in which sciaenids are important appear to be most widely distributed on the relatively extensive, and relatively muddy (compared with some other Indo-Pacific regions) continental shelves of the Indian subcontinent. From the Indus delta and the Gulf of Kutch, and the wide shelf off the Gulf of Cambar, at least as far as Orissa and Mysore, the demersal fish fauna resembles a richer version of the Atlantic sciaenid communities. The dominant sciaenids are *Pseudosciaena diacanthus* and *Otolithoides brunneus,* and these and other species of sciaeneids comprise 41.3% of the demersal landings at Bombay, from trawlers ranging from the Gulf of Kutch to Goa (Rao *et al.,* 1974). Obviously, this percentage varies regionally and with depth; off Karachi and the Sind, the greatest abundances are at 30–50 m, and sciaenids extend to at least 75 m. On the shelf near the Swatch-of-no-Ground, sciaenids form as much as 60% of the catches and on the Kori Great Bank, where breams (*Pomadasys hasta*) dominate, sciaenids form only 5–10%. Polynemids (mostly *Polydactyclus quadrifilis*) form less than 1% of the biomass along the whole west coast of India, but locally (Gulf of Cambray, Divarka) they may form as much as 20%; grunts (*Pomadasys hasta*) regionally form less than 5% of landings but locally, as on the Kori Great Bank, they may form up to 35% of biomass. Ariid catfish and rays are usually present, and form between 5–10% of biomass, though elasmobranchs as a whole often make up nearly half of biomass. On the wide shelf areas of the northeastern coast of India, these faunas extend with no consistent changes in abundance as deep as trawl survey data are available, usually about 75 m.

Southwards, community composition appears progressively to change, and off Mysore the fish community is very different, beginning to resemble that of the western Pacific, as recorded in the integrated catches of survey trawlers: *Leiognathus splendens* and other slipmouths comprise as much as one-third of biomass, followed in importance by the *Ilisha*-like clupeid *Opisthopterus tardoore;* these two groups usually comprise half of all trawl catches. Sciaenids comprised less than 10%, and *Lactarius lactarius* is almost equally abundant. Pleuronectiformes, Polynemidae, *Arius, Platycephalus, Trichiurus*—all predictable groups in this kind of fish community—are present at less than 5% of total biomass. In none of the data we have found for the west coast of India is it possible to establish the real differences between fish communities on sandy, harder deposits (presumably including genera like *Nemipterus*, seen in the landing statistics) and the sciaenid communities that appear to dominate the region.

In the Bay of Bengal, landings on the Andhra Pradesh coast resemble those off northeast India quite closely; once again, sciaenids are dominant, at about 20% overall and with local and regional variation of the same scale as in the Bombay–Karachi region. Leiognathidae form less than 10% and *Lactarius* and *Opisthopterus* each less than 5%, about the same abundance as *Saurida, Cynoglossus, Psettodes,* and individual groups of elasmobranchs. Off the Ganges mouths, at the head of the Bay of Bengal, Rajiidae and Ariidae consistently dominate biomass, with sciaenids usually in third place, followed by polynemids and flatfish of several genera.

Proceeding eastward once again, we have better information for the coast of Burma from the FAO/UNDP surveys of the late 1960s and later surveys, including the extensive Norwegian investigations of 1979–1980, all reviewed by Pauly *et al.* (1984a). In the data from the FAO/UNDP surveys, sciaenids form 27% and *Pomadasys hasta* about 11% of all fish biomass, with Polynemidae, Muraenesocidae, Ariidae, *Ilisha, Leiognathus, Trichiurus, Psettodes,* and *Saurida* all being significant components. These form the inshore fish fauna, representing the fish biomass peak that occurs at depths less than 50 m, as also occurs almost throughout the tropics. Also present, at somewhat greater depths (50–100 m), are significant (though <5%) abundances of hard-bottom genera—*Lutjanus argentimaculatus, L. sanguineus,* and *L. malabaricus* (especially off the Irrawady delta), *Nemipterus japonicus* (especially on the Arakan coast); and *Pennahia macrocephalus* (especially on the Tenasserim coast).

The Burmese trawl surveys by FAO/UNDP included 87 identified species, which is an unknown fraction of the total actually present: these included 16 species of Carangidae, 6 Lutjanidae, 2 Serranidae, 3 Nemipteridae, and 2 Lethrinidae among the hard-bottom genera. Three species of Rajidae, 2 of Ariidae, 2 of Polynemidae, 6 of Sciaenidae, 2 Drepanidae/Ephippidae, and one of Psettodidae comprise the soft-bottom specialists typical of the east–west coast off the Irrawady delta.

The Burmese coast has been subjected so far to relatively light mechanized fishing pressure, and the relatively long period from the first surveys of the mid-1950s to the most recent in 1980–1983 are useful, as Pauly *et al.* (1984a) point out in confirming the relative taxonomic stability over decadal time scales of fish stocks on a tropical continental shelf in the absence of fishing stress. This is useful confirmation of what is really obvious, anyway: that the major taxonomic changes seen in the Gulf of Thailand (see Fig. 6.2) and other places are, in fact, the result not of natural community dynamics, but of heavy fishing pressure. On this coast, there are also interesting differences in the size-structured depth distribution of some genera, such as *Nemipterus,* compared with other parts of their range. These differences suggest that the low oxygen water that may occur over the deeper parts of the shelf results in shoreward crowding of fish stocks and restricts the normal ontogenetic bathymetric distribution of size classes. Nevertheless, there is in general a normal bathymetric progression of ecological types, from forms such as *Pomadasys, Arius, Dasyatis,* and so on in shallow water, and *Lutjanus* and *Nemipterus* at midshelf depths, through to *Priacanthus, Peristedion,* and deep prawns and rock lobsters at depths exceeding 100 m.

Off the Thai coast of the Andaman Sea, Leiognathidae dominated the virgin stocks, followed by Mullidae and Sciaenidae. The Malaysian west coast surveys confirm once again the general composition of the Bay of Bengal fauna, with sciaenids forming 15–25% of landings, followed by the usual characteristic group of other genera. As with all these resource-oriented surveys, great care has to be taken in drawing assumptions: the Malaysian surveys grouped almost half the biomass in just two categories, for instance, manure fish (presumably not for human consumption) and mixed fish. These categories probably conceal an abundance of such forms as *Opisthopterus,* recognized in the Mysore surveys.

Farther to the south, in the Malacca Straits, and the South China Sea, we can turn to the reports of Green and Birtwhistle (1927) and Ommaney (1961). For this region, resource survey data are well stratified, more exploratory, and less restricted to the assessment of grounds already known to be capable of supporting a commercial fishery. Menasveta (1970) is a very useful historical document concerning the fisheries exploration of this region and includes descriptions of many trawl surveys not discussed here, where we are especially interested in the communities of demersal fish on which the commercial surveys were made; very few sets of exploratory fishing data from the western Pacific were reported in a manner compatible with our purpose.

Based on more than 150 trawl hauls dispersed over the whole of the continental shelf of the South China Sea within the 200-m line, Ommaney showed that there were two distinct fish faunas. In one of these faunas, on muddy inshore grounds shallower than 40 m, mostly off Malaysia, the following typical sciaenid community families dominated the catches: Trygonidae, Ariidae, Lactaridae, Drepanidae, Leiognathidae, Gerreidae, Polynemidae, Sciaenidae, and Pleuronectiformes. The existence of this fauna is confirmed along the east coast of Malaysia by Thai–German trawl surveys (Menasveta, 1970) which found sciaenids, *Leiognathus,* and rays inshore of a more diverse fauna. We shall return later to Ommaneys second, offshore fish fauna.

The inshore fauna of sciaenids also extends along the western coast of Borneo, where it is typical of the region and data from seven coastal sites from east of Kuching to Sandakan on the north coast, facing the Sulu Sea, suggest that there is mutual exclusion between sciaenids (and their associated genera) and nemipterids and other fish of more sandy deposits.

For the Gulf of Thailand, which is the site of a large tropical trawl fishery that underwent explosive growth once its potential was understood and a suitable, light-bottom trawl was developed, good landing statistics exist for 40 categories from each of nine subareas over a period of 20 years; some depth stratification of the data also exists. These data enable us to understand the broad outline, if not the detail, of the fish communities originally present, and also how they have been modified by a decade of heavy trawl fishing. In 1961, soon after the beginning of trawling, the catch composition was clearly different from the Bay of Bengal and western India, and more closely resembled what seems to be the situation on soft bottoms in the Philippines region. Though the basic group of families (rajids, ariids, sciaenids, nemipterids, *Pomadasys, Cynoglossus, Psettodes*) were present in the Gulf of Thailand, the Indo-Pacific specialists *Leiognathus, Lactarius, Scopelopsis,* and *Saurida* were far more relatively abundant than in the other regions we have discussed; species of *Leiognathus* alone comprised 28.7% of all landed biomass in 1963. The fish fauna along the 500-km coastline of the western Gulf of Thailand is quite uniform. In only one of nine subareas were *Leiognathus* not ranked first in abundance, and throughout the whole Gulf the same 10 species groups occupied the first 10 places in abundance ranking of 40 categories. Only in the inner Gulf and off the delta of the Mae Nah Chao Phraya, is the ranking significantly different from elsewhere; here, rays, Mullidae, and *Scopelopsis* were about as abundant as *Leiognathus,* while carangids were the most abundant group.

Table 6.2 shows the striking changes that occurred in the catch composition in the Gulf of Thailand, paralleling but different from the situation in the Gulf of Guinea; already by 1972, the total abundance of fish and large invertebrates had dropped to 17% of the original catch-rate, and the largest component was now of squids (*Loligo*), which had previously been ranked lower than 10th in abundance. The apparent abundance of *Leiognathus* had dropped to about 7% of its original level, though still ranking second overall. This truly fundamental change in the biological resource base of a major tropical fishery has been sustained to the present time (Table 6.2) and has become a cause for concern by fishery managers

TABLE 6.2. Changes in Ranked and Relative Abundance of Demersal Fish in the Gulf of Thailand with the Development of Commercial Trawling[a]

	1963		1972		1982	
Rank	Fish	%	Fish	%	Fish	%
1	Leiognathidae	24.10	*Loligo* spp.	27.40	*Loligo* spp.	19.80
2	Carangidae	6.60	Leiognathidae	9.40	*Priacanthus* spp.	8.70
3	*Nemipterus* spp.	6.20	*Nemipterus* spp.	9.00	*Nemipterus* spp.	8.70
4	Sciaenidae	6.20	Carangidae	7.30	Leiognathidae	7.60
5	Mullidae	5.40	*Saurida* spp.	6.30	*Saurida* spp.	6.20
6	Rajidae	4.90	*Sepia* spp.	5.80	*Sepia* spp.	2.90
7	Gerridae	4.30	Mullidae	3.70	Carangidae	2.90
8	*Saurida* spp.	3.80	*Priacanthus* spp.	3.70	Lutjanidae	2.10
9	*Scolopsis* spp.	2.50	Brachyuran crabs	3.10	Ariidae	1.60
10	Ariidae	2.50	*Scolopsis* spp.	2.70	*Scolopsis* spp.	1.10
	Rel. abundance	100.00	Rel. abundance	20.88	Rel. abundance	17.42

[a] From Ritrasaga (1976) and M. Boonyubol (personal communication).

and ecologists everywhere. It is clearly of comparable ecological importance to the Gulf of Guinea *Balistes* explosion.

For the complex coastal areas of the Philippines, we have a number of surveys of demersal fish faunas, starting with the immediately postwar investigations of the U.S. Fish and Wildlife Service (Warfel and Manacop, 1950). This survey, though it concentrated mostly on the Lingayen Gulf, Manila Bay, and Guimaras Strait, was able to show the extent of soft-bottom fish faunas throughout the archipelago, stratified by depth. It is useful to note in the results of this survey of largely untouched fish faunas that the same families of fish are abundant right across the continental shelf from shallow water to 75 fathoms, the limit of trawling with small boats. Over this depth range, and for the Philippines as a whole, there was remarkably little change in the dominance of Leiognathidae (36% biomass overall), Pomadasyidae (8%), Dasyatidae (7%), Trichiuridae (4%), Lutjanidae (4%), and Sciaenidae, Mullidae, and Nemipteridae (each 3%). In fact, this is a typical tropical fish fauna on muddy and sandy bottoms, where fishing on coral has been avoided, and the lack of bathymetric change is almost certainly related to the great depth of the tropical water in this part of the Pacific. In some areas of the Philippines, such as the predominantly muddy Panay Gulf, the survey found special species assemblages—here the fauna was dominated by *Trichiurus* (30%), sciaenids (19%), *Nemipterus* (10%), and other mud-bottom forms that occurred significantly (e.g., Polynemids at 1%). Only in this area were slipmouths, lizardfish, and mojarras present in insignificant numbers.

More recently, a detailed study of the fish assemblages of the Samar Sea, from 20 to 90 m depth (McManus, 1985), has demonstrated that there is a depth-mediated faunal distinction between 30 and 40 m, and that this is independent of season and location (and hence substrate) within the Samar Sea. However, it appears that this faunal boundary does not separate faunas as different from each other as the boundary that occurs at about the same depth in the Gulf of Guinea.

Southward from the South China Sea, there are also indications of the existence of a soft-bottom community in the Gulf of Carpentaria and along much of the tropical coast of northern Australia in depths of less than 20 m that seems to include quite typical species; sciaenids are represented by *Johnius coiter* and *Johnieops vogleri,* and *Polynemus nigripinnins, Cynoglossus* spp., *Ilisha* spp., *Thrissa hamiltoni,* and *T. setirostris,* as well as *Harpodon translucens* also occur.

Finally, at the two far ends of the Indo-Pacific Region we also have some evidence of a similar fauna. Off Mozambique, trawling surveys on Sofala Bank and in Delagoa Bay appear (though the evidence is not very clear) to have sampled analogs of both sciaenid- and sparid-dominated communities on these continental shelf areas (Saetre and de Paula e Silva, 1979; Brinca *et al.,* 1981). Inshore, at <25 m, Sciaenidae (8–48% biomass) and Pomadasyidae (6–36%) dominate the fauna, together with Cynoglossidae, Ariidae, Mullidae, Polynemidae, Synodontidae, and other soft-bottom families. The dominant sciaenids are *Johnius belangeri, J. dussumieri,* and *Otolithes ruber;* the main grunts are *Pomadasys hasta* and *P. maculatus.* Offshore, from 25 to 50 m, goatfish (*Upenaeus bensasi, U. vittatus*), lizardfish (*Saurida undosquamis*), grunt (*Pomadasys maculata*), and sparids (*Nemipterus delagoa*) dominated the shelf fauna. A similar fauna, but with some species replacement, occurred deeper than 50 m. The general distribution of this fauna corresponds with increasing amounts of riverine muds from the Save and Zambesi inshore, and harder, sandier deposits offshore. Of course, as with other trawl surveys, area of high sand waves or coral heads were avoided, and this must bias the overall faunistic description.

In the eastern tropical Pacific, obviously isolated by many thousands of miles of open ocean from the western Indo-Pacific soft-bottom fish faunas, the same mix of genera appears to occur; in the few muddy bights and gulfs (Tehuantepec, Panama, Guayaquil) that occur along this coast with its very narrow continental shelf there is a fauna comprising 22 significant species of sciaenids, 13 ariids, 2 polynemids, 5 flatfish of various families, 7 pomadasyids, and so on. But it should be noted that *Leiognathus, Lactarius, Saurida,* and *Nemipterus* do not occur; this constitutes an important difference between the eastern and western Indo-

Pacific regions. This fauna appears to be not well investigated outside the coastal lagoons, perhaps because the demersal fisheries of the region are dominated by the imperative of the North American market for penaeid shrimp; fish are once again a by-product.

Indo-Pacific Hard-Bottom Communities

Turning now to the fish communities of hard, sandy, and somewhat rocky sea bottoms, so far as they can be described from the few surveys where such grounds can be distinguished in the data from the more trawlable, softer deposits, we shall refer especially to the surveys of the Indian Ocean Banks, to landing statistics from Oman, to some offshore surveys of the South China Sea, and to surveys along the north coast of Australia.

North of the Indus mouth, along the sandy coast of Baluchistan and westward through the Persian Gulf, we know that the sciaenid association is replaced with a fauna dominated by snappers and groupers. On the Omani coast, lutjanids, serranids, and lethrinids form 51% of landed demersal biomass, followed by sparids (9%) pomadasyids (5%), and nemipterids (3%)—exactly the dominant groups we would expect to see in a "lutjanid" community as defined in the Atlantic Ocean.

On the North Kenya Banks, which are the largest area of continental shelf of the East African coastline, line-fishing surveys by Morgans (1964) showed that the fauna of useful fish above thermocline depths is dominated by 14 species of the same families: *Lutjanus rivulatus, L. bohar,* and *Lethrinus crocineus* each alone contributing more than 10% to the apparent biomass; below the thermocline, the fauna of *Lutjanus, Epinephelus,* and *Lethrinus* species gives way to a fauna of smaller fish, which Morgans (1964) called the "pink-fish fauna," with *Cheimerius nufar, Pristipomoides microlepis, Polysteganus caeruleopunctatus,* and *Argyrops spinifer* being the most abundant species. The FAO/UNDP and Norwegian trawl surveys reviewed in Food and Agriculture Organization (1982) described Morgans' shallow-bank fauna also on the North Kenya Bank, but found a fauna dominated by small sharks and rays on the trawlable grounds of the narrow continental shelf off Ungama Bay. Southward, the banks around Pemba Island and the Malindi Bank, of much smaller dimension than the North Kenya Banks, have a reduced fauna of large Lutjanidae and a higher relative abundance of unfortunately undefined small noncommercial species. Still farther to the south, on St. Lazarus Bank off Mozambique, *Lutjanus bohar* dominates a typical rocky bank community of snappers, groupers, and sea bass (Saetre and de Paula e Silva, 1979).

Fisheries surveys were performed with hand-lines in 1948–1949

(Wheeler and Ommaney, 1953) of the Mauritius and Seychelles banks and the great arc of shallow banks between, including Nazareth and Saya de Malha, as well as Chagos 600 miles to the east and Aldabra an equal distance westward; this comprehensive survey identified 44 commercially useful demersal fish species, almost all of which were lutjanids, serranids, and labrids; of these, *Lutjanus civis* and *Lethrinus ramak* comprised more than 75% by number of individuals. With the exceptions of Mauritius, Rodriguez, and the Hawkins Bank in the extreme southern end of the survey area, the distribution of species was remarkably uniform for such a great area; this fish community, known only by its larger members, occupies the sandy flats at 15- to 60-m depths on the tops of these oceanic banks, which are otherwise dotted with coral heads, though in their shallower parts there are sea-grass meadows. However, it is important to bear in mind that this survey was done with very selective gear, and a glance at the systematic listing of the ichthyofauna, assembled for taxonomic purposes, for this region puts these results in perspective; of 322 listed demersal fish species for the Seychelles, 138 were of the same families (Serranidae, Lutjanidae, Lethrinidae, and Labridae) as were taken in Wheeler and Ommaneys survey, while there also exist at the Seychelles 46 Pomacentridae, 13 Balistidae, 42 Gobiidae, 24 Scorpaenidae, 37 Apogonidae, and 21 Chaetodontidae, all hard-bottom genera. In the same listing, there are no sciaenids, drepanids, or ariids, and only a single polynemid. More recently, the 1978 trawl survey of the Mahé Plateau by a Norwegian expedition (Anonymous, 1978b) found a typical bank fauna dominated by *Lutjanus sebae* and *L. lineolatus, Pseudopeneus seychellensis, Saurida undosquamis,* and *Nemipterus peroni*; these trawl catches obviously more closely reflect the whole fish fauna of the banks than the highly biased line catches of Wheeler and Ommaney. Trawl surveys of the inner Mahé Bank by the local fishery department recorded a similar list of species; soft-bottom genera are conspicuously absent.

Wheeler and Ommaney's survey of 1947–1948 itself demands some comment, apart from their findings. With a small wooden ex-Scottish herring drifter, about 20 m in length, without radar, and with a series of almost totally ineffective echo-sounders, these two scientists made 42 voyages from Mauritius, totaling 27,000 miles steamed, during which they studied dozens of individual banks and islands, some as much as 1000 miles from their home port, recording biological data and illustrating the living colors of their catches in watercolor paintings as they went along; of the responsibility that their skipper had for the safety of the expedition, Wheeler and Ommaney (1953) wrote

. . . while out of sight of land, no other ship of any kind was seen during the entire commission, and no contact could be established with Mauritius or Seychelles on the effective wave length of our radio telephone.

It is ironic that this well-meaning survey by two dedicated biologists, though it located and quantified rich demersal fish stocks, had little impact on the fisheries of the region. It was left to a Japanese distant-water operation using dories and motherships finally to exploit these riches, and later to introduce the technique to Seychellois and Mauritian fishermen.

In the Hawaiian archipelago, where demersal fish comprise less than 5% of all landings, the small production is based on a rather similar group of species, of which Lutjanidae dominate the landed biomass, especially in shallower areas; the dominant lutjanids are *Pristipomoides sieboldii* and *Aprion virescens,* and include six other common species. A serranid (*Epinephalus quernus*), some labrids (*Bodianus* spp.), and a scorpaenid (*Pontinus macrocephala*) are also of importance, especially deeper (Ralston and Polovina, 1982). Ranked second in group abundance to lutjanids in shallower water are another hard-bottom family, Mullidae, which comprise about 20% of landings. At mid-Pacific seamounts in the general region of Midway Island, there are significant populations of the pelagic armorhead *Pseudopentaceros wheeleri* and the alfonsin *Beryx splendens,* which have formed the basis of Japanese and Soviet fisheries, though the stocks are currently in decline; the existence of abundant small mesopelagic fish, mostly *Maurolicus* and the pelagic mysid *Gnathophausia,* are thought to be the principal diet of the stocks of armorhead and alfonsin.

On the Wadge Bank, which is a widening of the Indian continental shelf at the southern tip of the subcontinent off Cape Comorin, the fauna sampled by survey trawlers operating from Columbo was very similar to that sampled by lines on the Indian Ocean Banks. At least 40% of the indicated biomass comprised groupers and snappers (e.g., *Lutjanus argentimaculatus, L. sebae, Lethrinus nebulosus,* etc.).

In the South China Sea, in addition to the survey reported by Ommaney (1961) which demonstrated clearly the existence of a typical sandy ground fauna of lutjanids, mullids, nemipterids, and priacanthids, there are more recent but unanalyzed trawl survey data from 119 trawl hauls made northeast of Sarawak by training ships from Hokkaido University, between the mouth of the Rajang River and the Indonesian islands of Kepuluan Natuna Besar. This area, with sandy bottoms at 60–90 m, and with warm bottom water, bears a typical lutjanid assemblage in which Lutjanidae form 49% of biomass, followed by *Priacanthus* (10.2%), Pentapodidae (8.3%), carangids (7.1%), and triglids (2.6%). Lesser numbers of *Parapristipoma, Saurida,* sciaenids, nemipterids, and others comprised the remainder of the assemblage. The principal snappers are *Lutjanus sebae, L. lineolatus, L. lutjanus, L. vitta,* and *L. gibbus.* Associated species include *Nemipterus bathybus, N. personii, N. nematophorus, Epinephelus hata, E. diacanthus, E. areolatus, Upeneus bensasi, Psettodes erumei,* and *Argyrops spinifer.*

The most complete analysis of continental fish communities for the Indo-Pacific that we have seen is that of Okera (1982), who surveyed the Sahul shelf of northwest Australia, and the Arafura shelf, including the Gulf of Carpentaria, along the northern Australian coast. As we have already described, these regions are for the most part sandy; Okera noted that the Sahul shelf has complex topography (rendering trawling difficult) and the benthic epifauna includes many soft corals and large cup sponges. On both shelves, sand dominates the deposits shallower than 40–50 m; on the Arafura shelf, there is a middepth zone of silty clay, which grades once again into sands near the shelf-break. Middepths on the Sahul shelf are characterized by sandy ridges with silty basins between them, and at the shelf break there are numerous steep-sided banks. On the Arafura shelf, close inshore, and especially where conditions are somewhat brackish, there is a muddy zone.

Through this region, a survey of 266 trawl hauls was made in 1980–1981, each haul of 30 min duration; biological data were taken and identifications made of all 350 species taken in the catches. A subjective matrix technique clustered shallow species, deep species, and those with a eurybathic distribution. From this matrix, Okera extracted six recurrent species associations (Table 6.3). A coastal soft-bottom assemblage occurs in <15–20 m and comprises sciaenids, polynemids, *Harpodon,* and *Cynoglossus,* together with a series of bentho-pelagic clupeids. From other evidence, we know that this community changes its distribution seasonally; during the wet season, the whole Gulf of Carpentaria becomes "estuarized" and the inshore brackish water fauna extends very widely. In this region it appears to comprise five polynemids, three species of *Johnius,* and one of *Otolithus,* plus one species each of *Cynoglossus, Ephippus,* and of *Arius.* It is associated with a few species of *Nemipterus, Saurida, Sillago, Lates,* and *Platycephalus.*

An inner-shelf assemblage occurs below this, down to about 50 m, and comprises grunts (*Pomadasys*) and bentho-pelagic species (*Therapon theraps,* etc.) and large aggregations of *Leiognathus, Rastrelliger,* and smaller bentho-pelagic species; on harder grounds of coralline rubble and sand, *Lethrinus, Pristotis, Scolopsis,* and *Siganus* joined the other species and the community resembled a fauna at similar depths with which Okera was familiar off East Africa. Deeper, from 60 to 120 m, a typical midshelf assemblage, with some internal zonation (*Upeneus* above, *Priacanthus* below) contains many species of groupers and snappers, as well as *Saurida, Nemipterus,* and *Lepidotrigla*—in short, recalling what we have called a lutjanid community in the Atlantic. A shelf-break assemblage mingles some of the deeper-penetrating species of the midshelf assemblage with slope forms (recalling those recorded off tropical West Africa) such as *Chlorophthalmus, Antigonia, Chaunax, Peristedion,* and other

TABLE 6.3. Recurrent Species Groups of Demersal Fish on the North Australian Continental Shelf[a]

Neritic mud assemblage (Coastal mud, <15–20 m)	Inner shelf assemblage (Muddy sands, 10–50 m)	Midshelf assemblage (Mixed deposits, 60–110 m)
Principal species	Principal species	Principal species
Euristhmus nudiceps	*Caranx bucculentus*	*Apolectis niger*
Harpodon translucens	*Pomadasys maculatus*	*Lepidotrigla spilotera*
Ilisha melastoma	*Selaroides leptolepis*	*Lutjanus malabaricus*
Johnieops vogleri	*Therapon theraps*	*Lutjanus vitta*
Johnius coiter	Secondary species	*Nemipterus hexodon*
Polynemus nigripinnis	*Gazza minuta*	*Pristipomoides multidens*
Setipinna papuensis	*Herklotsichthys koningsbergi*	*Psenes arafurensis*
Secondary species	*Herklotsichthys macullochi*	*Saurida filamentosa*
Cynoglossus spp.	*Leiognathus leuciscus*	*Trichiurus lepturus*
Rhinoprenes pentanemus	*Leiognathus splendens*	*Upeneus moluccensis*
Thryssa hamiltoni	*Paramonacanthus japonicus*	Secondary species
Thryssa setirostris	*Pelates quadrilineatus*	*Acropoma japonicum*
	Secutor insidiator	*Nemipterus bathybus*
Shelf break assemblage (Mixed deposits, 12–220 m)	*Secutor ruconius*	*Nemipterus virgatus*
	Trixiphthichys weberi	*Priacanthus cruentatus*
Principal species	Tertiary species	*Psenopsis humerosus*
Acropoma spp.	*Eleutheronema* spp.	*Siphamia* spp.
Nemipterus bathybus	*Ilisha* spp.	*Synagrops philippinensis*
Nemipterus virgatus	*Polynemus* spp.	*Upeneus sulphureus*
Psenoposis humerosus	*Thryssa* spp.	
Saurida filamentosa		Offshore sand assemblage (Sand, 80–90 m)
Siphamia spp.	Hard bottom assemblage (Boulders and reefs, various depths)	
Synagrops spp.		Principal species
Trichiurus lepturus		*Pseudorhombus dupliciocellatus*
Upeneus moluccensis	Principal species	Secondary species
Secondary species	*Drepane punctata*	*Adionyx ruber*
Antigonia spp.	*Lutjanus argentomaculatus*	*Alutera monoceros*
Ariomma spp.	*Lutjanus erythropterus*	*Aptochtrema rostrata*
Chaetodon modestus	*Lutjanus russelli*	*Fistularia petimba*
Chaunax spp.	*Pomadasys hasta*	*Lethrinus lentjan*
Chlorophthalmus spp.	Secondary species	*Nemipterus tambuloides*
Coelorhynchus spp.	*Abalistes stellaris*	*Parascolopsis inermis*
Epinnula spp.	*Argyrops spinifer*	*Parupenaeus pleurospilus*
Hydrolagus spp.	*Lethrinus lentjan*	*Pentaprion longimanus*
Monocentrus japonicus	*Lethrinus nematocanthus*	*Priacanthus hamrur*
Peristedion spp.	*Lutjanus sebae*	*Saurida undisquamis*
Saurida elongata	*Lutjanus vitta*	*Upeneus* spp.
Zeus faber	*Parachaetodon ocellatus*	
	Plectorhyncus pictus	
	Pristipomoides multidens	
	Zabidius novaemaculatus	

[a] From Okera, (1982).

cold-water genera; but the dominant species that characterize the assemblage are small *Siphamia, Acropoma,* and *Synagrops.* At a few isolated patches of reef, a different assemblage of groupers occurs from those of the typical shelf assemblages; again, different species are specific to offshore reefs in the Gulf of Carpentaria, and to inshore reefs within the influence of brackish water. Finally, a special and anomalous assemblage occurs only on the Sahul shelf, on very clean sandy bottoms; the genera are similar to the typical shelf assemblages, but some species replacement occurs. A second, smaller set of data is available for the extreme southern end of the Sahul shelf, for which Masuda *et al.* (1964) describe the distribution of some of the larger trawl-caught species off Cape Thouin; *Lethrinus ornatus* is the most important species in this region. Some species replacement was found on each side of the warm-water front which cuts across the shelf in this region. On the warm side *Lutjanus janthinuroptis, Argyrops spinifer,* and *Scolopsis temporalis,* and on the colder, western side *Nemipterus peronii, Lutjanus vitta,* and *L. sebae.*

Okera was able to relate the general distribution of these assemblages with the geographical and substrate features of the northern tropical coastline in very much the same way as has been done for the Atlantic; for example, at the eastern end of the survey area, the shelf of the Gulf of Carpentaria is broad and shallow, so that the midshelf assemblage is attenuated both spatially and in its species composition, and neritic species extend beyond their normal boundaries. It is particularly fortunate that this description exists (even if unpublished) since it is a unique record of the distribution of fish in relation to the geography of the continental shelf for part of the Indo-Pacific region in a near-natural state; more than a decade of regulated trawling does not appear to have had major impacts on the species composition at the time the survey was performed. However, there is now some evidence (K. Sainsbury, personal communication, 1986) that some faunistic modification has occurred, including a second (after the Gulf of Guinea) outburst of a balistid, this time involving *Abalistes stellarius.* It is also particularly useful to have a description of the relationship between the lutjanid communities of a tropical shelf and the fauna of coral reefs, a relationship that has been refered to in the tropical Atlantic but not so clearly observed. The relationship is important since the nonpelagic species potentially of fisheries significance in the two end-members of a series are always dominated by large snappers and groupers.

Transitions to the Pacific Subtropical Faunas

We complete this discussion of the tropical demersal communities by considering the evidence for a transition to subtropical assemblages to the north and south in the Pacific Region resembling the situation described for the Atlantic. Geographically, the situation is simplest in the eastern tropical Pacific, along the relatively straight coast of the American continent. On the coasts of Mexico, Baja California, and southern California, the tropical fauna occurs at least as far north as Cape San Lucas at the tip of Baja California, where there are coral reef outliers and a few central Pacific species. The continental shelf northwards from here is narrow and includes many deep basins; this shelf lies above an active tectonic margin. Though sparids are absent, the transition zone is marked by the occurrence of a few very large species of sciaenids (notably the endemic *Cynoscion macdonaldi,* which is now in danger of extinction and is actually one of the few commercial species of marine fish actually to achieve that distinction!), and a flowering of endemic scorpaenids that extend into the temperate and subarctic faunas.

The southern transition to subtropical fish in the eastern Pacific is, of course, profoundly influenced by the upwelling anomaly at 5–15°S off Peru. Not only is this zone unique at such latitudes for the magnitude and continuity of coastal upwelling, but also for the nature of its decadal-scale variability in the occurrence of El Niño conditions which, as we have seen in Chapter 3, significantly modify the north–south distributions of many marine organisms, including demersal fish. Under normal conditions, the demersal community of the Peruvian coast is characterized by such forms as hake (*Merluccius gayi*), the gurnard *Prionotus stephanophrys*, a series of sciaenids (*Cynoscion analis, Paralonchurus peruanus, Micropogonias* spp., etc), grunts (*Isacia conceptionis, Paralabrax humeralis),* and breams (*Cheilodactylus variegatus, Seriolella* sp.). During El Niño conditions, tropical Panamanian species such as *Polydactylus approximans, Chloroscombrus orqueta,* and atherinid silversides (to a total of 51 species in 1982–1983) appear in the Peruvian demersal ecosystem, and complementary changes occur in the demersal ecosystem in the Gulf of Guayaquil to the north.

In the western Indo-Pacific, on the wide continental shelves of the east China Sea and Australasia, the situation much more clearly resembles what occurs on both sides of the Atlantic. The tropical fauna is replaced by subtropical fish not far to the south of Hong Kong, and near Carnarvon in Western Australia and Cooktown in Queensland. The subtropical fauna is dominated heavily by sea breams (Sparidae, Nemipteridae) and sciaenids associated also with triglids and latilids. The dominant species are the breams *Taius tumifrons, Pagrosomus niger, Nemipterus virgatus* and *Gymnocranius orbis,* some large croakers (*Nibea, Pseudosciaena*), and lizardfish *Saurida tumbil*; this association of fish resembles very closely in its ecology the fauna of the northern and southern subtropical regions in the Atlantic. As in the Atlantic, increasingly northward this fauna has important seasonal onshore–offshore migrations related to seasonal water temperature changes over the continental shelf. Menasveta (1970) describes very extensive Japanese trawl fisheries and surveys of the Gulf of Tonkin in the late 1930s that appear to have been uncommonly well documented. In the Gulf proper, the sea bream *Evynnis japonicus* dominated the landings, of which they comprised almost 40%; outside the Gulf of Tonkin, south of Hainan island, only bull trawlers were licensed to fish, and the catch of these vessels was dominated by a different species of sea bream, *Taius tumifrons.*

The southern subtropical fauna of the Australian coast is somewhat special, though sparids form as much as 20% of fisheries-indicated biomass, and some sciaenids occur; the subtropical fauna, which also occurs on the northern coast of New Zealand, is dominated by the flatheads *Platycephalus* and *Neoplatycephalus,* the breams *Chrysophrys auratus* and *Dactylopagrus morwong,* the sciaenids *Atractoscion atelodus* and *Sciaena antarctica,* together with several Australian endemics like *Arripis trutta.*

FISH OF CORAL REEFS AND ATOLLS

The fisheries of coral reefs, prosecuted with hand-lines, traps, and monofilament gill-nets, are usually targetted at a relatively few high-level predators, which are usually large fish relative to the extremely diverse assemblage of reef fish that sustains them (Russ, 1984).

As we have noted above, in our discussion of demersal fish assemblages of continental shelves, the highly diverse fauna of coral reefs is the end-member of a series of increasing diversity, and perhaps forms a continuum with the lutjanid assemblages of rocky and sandy continental shelves, especially where the topography is broken. However, because of its extreme character, it warrants special discussion.

The coral reef ecosystem, and notably its fishes, has for long been a focus of study in biological diversity and the specialization of species to niches; like the tropical rain forest, the great number of species packed into a small spatial dimension has been related to the diversity of substrates offered by the organisms (trees, corals) that determine the general form of the biocoenosis. The number of reef-fish species in a single zoogeographic region varies from hundreds to thousands, including both demersal species (in the sense that they take their food from benthic organisms) and pelagic species (in the sense that their prey is planktonic), though the difference in body-form between the two groups is not quite as distinct as experience with only continental shelf fish might lead us to expect.

Most families of fish that occur in the tropical seas include species that occur in the coral reef fauna, and some families are almost entirely restricted to reefs; almost all species of Chaetodontidae occur on coral reefs, as well as the majority of Scaridae and Labridae. Acanthuridae, Holocentridae, Balistidae and Ostraciodontidae, Pomacentridae and Serranidae, as well as Belennidae and Muraenidae are all dominant among the demersal component. The dominant families of pelagic fish associated with reefs, apart from the top predators such as *Sphyraena,* Carangidae, and sharks, are forms like the damselfish (Pomacentridae), silversides (Atherinidae), small lutjanids, such as *Caesio* and its relatives, and some Acanthuridae and Holocentridae.

However, as Hobson (1974) points out, there are some highly unusual characteristics of this assemblage of fish that must be understood if its structure is to be interpreted correctly. Not only are most of the reef fishes among the more highly (and recently) evolved teleosts, but most of the total diversity that exists today among higher teleosts has evolved specifically among the species that occupy the reef habitat. This might seem strange in so ancient an ecosystem as the coral-algal reef that has existed, as an ecological unit, since at least the Cambrian. However, as Hobson reminds us, and more recent evidence of the occurrence of biotic crises during geological time make more certain, reef ecosystems have existed as stable ecosystems only between biotic crises on a planetary scale; further, the composition of reefs during each stable period of their existence has been progressively different. Scleractinian corals, which dominate modern reefs, and teleost fish, which dominate their fish fauna, both commenced their radiation during the Mesozoic, principally during the Jurassic, about 150 my BP. During the Cretaceous, reef fish were very different from today; the primitive beryciform acanthopterygian fish had radiated into a highly diverse reef fauna but quite different from that of the later Tertiary and Quaternary reefs. Today, only the nocturnal holocentrids represent the remnants of the Mesozoic beryciform radiation on reefs, and all other modern beryciforms are bathyal fish of the deep-sea ecosystem.

The Tertiary radiation of the tropical marine fauna included a massive radiation of scleractinian corals subsequent to the collapse of the coral ecosystem at the time of the Cretaceous–Tertiary biotic crisis—the "Gubbio event" of dinosaur extinction fame. The modern Perciformes radiated subsequently to this event and now almost totally dominate the coral reef fauna; as Hobson notes, 98% of the coral reef fish species at Hawaii have reached the perciform level of organization, and 75% actually belong to the order Perciformes. This radiation has been of a group of fishes that developed a complex jaw mechanism with a protrusible premaxillary, and pectoral fins stiffened with hard rays and optimally positioned for high maneuverability.

As many authors have described, there are two rather distinct fish faunas on a coral reef: the diurnal fish and the nocturnal fish. At night, the diurnal fish use a variety of techniques to conceal themselves in holes and under ledges even, as in some labrids, secreting a mucous net for protection around the entrance to their resting place. The diurnal species tend to be larger, more boldly patterned, and brighter in coloration than the smaller, darker, more cryptic nocturnal species. Most species of clupeids, holocentrids, apogonids, kuhliids, and lutjanids are nocturnal, as are many individual species of other families; these are mostly generalized, less highly evolved groups. On the other hand the diurnal species are the more advanced, specialized groups such as chaetodonts and tetraodonts. Though there are diel differences in the behavior of continental shelf fish species, they are not as profound as the change that occurs on a reef at twilight; the total diversity on a coral reef owes much to this dual nature of its fish assemblage.

The richness of fish assemblages on coral reefs is determined by many of the same factors as we have already discussed for the invertebrates of the reef ecosystem; coral reefs in the western Indo-Pacific support the most diverse fish communities, with the greatest diversity occurring in the Philippines, while the least diverse are in the Atlantic. Coral reef fish follow the general pattern of diversity of tropical marine organisms; Sale (1980) shows that the numbers of species of fish in regional reef faunas ranges from more than 2000 in the Philippines to less than 200 in the northern Caribbean. Interestingly, there appears to be no latitudinal component in this variation, even after the relative distance of each data set from the center of highest diversity in the Philippines has been taken into account. However, diversity of reef fish cannot simply be explained by the zoogeographic history, discussed earlier, of the individual tropical oceans, nor by the vicariance effect suggested by Rosen (1981) for reef corals; not only is there a relationship between total sample size in the available surveys and indicated species richness (Sale, 1980), but also with the complexity of the topography and of the other ecological elements of each reef (Talbot, 1965).

Studies of the diversity of reef fish by Whittaker (1960) led him to suggest that two forms of diversity may exist independently. Similar-sized patches of similar habitat within the same general region may exhibit a range of diversity caused by the occurrence of more or fewer of a single set of species adapted to this habitat. On the other hand, similar habitats occurring in different geographical regions may exhibit different diversity because different fish communities are available to populate the habitat in each region. Whittaker terms within-habitat diversity alpha-diversity, and between-habitat diversity beta-diversity. The two form the whole diversity pool of coral reef fish assemblages.

Of more concern to fisheries managers is the extent to which the fish community of coral reefs (as an end-member of the diversity series within fish communities generally) is naturally stable, and can be depended upon as a predictable resource.

Coral reef fish communities have some very special characteristics that are not exhibited by the fish communities of tropical continental shelves. Individuals may have a very strong affinity for a particular home site, where space is defended against newcomers (recognized as such) as well as affording protection from predators; intraspecies affinities are common, with pair-bonds being formed and maintained; mutualistic interspecific associations may exist, and Batesian mimicry (by which one inoffensive species mimics another noxious or dangerous species) is common. Thus, intra- and interspecific relationships of all kinds are extremely numerous, and have been a rich source of material for ecological research ever since the availability of underwater breathing equipment for biologists. The kinds of relationships that have been revealed by such studies do occur elsewhere in the ocean, but in the coral reef ecosystem are much more developed than anywhere else.

It has become a paradigm for this and other ecosystems, originating in the concept of Elton (1927), that the very diversity and richness of interspecific relationships is a guarantor of stability; many investigations, proceeding from this assumption, have seemed to find it so. However, the real facts may be otherwise, as Sale (1980) demonstrates with some conviction. He suggests that three types of studies have led their authors to assume stability in coral reefs; colonization of new or denuded habitats, long-term monitoring of species groups at a site, and long-term monitoring of an entire reef assemblage. Sale suggests, after reanalysis of 16 previously published studies, that only one appeared to demonstrate stability of a reef fish assemblage over a substantial period of time. In all other cases, he suggests convincingly, there are significant changes in composition and in relative abundance between species that are present throughout the sampling period. Sale, however, did not address the question of whether these changes in species composition are greater or smaller, or about the same, as occur in other tropical fish communities. This is a key question that still needs to be addressed.

What is known of the recruitment mechanisms of reef fish make it inevitable that some level of between-year changes must be expected, but nothing tells us the magnitude of these year-to-year differences in recruitment. Almost without exception, reef fish have planktonic eggs and larvae, and their reproductive mechanisms have evolved to enhance the distribution of their reproductive products away from the home range of the parents, or even from the parent stock; indeed, given the insular and fragmented nature of the reef ecosystem, such a mechanism must be essential to its distribution and survival. Since patchiness of planktonic larvae must occur at sea, and recruitment of each species contains some degree of seasonal signal, the availability of larvae in the water over the reef at their critical stage for settlement may not include the full range of species able to find suitable habitat on that particular reef. Therefore, it is to be expected, as several authors have suggested, that recruitment should be a stochastic process (e.g., Sale, 1978; Sale and Dybdahl, 1975), and there appears to be some evidence that this is correct. For instance, Luckhurst and Luckhurst (1977) found differences between recruitment levels for many species over marked quadrats during the same season in 2 successive years. However, the experiments performed by Sweatman (1985) on a matrix of artificially placed coral heads showed clearly that in some species of planktivorous reef fish there was increased settlement of larvae in sites already occupied by residents of the same species. Conversely, there was exclusion by resident fish or larval forms of some other species. As Sweatman suggests, such a mechanism could account for the clumped or overdispersed distributions observed in some species of reef fish.

It is possible that the recruitment mechanism in individual reef fish species will be found to resemble what is described for the more accessible shallow-water rock-fish *Semicossyphus pulcher* of the Californian coast; the between-year consistency of recruitment in this species is related to the presence and abundance of upstream populations as a source of larvae ready to settle. This factor, together with between-year differences in circulation patterns in the California Current, determine whether recruitment is approximately constant each year, or whether it is stochastic (Cowen, 1985).

These, then, are the principal characteristics of reef fish communities or species assemblages that form the resource base for many island fisheries in the Caribbean and Indo-Pacific; the community aspects and the interspecific relations have been stressed in this brief review, for it seems to us absolutely essential to have at least a preliminary understanding of the likely consequences of disturbing such an internally complex ecosystem by selective removal of some species, or some size classes of some species. Our present level of understanding of the dynamics and interactions of coral reef fishes is still very scanty. Progress may come through concerted, coordinated efforts and rigorously defined experiments together with computer simulations. We may also hope to learn much from the analysis of unplanned experiments through the impact of fisheries on reef ecosystems as Munro (1983) and co-workers were able to do for Jamaican reefs.

TROPICAL ESTUARINE FISH ASSEMBLAGES

Relatively few fish are wholly adapted to life within estuaries *sensu stricto,* yet estuaries appear to be important in the recruitment of at least some of the fish communities of warm-water continental shelves. As Day *et al.* (1981) point out, most species restricted to midlatitude estuaries are small fish, such as gobies, syngnathids, atherinids, and some clupeoids. Definition cannot be precise, but the number of such species is commonly less than 10% of all species occurring in an estuary.

The great majority of fish that are found in estuaries are juveniles and adults of species occurring also on the adjacent continental shelf, if one discounts the passage of anadromous and catadromous fish through the estuary. There is also, in the tropics, usually some descent into estuaries by freshwater fish during the rainy season. In tropical seas, and especially in monsoon areas where estuarization of the continental shelf occurs, the distinction between estuaries and the neritic sea is slighter than in non-monsoon areas.

However, there are also sufficient descriptions of specialized estuarine fish faunas for it to be clear that this is often a reality in the tropics, probably where the size of the estuary and its tidal regime permit a relatively long flushing period. Such estuarine communities comprise genera that occur there globally, such as the sea catfish, or regionally, such as *Centropomus* and *Diapterus* on the American tropical coasts. They may, as in Singapore and Papua New Guinea, comprise genera with large scales or mucilaginous skins (*Gerres* and *Leiognathus,* respectively) and able to support salinity changes. Finally, in the West African estuaries, they comprise simply a species replacement within genera compared with the inshore fish community of the inner continental shelf, and sampling in coastal lagoons does not reveal a population of juvenile offshore species (Longhurst, 1965), but a resident population of just a few species instead (Pillay, 1967a; Pauly, 1975).

It is significant that on subtropical coasts, such as those of southern Africa or the south eastern United States, the species that regularly enter estuaries as juveniles are species of families having most of their members in the tropics, such as sciaenids. As Blaber (1980) suggests, and as we shall discuss below, the turbid waters of subtropical estuaries may be used by these fish in their juvenile stage as essential analogs for the turbid, inshore habitats of tropical continental shelves in which their groups presumably evolved, and which still, for their tropical congeners, act as nursery grounds. This suggestion leads us on to investigate to what extent estuaries and coastal lagoons determine the abundance and kind of demersal fish on the adjacent tropical continental shelf. The question we have to address is to what extent tropical estuaries and lagoons have a special brackish-water fish fauna, and to what extent they principally serve as nursery grounds for the larvae and juveniles of continental shelf species. Further, to what extent is the ecology of the well-studied warm-water estuaries of the eastern coast of North America relevant to tropical estuaries and lagoons? Classical studies, such as those of Gunter (1967) and Pearson (1929) and many modern surveys, make it quite clear that in the southeastern United States the most abundant demersal fish species of the continental shelf are dependent on estuaries for the survival of their late larval and juvenile stages. As Bozeman and Dean (1980) show, juveniles of several species of sciaenids (*Leiostomus xanthurus, Lagodon rhomboides, Micropogon undulatus*), as well as of bothid flounders, clupeids, and other fish, enter South Carolina estuaries in vast numbers in autumn, and there is a mass emigration again in the spring. Musick (1987) describes a summer immigration to the same estuaries of juvenile sharks of several species from the continental shelf. Deegan and Day (1984) discuss various mechanisms by which these migrations might occur, and show that there are regional and specific differences in the ecology of juveniles. Observations such as these have caused much concern over the destruction of coastal lagoon and marsh habitat for industrial and recreational development, apparently to the cost of continental shelf fisheries resources. So there is now an extensive literature concerning this problem along many subtropical and tropical coasts, advocating wetland protection; many of these studies are less than objective. Their authors tend to sample only within estuaries and lagoons, and since they find there many juvenile fish of some continental shelf species, they can conclude that the offshore fisheries are heavily dependent on estuarine habitat. Seldom is the study properly designed, so as to search the whole habitat of each individual species and quantify the proportion of juveniles occurring in estuaries and on the continental shelf. In fact, sampling is seldom carried out beyond the threatened estuarine and lagoon itself, so the results are biased from the beginning.

We have briefly reviewed about 20 published studies of tropical estuarine fish faunas to try to determine to what extent the estuarine-dependency of Gulf of Mexico demersal fish is universal throughout the tropics. Undoubtedly, in some cases, and especially in subtropical western boundary current situations, the same phenomenon occurs: for instance, Chao *et al.* (1987) found that a barrier-beach lagoon, the Lagoa do Patos in southern Brazil, is used as a nursery area by massive numbers of juvenile sciaenids (*Umbrina, Micropogon, Macrodon, Pogonias*) along with mullets (*Mugil*) and silversides (*Atherinus*). The same may occur in South African estuaries (Day *et al.*, 1981). However, every other study we have reviewed discusses the existence of juveniles of offshore species on feeding migrations, but not one gives any evidence of very large numbers of small fish being present, as must be the case to recruit the continental shelf populations. In fact, some studies specifically indicate rather small numbers of each recorded species being present in the samples (e.g., Pillay, 1967b).

However, these studies do confirm that throughout the tropics some genera absolutely depend on estuaries for juvenile growth; mullets (*Mugil*) and some clupeids (*Brevoortia, Ethmalosa, Anchoa, Cetengraulis*) appear to fall into this category and their postlarvae and juveniles occur in great numbers in estuaries. Other genera, like *Caranx*, too frequently appear in descriptions of juvenile fish in mangrove creeks for this not to be their primary juvenile habitat.

Thus, we have been unable to find evidence for the relevance of the Gulf of Mexico model of estuarine-dependence generally throughout the tropics, and much to the contrary. We do not think that evidence presented to date for the lagoons of the Pacific coast of Mexico, Panama, the coast of Venezuela, Australia, Puerto Rico, or the Gulf of Guinea can be held to support an apparently current opinion among warm-water fish biologists; that the Gunterian hypothesis of estuarine-dependence of continental shelf fish stocks is valid for the tropical regions as a whole.

We suggest, as do Blaber and Blaber (1981), that the solution to this paradox lies in the relative estuarization of the continental shelf in much of the tropics, so that low salinities and muddy deposits do not have such a sharp boundary as on the coast of the southeast United States. Several studies suggest that the neritic regions of the continental shelf act as juveniles habitat for many offshore species. Lowe-McConnell (1962) found good evidence for this in her trawl surveys off Guiana, encountering localized populations of juvenile sciaenids and other fish on the inner continental shelf, and Ranier (1984) reports that juvenile *Polydactylus, Secutor,* and *Johnius* occur inshore in the Gulf of Carpentaria, while adults occur offshore. To what extent the fauna typical of sandy, shelly deposits of outer continental shelves (Sparidae, Priacanthidae, Nemipteridae, etc.) follow this habit is not clear, and there is no clear indication that they move consistently inshore onto neritic muds as juveniles. Indeed, even in the southeastern United States the porgies (Sparidae) of the outer continental shelf do not appear in the lists of abundant estuarine juvenile fish in studies such as that of Bozeman and Dean (1980) in South Carolina marshes.

Chapter 7

Pelagic Fish of Tropical Oceans

There is a great wealth and diversity of pelagic fish in tropical seas, and important fisheries are based upon them. These fisheries are driven by a strong global demand for tropical clupeoid products (meal and solubles) to supply the international protein market, and for tropical tuna for the Japanese table and for European and North American sandwiches, salads, and cat food. During the great period of fisheries expansion between 1950 and 1970, the proportion of the world fish catch that came to be used for industrial production of fish meal rose from less than 10% to more than 35%, and the tuna fisheries have come to be based on the most sophisticated catching techniques yet developed in any fishery.

Of the approximately five million tons of fish meal that entered the international trade annually in the early 1970s, more than three million tons were made from tropical clupeoid stocks, with high-latitude herring contributing less than one million (Longhurst, 1971a). Though the decline of the Peruvian anchovy fishery since then has changed the pattern considerably, there is no reason to suppose that the demand for fish meal, oil, and solubles will decrease as the variety and production of formulated human foods continues to increase world-wide, nor that the production of tropical clupeoids cannot return to the situation prior to the collapse of the Peruvian stocks. And at the other end of the scale of values in fishery products there is no reason to anticipate a substantial decline in the world-wide long-term demand for high-value meat from tropical tuna.

Tropical pelagic fisheries are based on a variety of different taxonomic groups, some of which have pantropical distributions; as we have noted, the zoogeography of pelagic fish species more closely resembles that of plankton than the benthos, though there is a greater diversity of species in most groups in the western Pacific region than in the Atlantic or the eastern Pacific. Pelagic fish belong to several Orders: the relatively primitive and herring-like Clupeiformes usually occurring over continental shelves; Atheriniformes, including the exocoetid flying fish; the small, meso- and bathypelagic Myctophiformes of the open oceans; and two highly evolved families of Perciformes, the carangid jacks and trevallies of continental shelves, and the scombroid mackerels, seerfish, and the oceanic tunas and billfish.

Our discussion of the pelagic fish assemblages will take two directions; we shall review the groups of fish that comprise these assemblages and their principal biological characteristics, and then we shall discuss some regional species assemblages and some of the biological interactions within them. We shall avoid so far as possible further discussion of the bentho-pelagic forms that have already been reviewed in the previous chapter. One of the characteristics of the tropical fish fauna, generally, that sets it apart from temperate faunas is the wealth of clupeoids and scombroids that are bentho-pelagic and are therefore frequently caught in demersal trawls.

CLUPEOIDS OF TROPICAL SEAS

With the exception of a few genera of freshwater Osteoglossiformes that represent the remnants of a Mesozoic fish radiation, the Suborder Clupeiodei are the most primitive of Holostean fish. Besides the more important and familiar herring and anchovy-like genera (e.g., *Sardinella, Sardinops, Engraulis*), there exist in tropical seas several that are quite different in morphology and ecology from these more typical forms. Six families of fish concern us here: three families of Elopiformes, Elopoidae, Megalopidae, and Albuloidae, each with a very small number of species of large or even very large, carnivorous fish restricted to the tropical seas; and three families of Clupeiformes, the predatory Chirocentridae, the Clupeidae, and Engraulidae, with many genera and species of small microphagous fish occurring both in the tropics and in high latitudes.

Elopiformes are a rather consistent group of silvery, highly active predatory fish, which occur only in tropical seas and sometimes enter lagoons and other brackish environments; they all have leptocephalus larvae resembling those of eels. There are from one to five species of *Elops,* depending on the authority, distributed in all three oceans; these are rather slender, fast fish not usually reaching 50 cm, but, like the other Elopiformes, not highly regarded as a market fish, because of the numerous small bones in their muscles. Two species of *Albula,* one ubiquitous and one restricted to the central American region of both oceans, are similar to *Elops,* though rather less active and with a smaller, inferior mouth, which suggests feeding on benthic organisms. Two species of tarpons (*Megalops*) inhabit the Indo-Pacific and Atlantic, respectively: these are very large fish, reaching 2 m in length, with extremely large, hard scales. Tarpon enter brackish water much more habitually than *Albula* or *Elops*.

The wolf-herring (Chirocentridae) is anomalous among the Clupeiformes: reaching 3.5 m in length, long and compressed in the body, and with fang-like teeth, it occurs only in the Indo-Pacific, where it forms the basis of a small fishery in the Bay of Bengal; only a single species (*Chirocentrus dorab*) exists and it has been suggested that this is, in point of fact, not a herring at all but another remnant of the osteoglossomorph fauna of the Mesozoic. *Chirocentrus dorab* occurs around the Indian Ocean from southern Africa to the Red Sea, and from Japan to New Zealand.

The remainder of the Clupeiformes comprises 25 genera in the tropics, 10 in the subtropical regions, and a single boreal genus. The east–west diversity gradient exists only in tropical regions for these families, but here it is very strongly expressed: 10 species occur in the eastern tropical Atlantic, and 58 in the western Atlantic; 26 species are found in the eastern tropical Pacific, 77 in the western Indo-Pacific. Within single genera, the same distribution is also expressed; there are 3 species of *Sardinella* in the Gulf of Guinea, 6 in the western Atlantic, and 12, 14, or 16 (depending on the authority) in the western Pacific. The distribution of genera and species of clupeoids in the tropics is illustrated in Table 7.1.

The 30-odd genera that comprise the Clupeiformes are rather consistent in morphology and ecology. They are all of relatively small size (10–30 cm total length), they are generally soft-bodied with rather deciduous scales, and they are typically shoaling fish, the shoals sometimes being of very great size; these shoals occur mostly over continental shelves, though migration across deep water occurs in some species. Shoals may occur at the surface, in midwater, or near the bottom, and typically break up at

TABLE 7.1. **Approximate Relative Geographic Distribution of Numbers of Species of Clupeoid Fish**

	Atlantic Ocean		Pacific Ocean		Indian Ocean
	West	East	West	East	
Boreal	1	1	1	1	—
North temperate	5	11	3	2	—
Tropical	10	58	26	77	17
South temperate	2	7	4	2	2
Antiboreal	—	1	1	—	—

night into smaller groups of feeding fish; this dispersion may be complete, with shoals only reforming at dawn. Clupeiformes occur in all seas, very frequently enter brackish water, and some are anadromous. The fish are usually microphagous, feeding either by snatching individual particles or by gill-raker filtration during forward swimming; the kind of feeding used may depend on the size and abundance of food particles available. Associated with their planktonic food, which may comprise either zooplankton or, in a few species in each region, phytoplankton, the flesh of almost all species is highly oily. A few small species are predatory, feeding usually on large planktonic crustacea and very small fish.

With few exceptions, Clupeiformes occur over or near continental shelves and often they are truly coastal fish, associated with which there is a widespread tolerance of low salinities, though not all species are euryhaline. Within a single genus, such as *Sardinella,* there may be a variable tolerance to low salinities; the tropical Atlantic *Sardinella aurita* is extremely stenohaline (or at least it is not known to occur in water lower than about 35‰), though whether this is because its other ecological requirements keep it in high salinity water or because it is physiologically incapable of tolerating low salinities is not clear. On the other hand, *S. maderensis,* also from the Gulf of Guinea is regularly encountered in estuarine conditions down to at least 20‰. Obviously, the anadromous genera (*Hilsa, Macrura*) are tolerant of quite fresh water; though temperate and subtropical anadromous clupeids (*Alosa, Dorosoma, Pomolobus*) occur in both Atlantic and Indo-Pacific Oceans, there do not appear to be any river-ascending clupeids in the tropical Atlantic, nor in the eastern tropical Pacific. A few nonmigratory, small clupeids occupy coastal fresh waters, such as the small West African *Pellonula vorax* that occurs high in estuaries but does not appear ever to descend to the sea. Similarly, *Rhinosardinia* behaves the same way in South America, and *Microthrissa* and *Pellonula* in the Indo-Pacific occur only at the heads of estuaries.

It may be useful to review the genera of Clupeiformes (apart from *Chirocentrus*) that occur in the tropical seas according to the feeding habits of their adults. In practical terms, four groups can be recognized: predators, zooplankton feeders, omnivores capable of taking both phyto- and zooplankton, and herbivores that specialize in phytoplankton. Obviously, larval and juvenile forms of all species utilize very small particles, often phytoplankton and protists.

The most common small predatory clupeids resemble *Ilisha* in general morphology; they are laterally flattened, with rather small tails, and a large upturned mouth, and are typical members of the bentho-pelagic clupeids that we have already noted as being a common constituent of tropical trawl catches. The genera comprising this group are *Ilisha* itself, *Pristigaster, Opisthopterus, Raconda, Pliosteostoma,* and *Odontogna-thus,* and all of these are restricted to very low latitudes. In addition, some curious large-toothed anchovies of the tropics must be included in this group: *Lycengraulis, Lycothrissa,* and *Coilia.* Some species of the thread-herring *Opisthonema* are also usually grouped here.

The great majority of clupeoids, including tropical forms, belong in the zooplankton-feeding group of genera; these fish occur more often than their predators in organized, dense shoals. They resemble the true her-rings of northern latitudes, and many species are important fishery re-sources in the tropics: Typical genera are *Sardinella* (but not all species), *Herklotsichthys, Clupeoides, Lile,* and *Jenkinsia.* The round herrings *Dussumeria* and *Etrumeus* that are so widespread in the Indo-Pacific feed mostly on zooplankton also, as do some of the small tropical anchovies, such as *Setipinna* and *Stolephorus.*

A few genera of tropical and subtropical clupeoids are characterised by their ability to utilize phytoplankton as well as zooplankton; not unnatu-rally, these tend to occur in places where blooms of phytoplankton occur at least intermittently. The sardines and anchovies of the upwelling re-gions of eastern boundary currents (*Sardinops, Engraulis*) are the most important example, as several of these are important in regions that are otherwise characterized as tropical. *E. ringens* seems to occur as at least two different forms, with long alimentary canals in the north and shorter in the south.

In each tropical region, however, some species of clupeoids specialize in direct consumption of phytoplankton, usually the blooms of large diatoms that occur in estuarine regions; periods between blooms may be bridged by these species by limited consumption of zooplankton, by filtra-tion of particulate organic material suspended in the water or even by browsing on diatom-rich superficial bottom sediments (van Thielen, 1977). Some of these herbivorous species are shad-like estuarine fish, similar in ecology to the very well-studied North American subtropical menhaden (*Brevoortia* spp.); *Ethmalosa dorsalis* of West Africa is an example of this type. Others are more typical continental shelf sardines, *Sardinella longiceps* and the thread herring *Opisthonema libertate,* both of the western Indo-Pacific, for example. In the eastern tropical Pacific, the little bay anchovy *Cetengraulis edentulus* consumes phytoplankton as the principal element of its diet. The subtropical and tropical gizzard shads *Clupanodon* and *Nematolosa* have a related, but specialized form, of feeding; they ingest the upper layer of mud in shallow parts of estuaries to utilize the benthic diatom layer that occurs there. Finally, some species of small tropical, neritic anchovies (e.g., *Anchoa starkSii, Thrissa* sp.) are also reported to be able to consume phytoplankton under some circumstances.

In the eastern tropical Atlantic, as we have noted for many biota, the diversity of clupeoids is relatively low; two species dominate the clupeoid fisheries of the region, the estuarine "bonga" *Ethmalosa fimbriata* and the sardine *Sardinella aurita* of the coastal upwelling regions. *S. maderensis* occurs the length of the coast, inshore, and is everywhere fished, but is nowhere landed in such large quantities as the two primary species; West African engraulids are not the object of a significant fishery, nor is there any evidence of unexploited engraulid potential in the tropical West African region, although there is some evidence that local engraulids have increased in abundance following declines in the abundance of *Sardinella aurita*.

Ethmalosa is both superficially and anatomically extremely similar to the western Atlantic menhadens (*Brevoortia* spp.), although Monod (1961) has suggested that this resemblance is to be regarded as evolutionary convergence. Be that as it may, almost all that is known of the ecology of the American menhadens may be said to apply, in general, to *Ethmalosa,* which is strong, active filter-feeding shad-like fish, attaining about 40 cm. Bainbridge (1960a) has shown, by comparing stomachs contents with plankton samples, that *Ethmalosa* obtains its food by passive filtration of phytoplankton (and, to a much lesser extent, zooplankton) with very little selectivity. Shoals of *Ethmalosa* are rarely found in clear water, or far from estuaries and lagoons. Major populations occur in the Senegal–Liberia and Niger Delta–Cameroons regions, but minor populations occur off every river mouth and lagoon in the tropical region. *Ethmalosa* spawns throughout the year at sea, off the river mouths, and within a few weeks the juveniles move into brackish water where they may be enormously abundant; at about 10 cm they join the adult onshore-offshore migration pattern which is related to the spring neap-tidal cycle. During neap tides, when diatom blooms occur in the estuaries, *Ethmalosa* enter to feed; during spring tides, when the estuaries are turbid with suspended sediments, *Ethmalosa* move to the inner continental shelf (Longhurst, 1960b). In some places, as in the western Nigerian coastal lagoons, populations of early-maturing, land-locked *Ethmalosa* coexist with the larger, migratory fish. These dwarf fish mature at <10 cm, compared with a normal length at first maturity of 21–22 cm; the lagoon-locked fish seldom attain lengths greater than 15 cm. A possible mechanism to account for this phenomenon, which occurs also in freshwater fish, is discussed in Chapter 9.

Sardinella aurita is unlike the other species of this genus that occurs on the West African coast; it is a long, herring-like fish, very round in section, whereas *S. maderensis* is a more typical tropical species, flatter, deeper, much less active, and apparently restricted to shallow water. Both in the eastern boundary currents of the subtropical regions, and in the tropical coastal upwelling off Ghana and the Ivory Coast, large stocks of *S. aurita* support important fisheries, particularly off Ghana. In this region, *S. aurita* alternates between periods spent as demersal schools at, or near, the edge of the continental shelf in the subthermocline environment, and as surface shoals closer to the coast (Boradatov and Karpetchenko, 1958). Normally, shoals of maturing *S. aurita* appear off the coast of Ghana at the time of seasonal upwelling, from June to late September; it feeds heavily during this period, as Bainbridge (1960b) has shown, largely upon *Calanoides carinatus,* a copepod which, as we have already seen, dominates the zooplankton of the tropical coastal upwelling ecosystem. Spawning occurs at the end of the upwelling season, after which the fish return to the deeper subthermocline waters offshore, where they again form dense demersal shoals which rise into midwater and disperse slightly at night. In the rare years when the upwelling does not occur or is very weak, such as in 1968, the fish do not appear at the coast, and spawning and the subsequent year-class fail.

In the subtropical regions of Senegal–Mauretania and Angola, the ecology of *Sardinella aurita* is compatible with the above description, though north–south seasonal migrations occur. Off northern Senegal, 1 to 2-year-old *S. aurita* aggregate at the frontal region from July–September with some daytime demersal schooling and nocturnal dispersion to the surface. Maturation occurs at this time and subsequently the fish move closer to the coast to spawn. In.the boreal winter, *S. aurita* moves to southern Senegal, where it displaces the neritic *S. maderensis* that migrate farther south to the tropical region. Comparable movements and concentrations of *S. aurita* occur in the Congo–Angola region in response to the austral seasons.

The western tropical Atlantic has a much more complex clupeoid resource; not only are warm-water engraulids available and exploited, but several species of *Sardinella* are important. Curiously, *Bevoortia* appears to be restricted to the subtropical regions, and there is no exact analog of the *Ethmalosa* fisheries of the eastern Atlantic.

The most important engraulid resources are based on about a dozen species: *Anchovia clupeoides* is common from the Caribbean to Brazil, and important in the local fisheries economies; *Anchoa mitchelli,* the bay anchovy, occurs from Cape Cod to Yucatan and has some industrial importance; *A. hepsetus* has a wider range, from Cape Hatteras to Uruaguay, and is abundant at many places in the Ceribbean; *A. tricolor, A. perfasciata,* and *A. elongata* have more central distributions along the eastern coast of South America and the southern Caribbean.

Two species of *Sardinella* are important in the western Atlantic, having rather similar ecology and some degree of geographical replacement. *S. anchovia* dominates along the northern coast of South America, and is especially abundant off Venezuela, while *S. brasiliensis* replaces it along the east coast of South America; the replacement occurs somewhat to the north of the Guianas. The major population of *S. anchovia* is concentrated along the eastern coast of Venezuela from Margareta Island to the Gulf of Carioco, where its life history is tied very closely to the upwelling cycle. Spawning peaks coincide with upwelling periods, when plankton abundance is greatest (Simpson and Griffiths, 1967). Because a modern purse-seine or midwater trawl fishery has not developed for this species, we are much less certain about its offshore ecology than for the West African *S. aurita*. The stocks of *S. brasiliensis* to the south in the Brazilian subtropical regions comprise a greater number of years classes (6–7) than the Venezuelan stocks (2–3) of *S. anchovia* and are relatively less abundant that the *S. aurita* stocks of the Atlantic eastern boundary currents.

The most important western Atlantic clupeid resources are the *Brevoortia* stocks that occur from the northern Gulf of Mexico (*B. patronus*) to the eastern coast of the United States (*B. tyrannus*), and off southern Brazil to northern Argentina (*B. pectinata* and *B. aurea*). The northern stocks have supported fisheries of world importance, reaching almost one million tons in the early 1960s, though catches have subsequently declined to about half this figure. The ecology and physiological energetics of *Brevoortia* are now very well known, due to intensive work on *B. tyrannus* (e.g., Durbin and Durbin, 1975). *Brevoortia* has an ecology very similar to that described above for the more tropical *Ethmalosa*; all species have a very fine gill-raker filter basket, and their diet is dominated by larger phytoplankton, principally estuarine diatoms. Long seasonal migrations maintain the stocks in areas of diatom and dinoflagellate blooms, and when these are not available the diet comprises microzooplankton filtered passively in the same manner as diatoms. Off eastern North America, spawning occurs rather early in the year in the Gulf of Mexico and progressively later as the stocks migrate northward; eggs occur only offshore, and the young fish move rapidly into estuaries where they spend the rest of the summer, leaving them to migrate south only at the onset of cold weather. Their dominant food during this period is estuarine diatoms. Migration of adult shoals takes place at the surface, but shoals occur deeper during the winter in the Gulf of Mexico and off Florida.

Though there are about 25 species of clupeoids in the eastern tropical Pacific Ocean, the resources are dominated by the Peruvian anchoveta *Engraulis ringens* of the eastern boundary current and the sardine *Sardinops sagax* which accompanies it. Along the coastal regions of intertro-

pical Central America, several engraulids form locally important resources; for many years, tuna bait-boats from the United States and
Central American countries relied almost entirely on *Cetengraulis myste-*
cetus for the live bait on which their operations depended; between 1939
and 1951 this species was entirely eliminated from the principal fishing
area in the Gulf of Nicoya, but is said to have been successfully reintro-
duced to the Gulf by the same bait-boats from other areas (Howard and
Landa, 1958). *C. mysticetus* has a diet in which phytoplankton is extremely important; as Bayliff (1963) showed, the stomach is muscular and
gizzard-like, a character it shares with phytophagous clupeoids, such as
Ethmalosa, and with mullets (*Mugil*). *Coscinodiscus* and other centric
diatoms dominate the diet of juveniles, while adults tend to feed on benthic diatoms such as *Melosira,* in proportion to the floristics of the benthic
diatoms of their habitat. Bayliff (1963) and Howard and Landa (1958)
showed that spawning occurs at the upwelling season, and that the
strength of the stock of *Cetengraulis* was sensitive to upwelling variability. Elsewhere along the coast, local fisheries utilize a number of other
species: *Anchovia starksi,* from El Salvador to Ecuador, *A. naso* from
Baja California to Peru, as well as the thread herring *Opisthonema liber-*
tate, especially in the Gulf of Panama.

The anchoveta (*Engraulis ringens*) of the Peru Current is anomalous in
the genus *Engraulis* for its very low-latitude distribution, which is a consequence of the great intensity and sustained nature of the upwelling
along the Peruvian coast. *E. ringens* differs from its higher latitude congeners in its rather fine gill-raker filaments, which also appear earlier in its
development than in other species of *Engraulis*. This is related to the
ecology of *E. ringens,* whose diet is based upon diatoms to a greater
extent than other species of *Engraulis*.

In the western Pacific there are many clupeid stocks of major fisheries
importance, of which the most significant are the various populations of
the Indian oil sardine (*Sardinella longiceps*), a variety of stocks of about a
dozen other species of *Sardinella,* and the river ascending shads, especially *Hilsa ilisha*. We shall discuss the biology of the most important
species, the oil sardine, in some detail below as a tropical example of a
clupeid stock that exhibits much between-year variability in abundance.

Other than the oil sardine, the sardine stocks of the western Pacific
resemble *S. maderensis* of West Africa—zooplankton feeding, coastal
species, with limited regional migration patterns, and not dependent on a
major seasonal upwelling system with its associated phytoplankton
bloom. Some fisheries have developed for small engraulids in this region
(Tham, 1968), perhaps the most important being for *Stolephorus,* of which
Wongratana (1983) recognizes 19–20 species in the region of the South

China Sea and the Philippines. *S. amboiensis* dominates in the tuna bait fishery in parts of the western Pacific (Dalzell, 1984b) and most are coastal, shoaling fish, though *S. punctifer* is a stenohaline oceanic species occurring >500 km from the coast (Dalzell, 1983). There is a possibility of some interisland migration in the archipelagos but no suggestion of any major interregional movements. Some species (*S. commersoni, S. indicus*) are continental shelf species that enter brackish water only occasionally; others (*S. bataviensis, S. heterolobus*) are more strictly coastal and neritic, while yet others (*S. tri, S. macrops*) are most abundant near river mouths and enter brackish water more freely. Some species (*S. indicus, S. batavius*) feed almost solely on zooplankton, while others (*S. heterolobus*) are more eclectic and are able also to consume phytoplankton. As Tham (1950) demonstrated, there is a positive correlation between zooplankton abundance in the Singapore Straits and the abundance there of *Stolephorus*, which forms an important part of the whole pelagic fish biomass. Li (1960) describes local fisheries for *Sardinella aurita, S. sirm, S. fimbriata,* and about five other species right through the tropical western Indo-Pacific from the African coast to the Philippines. To judge from his listings of local names, about four to six species are of significance in each region.

There are also data on the stocks of clupeoids and other pelagic species from the modern acoustic surveys done by the Norwegian survey ship *Fridtjof Nansen* along the coast from Somalia to Pakistan (Kesteven *et al.*, 1981). Pelagic fish were especially abundant on the northern part of the eastern coast of Somalia, and *Sardinella longiceps* and *Etrumeus teres* were found in spawning concentrations at the upwelling centers of Ras Hafir and Ras Hafun. Few fish were found on the northern Red Sea coast of Somalia, but there were other concentrations of the same two species along the southern Omani coast from 16 to 22°N. Off Somalia and Oman, the sardines were accompanied by *Trachurus, Decapterus* and *Scomber.* Off both Somalia and Oman, stock assessments for all pelagic stocks were of the order one million tons.

The stocks of *Hilsa* and *Macrura,* river-ascending shads of the Indian Ocean and South China Sea, also command attention. Species of these two genera occur importantly in the estuarine fisheries of Madagascar, of the eastern coast of Africa, of India and Peninsular Malaysia, and of the Indonesian archipelago. *Hilsa ilisha* has been well reviewed by Pillay and Pillay (1958) for the Indian region. As might be anticipated, this species exists as separate races associated with the various major river systems of the subcontinent; the Pillays remark that meristics and morphometrics confirm the reported abilities of Indian fish traders to judge the provenance of fish from various Indian rivers by their taste, which as the Pillays point out, is largely dependent on oil content and hence environmental factors. *H. ilisha* is a very important species from the Arabian Seas (Indus system) to the South China Sea (rivers of Vietnam) and it occurs in brackish waters throughout this huge area.

Outside the spawning season, *Hilsa* is a fish of coastal regions and the lower reaches of estuaries, where its ecology resembles *Alosa* of the Atlantic; its distribution follows the 22–28% isohalines (Pati, 1982) and coastal fisheries occur during the season of continental shelf estuarization. The stimulus for the annual spawning migration upstream is the regional discharge level of fresh water, whether caused by the local monsoonal regime or the spring thaw of the snowpack on the Himalayas. Fish may move several hundreds of kilometers upstream to spawn their planktonic eggs, and larvae and juveniles progressively descend back to the estuaries. Juveniles reach the sea when they are about 1 year old.

The only clupeoids of fishery significance in the central Pacific archipelagos are small local stocks of such forms as *Stolephorus purpureus*, the nehu of Hawaii, which is mostly of significance as a source (albeit insufficient) of live-bait for the small pole-and-line boats that fish for skipjack out of these islands.

It is probably sensible to discuss briefly in this section the exocoetid flying fish resources of tropical seas. The Exocoetidae are, like the Clupeiformes, relatively primitive teleosts that have a singular evolutionary specialization. They resemble herrings in their general form and appear to have originated relatively recently from a neritic origin: indeed, the neritic Belonidae, or needle fishes, are probably another divergence from the same stock. The more evolved genera (*Exocoetus, Hirundichthys*) are found in the open ocean, while some genera less apt for flight are restricted to a more neritic habitat (*Fodiator, Parexocoetus*). About 40 species of exocoetid flying fishes are distributed in tropical oceans and coastal seas. In some places, as in the Caribbean and Indonesia, they are sufficiently abundant to constitute a fishery resource of at least regional significance.

Many species of Exoceotidae, especially during their juvenile stages, have an extended, protuberant lower jaw. Exocoetidae reveal their neritic origin by the fact that all species, even those occurring in the open ocean, deposit their eggs on demersal substrates or some other solid object, such as a frond of *Sargassum* weed. This habit is the basis of one of the most important techniques in the Caribbean fishery for *Hirundichthys*; attracted to the boat by chumming with finely chopped fish, the *Hirundichthys* are induced to aggegate below rafts of palm fronds, in which they commence (or can be induced to commence by suitable strategems) to spawn. They are then boated with hand-nets. Off Sulawesi (Indonesia) it is the egg-laden palm fronds that are collected for the sale of the adhering egg masses of *Cypsilurus,* a local exocoetid.

Of course, the principal adaptation of Exocoetidae is to flight. This is achieved by planing close to the sea surface on a pair of wing-like pectoral fins which, when folded, extend to about the base of the caudal fin. In some genera (e.g., *Hirundichthys*) the pelvic fins are also modified as aerofoils, so that two pairs of wings are used for planing. The motive force for planing is achieved underwater, prior to launching into the air, though under calm conditions secondary thrust may be obtained from the lower, elongated fork of the caudal fin along the water surface; prior to take-off, and during the period of secondary thrust, the caudal fin beats at a frequency of about 50 cycles/second. This mechanism is the same as used by wingless needle fishes that are able to "run" along the water surface on their lower tail lobe. In both Exocoetidae and needle fishes, the flight mechanism is just that: an escape from a threatening underwater predator. In the largest and most highly evolved exocoetids, flight may extend for 100–200 m, with a duration of at least 10 sec; it seems probable that the ground effect, by which an aeroplane flares out just before touching down, is also effective in maintaining long flight in Exocoetidae, at least over a very calm sea. The geometry seems about right.

Flying fish are generalised pelagic predators, and their diet (Parin, 1970; Mahon *et al.,* 1986) comprises macroplankton (copepods, larval decapods, molluscs, salps, and even siphonophores) as well as fish larvae and small fish, including postlarvae of flying fish. The related half-beaks (*Belone*) are of very similar form and obviously (to any sensible biologist) evolved to feed actively on macroplankton, but one Caribbean species has the apparently odd habit of ingesting relatively large floating leaves of sea grass, perhaps for the sake of their epibiota. This is done by day, and the species is said to feed more normally at night.

VARIABILITY OF TROPICAL CLUPEOID STOCKS

It is now widely understood that high latitude clupeoid stocks are more variable on a decadal or secular time scale than demersal stocks, and even apparently rational management appears unable to prevent the fisheries from following a "boom and bust" cycle. We will now discuss between-year variability in two tropical clupeoid stocks in which recruitment variance appears to be driven at least partly by environmental variability.

We have touched on the first case already in Chapter 3, in relation to the biological consequences of El Niño–Southern Oscillation (ENSO) events. The anchoveta, *Engraulis ringens,* occurs from northern Peru to central Chile, and seasonally further to the south; meristic analysis indicates three separate stocks, two normally off Chile and one off Peru. South of Punta Aguja, at 6°S, spawning continues through the year, though there are spawning peaks in October and (more importantly) in February–March. The principal period of recruitment (of pelladillas, at about 8 cm L_t) to the fishery is in July–September, of fish spawned earlier the same year. Thus, in the period June–September, 2-year classes are in the fishery, though availability may not be highest at this season. Few fish survive beyond their second year of life.

The great fishery for anchoveta that boomed in the 1960s and reached an official total catch of about 12 million tons (which may really have been closer to 15 million tons) before collapsing, had until recently drawn attention away from the fact that in the Peru Current there existed a classic anchovy–sardine pair, such as dominates the clupeid fisheries of the Benguela, Canary, and California and other upwelling regions at much higher latitudes and which show a high degree of variance in the relative stock sizes of the two species. The Peruvian sardine (*Sardinops sagax*) has a very similar ecology to those of the other eastern boundary currents; it is large herring-like fish, feeding almost entirely on zooplankton, and having a somewhat different distribution from anchoveta. In all sardine–anchovy pairs, there are clear differences between the habitat requirements of the two species. In these pairs, anchovies usually spawn in water a few degrees cooler than sardines, and more coastally; in addition, anchovies may be more fecund than sardines. In other anchovy–sardine pairs, there is evidence of natural fluctuation in relative dominance of each member of the pair on both decadal and secular time scales; Soutar and Isaacs (1969) have interpreted sediment core material from the anoxic Santa Barbara basin to show (from the abundances of scales therein) how the California sardine–anchovy pair has fluctuated in relative abundance through the last two centuries. Until the industrial fishery intervened, it appeared that the species pair in the Peru Current was more permanently dominated by *Engraulis*; it was the anchoveta that supported the population of over 15 million sea birds that existed along this coast prior to the fishery, and had existed there for many centuries, furnishing guano deposits on the offshore islands that were used historically to fertilize pre-Columbian agriculture in coastal oases.

It is not possible to discuss variability of the clupeoid stocks of the Peru Current without considering the multiple effects on them of fishing pressure and ENSO events. In an unexploited state, dominated by anchoveta and preyed upon principally by bonito (*Sarda chilensis*) and sea birds, the Peruvian species-pair may not be much affected by an ENSO event except that the centers of their populations may shift temporarily southward, and recruitment may be enhanced by reduced bird predation. One of the consequences of El Niño seems always to have been a population crash of cormorants and boobies due to the unavailability (further south, deeper) of anchoveta. However, the 1972–1973 ENSO event was extremely strong (perhaps the strongest of this century until then) and anchoveta recruitment failure occurred in 3 years consecutively, 1971–1973. The reproducing stock was reduced drastically by continued fishing during this El Niño, when anchoveta was concentrated inshore by warm water. The anchoveta stock began to recover subsequently, and by about 1975 had probably reached as much as 4 million tons once again. However, subsequent ENSO events in 1975 and 1977–1978 caused further recruitment failure and reduction of the stock to a very low level. During the 1982–1983 ENSO event, which was certainly the most powerful yet studied, anchoveta occurred only as very small patches along the coast of southern Peru; spawning was reduced from September 1982 and did not occur off Peru during the normal spawning season of February–March 1983. It is thought that the total stock biomass was reduced to its lowest recorded levels by the end of the 1982–1983 ENSO event (Santander and Zuzunaga, 1984).

After the 1972–1973 ENSO, at the time of the initial collapse of the anchoveta biomass and of the fishery from its late 1960s levels, the stock of *Sardinops sagax* off Peru began to increase significantly, and an important fishery developed over the next 10 years; during the 1982–1983 ENSO the sardine schools moved southward from their normal distribution at 6–14°S, and catches were very reduced as fish moved deeper than normal in response to the warm surface layer that invaded the Peruvian region (Fig. 7.1). By February 1983 the sardine spawning index was the

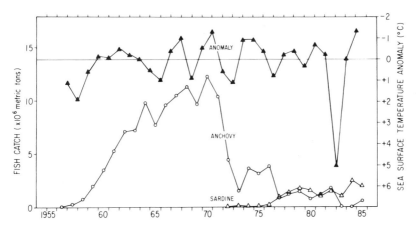

Fig. 7.1. Temperature anomalies (note inverted scale) and abundance of *Engraulis* and *Sardinops* off Peru, to show effect of ENSO events. (From R. C. Barber, personal communication.)

lowest recorded since 1972, with subsequent impaired recruitment; nevertheless the total biomass of sardines appears not to have been reduced by the events. It is noteworthy that this was in contrast to the spawning sucess of *Trachurus* and *Scomber,* both of which appear to have been enhanced by this ENSO event off Peru.

Turning now to the population biology of the Indian oil sardine, we note that this species has been studied much more extensively than the remaining Western Pacific species of *Sardinella.* In fact, there seems no real agreement even on how the different *Sardinella* species are actually distributed; each species probably comprises several individual stocks, and these are known to exist in *S. longiceps* from meristic evidence. Raja (1969) has suggested a rather restricted distribution for *S. longiceps,* rejecting the occurrence of this species east of the Bay of Bengal. He believes that the records of *S. longiceps* from Indonesia and the Philippines actually refer to *S. aurita,* a very similar species, though this is not a view accepted by other workers, who still continue to distinguish between *S. longiceps* as the oil sardine of the fisheries and canneries of the Sunda Islands and *S. aurita.* Be that as it may, it is from the west coast of India that we have the best information about *Sardinella* biology in the Indo-Pacific. As Raja reminds us, concern for the management of the oil sardine on the Indian coast goes back more than a century; Francis Day, in his 1865 "Fishes of Malabar," expressed concern that stimulation of the fish-oil trade might result in the fisheries for *S. longiceps* being "overworked." Already by 1910, James Hornell and his collaborators were seeking a means of predicting the highly variable appearance of oil sardines along the Indian coast each year (Raja, 1969); as we shall see, that search is still going on.

Sardinella longiceps is one of the major fishery resources of the western Indo-Pacific, but is highly variable in its yield, though less so in recent decades than formerly. Its yield to the coastal, still largely unmechanized, fisheries has fluctuated from less than 100 tons to more than 300,000 tons annually over the 60 years for which we have seen records. The fishery is concentrated along the western coast of India though *S. longiceps,* even in the restricted view of Raja, occurs from the Bay of Bengal to Kenya, excluding the Persian Gulf and the Red Sea.

Shoals appear seasonally at the coast, progressively northward from Calicut and Cochin as far as Vengurla, at the start of the southwest monsoon; these shoals are thought to move inshore from deeper water. The inshore movement begins in June, when populations of adults with mature gonads appear at the coast and spawn soon after their arrival. As the season advances, a second wave of shoals of immature O-group fish arrives whose availability peaks from October to the end of the year. By this time the original mature adults have left the coast, and the fishery continues for a few more months as the younger fish disperse back to deep water. There is unanimity amongst those who (from Hornell onwards) have studied the oil sardine that it is the seasonal bloom of phytoplankton, dominated by diatoms, occurring during and immediately after the monsoon that draws the fish to the coast. Raja (1969) reviews the many studies of feeding in *Sardinella longiceps* that make it clear that, though juveniles <15 cm are heavily dependent on small crustacean zooplankton, beyond this size the diet becomes increasingly dominated by phytoplankton as the fine filaments on the gill-rakers progressively develop. The phytoplankton diet is dominated by the larger diatoms: *Fragillaria, Coscinodiscus, Pleurosigma,* and *Biddulphia.* Again, starting with Hornell, there is some suggestion of feeding on benthic organic detritus during the period October–December; however, we note that this is also the period when the "chakara" along this coast often become suspended as liquid mud in the water column, and there is much mobility and transport of organic detritus. It seems more plausible that at this season *S. longiceps* frequently encounter regions where the nonliving organic suspension has a sufficient content of viable or nonviable plant cells to make it a useful food. There seems to be no need actually to infer benthic feeding from the evidence available.

Figure 7.2 shows the fluctuations in the availability of *S. longiceps* (and Fig. 7.3 of *Rastrelliger kanagurta*) in the Kerala–Mysore section of the Indian coast over a period of 58 years: this coast, it should be noted, supplies 99% of all Indian oil-sardine landings, and supports a species pair (in this case, sardine–mackerel) recalling the species pairs of the eastern boundary currents. The variability of their landings requires explanation, and prediction would also undoubtedly be of great value to the fishery. A number of correlations have been made between environmental variables and the availability of oil sardines, starting (yet again) with the redoubtable Hornell, who suggested a connection with local rainfall. This idea has been renewed recently by Anthony Raja (1972b, 1973), who finds a positive correlation between the abundance of juveniles and the rainfall during the peak spawning period of the preceeding season; his statistical evidence is also linked with observations on the relation between the relative atresia of adult gonads during a spawning season and rainfall, and he claims successful prediction for at least 3 separate years. Rainfall is a component of the general seasonal development of the monsoon, and the development of phytoplankton blooms which, as Nair and Subrahmanyan (1955) suggested, are linked to the appearance of the sardines along the coast. An inverse relationship between abundance of oil sardines and mackerels has frequently been reported along the Kerala–Mysore coasts.

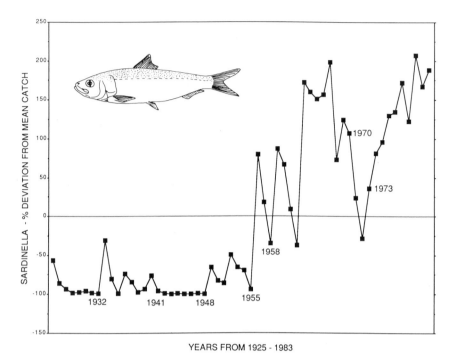

Fig. 7.2. Catch of oil sardine, *Sardinella longiceps,* on the southwest coast of India, 1925–1983, expressed as deviation from long-term mean. (Data from Raja, 1969, and FAO landing statistics.)

Rastrelliger has a more eclectic diet than *Sardinella longiceps,* not requiring phytoplankton blooms but able to utilize them. It is possible, Raja (1969) thinks, that one species may exclude the other from the coastal zone depending on the relative strength of the phytoplankton bloom.

We have assembled a data series (Fig. 7.4) for the landings of both species covering 58 years (1925–1983) which confirm the high variability noted as early as 1895–1910 by Hornell; this data series describes a fishery utilizing very similar gear throughout this long period. Our data clearly show the main features of the variability: that *Sardinella* are basically more variable in their availability than *Rastrelliger* and that two major events have peturbed the fishery since 1925, both apparently environmentally driven. There was an almost complete disappearance of *Sardinella* during the years 1941–1949, though *Rastrelliger* catches remained at a normal level. The year 1941 saw a catastrophic failure of the monsoon

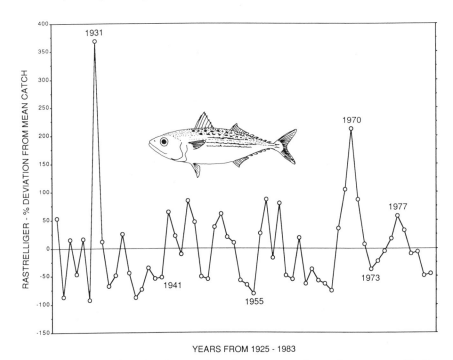

YEARS FROM 1925 - 1983

Fig. 7.3. Catch of mackerel, *Rastrelliger kanagurta,* on the southwest coast of India, 1925–1983, expressed as deviation from long-term mean. (Data from Raja, 1969, and FAO landing statistics.)

rainfall and the strongest ENSO event that had occurred since 1826. From 1954, rainfall on the west coast of India increased significantly over the average for the preceeding 25 years and in 1955–1956 the *Sardinella* catches suddenly increased from <50,000 tons annually to a new plateau of 100,000–300,000 tons, which was maintained until at least 1983 with all the appearance of a sustainable fishery (Fig. 7.5). During this period the availability of *Rastrelliger* has remained at about its pre-1955 levels. Embedded in these data are some events that seem to indicate that unusually high abundances of *Rastrelliger* (as in 1968–1972) is reflected in low availability of *Sardinella*. This will be a most fruitful subject for further study, but the impression gained from this data set is of a pair of interacting pelagic species whose relative abundances are strongly under environmental control and whose populations were well able to sustain continual fishing pressure from an unmechanized fishery.

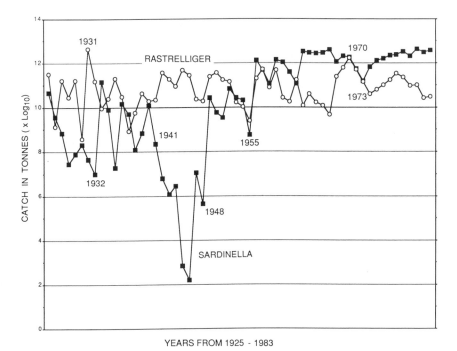

YEARS FROM 1925 - 1983

Fig. 7.4. Catches of *Sardinella longiceps* and *Rastrelliger kanagurta,* southwest India 1925–1983, to show detail of the population crash in the 1940s of sardines. (Data from Raja, 1969, and FAO landing statistics.)

CARANGIDS AND SCOMBROIDS OF COASTAL SEAS

As well as the oceanic tuna and billfish, which are reviewed in the next section, a great variety of genera of these two highly evolved perciform families occurs over continental shelves and in coastal waters: mackerels, seerfish, jacks,and trevallies are a prominent feature of the pelagic eco-system throughout the tropical seas.

Carangids are diurnally active, large-eyed fish having very small scales except along the lateral line where they form a series of armoured scutes; carangids are usually laterally compressed and most are strong, active swimmers. The basic carangid form is expressed in the jacks or trevallies (*Caranx*), which are strong predatory fish occurring in loose shoals in neritic regions; species of *Caranx* can attain about 60–75 cm TL, and are often golden or yellow in dorsal coloration. Shoals of *Caranx* frequently form feeding flurries right at the surface by forcing shoals of small fish surfacewards, and the splashing of a feeding group of *Caranx* is suffi-

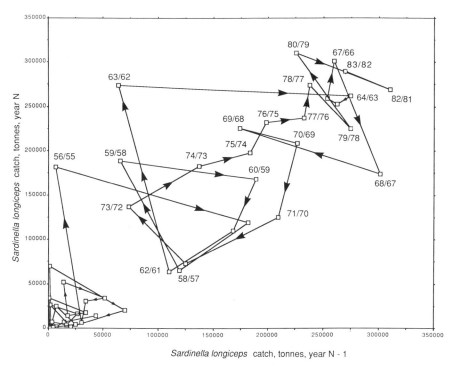

Fig. 7.5. *Sardinella longiceps* catches as a function of catches in previous year, to show the difference between the periods before and after 1955, and the cyclic trajectory taken by this function. (Data from Raja, 1969, and FAO landing statistics.)

ciently prominent to attract fishermen and fish-eating birds from considerable distances. Such feeding flurries, attended by a multitude of terns (feeding on the same prey) and sea eagles (feeding on the *Caranx*), are an unforgettable sight along the African coastline.

Of the same general shape, but less heavily built and not reaching so large a size as *Caranx,* are the genera collectively known as horse mackerel and pompanos (*Trachinotus, Decapterus, Trachurus, Selar*). These are less active fish than trevallies, forming a series from pompanos (*Trachinotus*), which most closely resemble trevallies to the smaller mackerel-like *Selar* of tropical coasts and *Trachurus* of cooler eastern boundary currents. These are all predatory fish, of whose life histories not a great deal is known in most regions; the tropical forms occur in the catches of bottom trawls on continental shelves along with some smaller carangids which depart from the *Caranx* body plan: very laterally flattened, with a very high forehead, and diamond-shaped in lateral plan, genera such as

Vomer, Chloroscombrus and *Scyris* occur throughout the tropics in the bentho-pelagic fish fauna.

In the eastern Atlantic, two species of *Trachurus* occupy the continental shelf from Morocco to Namibia (Boely and Fréon, 1980). Their abundance is much higher in the eastern boundary currents to north and south of the tropical area: the subtropical species is *T. trachurus* (with subspecific distinction between the northern and southern populations) which occurs only in the eastern boundary currents and does not, for instance, occur south of Cape Verde in Senegal, while *T. trecae* sparsely occupies the tropical Gulf of Guinea as well as maintaining populations in the more equatorward regions of the eastern boundary currents. Where the two species co-occur, it is reported that *T. trecae* occurs closer to the coast than *T. trachurus*. In the subtropical regions, both populations migrate with the intertropical front. In the Canary Current, *T. trachurus* reproduces off Mauretania (20–26°N) in winter, while *T. trecae* reproduces over a longer period, from February to June and farther south, from Cape Verde to Cape Timris (14–19°N). Commercial fishing concentrates on the spawning concentrations both here and in the southern region. *T. trecae* extends through the tropical Gulf of Guinea region in small numbers, with its principal populations near the bottom at the edge of the continental shelf.

In the western Pacific, roundscads (*Decapterus*) are important; indeed, two species (*D. russelli* and *D. macrosoma*) alone formed almost 40% of all commercial landings and 10% of all catches in the Philippines during the 1970s (Anonymous, 1978a), though, as Ronquillo (1974) describes, the fishery is based on juvenile age groups; landings are generally of fish of <20 cm in length, though they are an excellent market fish even so. The adults were not available to the fishing gear then in use, which were principally lights to attract fish to various capture devices at night. The Sulu Sea, and other passages between the islands of the Philippines archipelago are the principal population centers for *Decapterus*. Though it is entirely internal to the Philippine islands, the Sulu Sea is deep (<5000 m), but with shallow sill depths to the open Pacific Ocean, and the areas occupied by *Decapterus* stocks are over the shoaler parts on which are situated the island chains surrounding the Sulu Sea itself—on the south coast of Palawan, in the passages between Leyte and Panay, along the south coast of Mindanao, and on either side of the great southern peninsula of Luzon. Though the fish are present year round, there is a peak in landings during the dry season of April and May. There appears to be some general relationship between regions of high plankton abundance and *Decapterus* fishing grounds; roundscad have a diet almost entirely based on zooplankton.

Spawning of *Decapterus* occurs rather sporadically through the year over this large and very variable region, and reproductive seasons are not simultaneous everywhere. Added to the local variability in reproductive seasons are differences in relative abundance of the two species that form the bulk of the catches. Off southern Palawan, catches are mostly of *D. macrosoma,* while near Manila and along the north coast of Palawan they are mostly of *D. russelli*; no general explanation for differential distribution appears to have been advanced yet. However, it must be emphasized that current knowledge of the biology and distribution of this resource is based on very fragmentary knowledge of the adult distributions, migrations, and spawning sites, and almost entirely on what can be gleaned from the regional availability of juveniles to a single form of fishing gear.

As widely distributed as the carangids, the smaller scombroids, the Scombrini or mackerels, are less diverse and of greater fisheries importance, though principally in the Indo-Pacific. Though one genus (*Grammatorcynus*) is anomalous, being elongated and resembling the deep-sea snake mackerels, the Scombrini are a very homogeneous group. Resembling the temperate mackerel *Scomber,* though a little deeper and more laterally compressed, the western Indo-Pacific mackerel *Rastrelliger* is both a very important fishery commodity and a prominent and important member of the pelagic ecosystem of regional continental shelves.

Rastrelliger occurs throughout the western Indo-Pacific from the Mozambique Channel to the Red Sea, over most of the Polynesian archipelago from Queensland to Hawaii, and northward to Taiwan and the Philippines. Within this area, three species occur: *R. kanagurta* occupies the whole region, *R. brachysoma* has a distribution centered on the Sunda Shelf, and extending from the Andaman Sea to Fiji, while from the Gulf of Thailand to the Philippines a third species, *R. neglectus,* whose reality is actually in question, may be important. The species are distinguished largely on allometric proportions, *R. brachysoma* being relatively shorter and deeper in the body than the others, and *R. neglectus* being the thinnest and longest. Where the two species coexist, *R. kanagurta* occurs offshore where it feeds on macroplankton in water of at least 32%, while *R. brachysoma* is a neritic form and utilizes smaller food particles than the other species. In fact, *Rastrelliger,* like other mackerels, are eclectic feeders and stomach content analyses throughout their area of distribution (e.g., as reviewed by George, 1964) show that they are capable of utilizing a great range of food; records exist of stomachs being filled with green, microscopic algal material, with copepods and other small planktonic organisms, including medusae, and with macroplankton such as pelagic penaeids and larval and juvenile fish. Though there may be an ontogenetic progression from a microphagous to a macrophagous diet during growth, adults do not seem to be such predatory, piscivorous fish as, for instance, the high-latitude mackerels (*Scomber*). Further, there is some evidence that it is only the larger fish that are really capable of substantial filtration of diatom blooms to the extent that stomachs are filled with plant material.

Both species undergo extensive migrations, to which we shall return later in discussing interactions between pelagic species on the continental shelf, and this appears to occur by the movement of large shoals, numbering up to several million fish. Seasonally, *Rastrelliger* migrations bring a very significant resource to certain tropical coasts as we have discussed for the west coast of India, and the degree of uncertainty associated with their appearance may therefore be of great regional importance. Migrations are obviously linked to oceanographic cycles, though the patterns of migration are not at all clearly understood as yet, being mostly observed at the places where fisheries actually occur. Thus, in the Java Sea, it is known that as the current reversal occurs with the onset of the northeast monsoon, the *Rastrelliger* stock (together with a stock of *Decapterus*) moves off westwards with the flow of water and disappears from the fishery; after some weeks interval, a different stock enters the Java Sea fishery from the east. As the monsoonal currents reverse again at the end of the season, two stocks enter the Java Sea from the west, one apparently from the Indian Ocean the other from the South China Sea. Off western India, onshore movement of *Rastrelliger kanagurta* occurs with the northeasterly monsoon, and the fisery there is based on this fact.

As Dhebtaron and Chotiyaputta (1974) remark, it is somewhat surprising that the most important fish landings from the Gulf of Thailand comprised *R. neglectus,* rather than demersal fish, and reached almost 120,000 tons by 1966. The fishery collapsed in 1968, but a strong stock reappeared in the Gulf in 1970, after which it again contributed almost one-third of the total fish from the Gulf. The Gulf stock spawns in a relatively restricted band along the western coast (Fig. 7.6); within this region there are two spawning peaks, February–March (major) and July–September (minor). As with many pelagic fish in the tropics, spawning occurs only in the early part of the night, final maturation being very rapid during the evening. Larval diet is largely of very small crustacean plankton, while adults are capable of utilizing both zoo- and phytoplankton. Adult migration between the feeding grounds of the Inner Gulf and the spawning grounds along the west coast has been demonstrated by tagging (Somjaiwong and Chullasorn, 1972), though multisite tagging demonstrated a great deal of interchange of individuals throughout the western half of the Gulf.

A few larger species, the Spanish mackerels or seerfish (*Scomberomorus*) are also widespread on tropical continental shelves, and may be

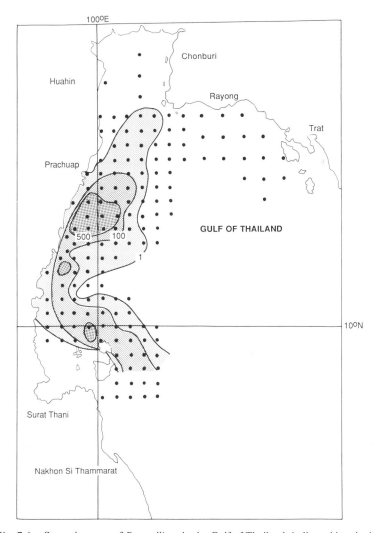

Fig. 7.6. Spawning area of *Rastrelliger* in the Gulf of Thailand, indicated by planktonic egg survey. Eggs m^{-2}. (From Dhebtaron and Chotiyaputta, 1974.)

extremely abundant at times. Though they are nowhere the single object of a major fishery, they are a consistent minor component of the landings on most tropical coasts and form 12% of Indian landings. Like *Rastrelliger* and *Scomber,* these are shoaling fish, sometimes occurring in very large migratory shoals, but are also (and perhaps more frequently) encountered as single fish or small groups of fish. Tagging studies off Queensland have shown that *S. commerson* undergoes both onshore–

offshore ontogenetic migrations (between spawning grounds along the inner side of the Great Barrier Reef and the coastal nursery ground), and seasonal north–south migrations of the adults (Mcpherson, 1981); however, it appears that on the western coast of India, spawning grounds are close inshore, reproductive products being recorded in the Gulf of Mannar and along the Coromandel coast (Chacko *et al.*, 1967). On the western Atlantic coast, *S. maculatus* performs similar seasonal migrations from the Gulf of Mexico and Caribbean north and south along the east coast of the United States and of South America during the summer.

It is appropriate to note here the existence of other predatory fish of continental shelves that may regionally have some significance, especially in the artisanal fisheries. Perhaps most important are the barracudas (*Sphyraena* spp.) that occur on all tropical continental shelves, particularly in clear, high salinity water (de Sylva, 1963). About 20 species exist in this group, and some are present in all three oceans. Adult size varies from less than 1 m (in most species), to almost 3 m in the great barracuda (*Sphyraena barracuda*). Young individuals and smaller species often aggregate into shoals, while older individuals and larger species are more usually solitary. Barracudas are probably "sidle and dash" predators, having a very streamlined form and a very large caudal fin relative to their muscular mass.

There is some indication from acoustic surveys of the Mozambique continental shelf that barracudas may at times associate into very large shoals. Saetre and de Paula e Silva (1979) recorded shoals of *S. obtusata* and *S. japonica* at the 200- to 250-m line on the continental edge off Mozambique that were locatable by echo sounder, giving characteristic traces near the bottom. Catch rates of <2 tons/hr were obtained from such aggregations.

During ontogeny, there is some seasonal migration of individuals between neritic habitats and the open continental shelf, and some species are specialized to the coral reef ecosystem. *Sphyraena* are almost entirely predatory upon other fish, their diet changing according to the fish they are associated with seasonally. They are a fishery resource of secondary importance, like the larger carangids, and like them may carry dangerous loads of ichthyotoxins.

THE OCEANIC SCOMBROID RESOURCE: TUNAS AND BILLFISH

A few species of three families of scombroid fish comprise the oceanic tuna and billfish resources of the tropics; swordfish (*Xiphias*) of the family Xiphidae, the sailfish and marlins (*Makaira, Istiophorus,* and *Tetrap-*

turus) of the Istiophoridae, and tunas (*Euthynnus, Katsuwonus, Thunnus*) of the Scombridae. Eleven species of tunas, swordfish, sailfish, and marlins dominate the large predatory fish component of the oceanic ecosystem, and the commercial catches from open tropical oceans; a few species of these groups are rarer or local. In addition, a further dozen species of smaller tuna-like fish, many of which are known as bonito, also occur in the same regions, though some are more important in continental shelf ecosystems than in the open ocean; these fish are of the genera *Sarda* (bonito), *Auxis* (frigate tuna), and *Orcynopsis* of the tropical Atlantic, *Gymnosarda* of the Indian Ocean, and *Cybiosarda* of the northern Australian region.

Many species of these open ocean scombroid genera are cosmopolitan and occur in all oceans within the tropical or subtropical zones; regional distributions are related to latitude or oceanographic regimes. Thus, yellowfin tuna (*T. albacares*), bigeye tuna (*T. obesus*), and skipjack (*K. pelamis*) occur in each ocean throughout the regions that are occupied by tropical surface water, but do not extend far into regions defined oceanographically as subtropical. However, within their general area of distribution, each of these species is specialized. Yellowfin is the tropical tuna *par excellence* and its poleward distribution is bounded by the 18–21°C isotherms, though temperatures in excess of 30°C may be avoided, as in the region off the west of Central America; longline catch data show that yellowfin and skipjack inhabit the tropical mixed layer above the thermocline, but are capable of penetrating down into cooler water.

Skipjack have a very similar distribution but have greater temperature tolerances, occurring in water as cool as 16°C and not seeming to avoid the warmest pools of surface water. Until recently, it was thought from fishing gear that yellowfin and skipjack were exclusively warm-water, near-surface species and indeed both species may frequently be observed feeding just below the surface, forming the "breezing schools" so familiar to tuna fishermen. Now, however, we have evidence from acoustic tags that transmit data on depth (or temperature) that both species make regular, short-term feeding dives to 200–400 m, reaching into water <10°C. Both species apparently spend much time at midthermocline depths, with some suggestion that skipjack spend more time there than yellowfin. From this common depth, forays down into cold water and up toward the surface are made. Day–night bathymetric differences are also recorded in these data. Hunter *et al.* (1987) review the results so far available from this technique.

On the other hand, bigeye tuna that have a very similar geographical distribution to yellowfin and skipjack have a markedly different bathymetric distribution; bigeye seldom appear at the surface, but are abundant at thermocline depths, whatever these may be regionally, throughout the

tropics. Albacore (*Thunnus alalunga*) are subtropical fish with a temperature tolerance of 14–23°C and a surface distribution in all oceans poleward
of the three principal tropical species; however, like many species of
plankton, albacore do occur right throughout the tropics, almost as stere-
otypes of the tropical submergence phenomenon. Especially in the western basins of oceans, they occur at approximately the same depths as
bigeye; apparently, their preferred temperatures occur in the eastern
parts of tropical oceans over too narrow a depth range for them to occupy
it successfully.

Finally, to carry this series to its conclusion, the bluefin tuna (*T. thynnus*) of temperate seas also has a deep distribution through parts of the
tropical oceans, especially on the western sides of the Atlantic and Pacific. Bluefin are extremely large fish, weighing as much as half a ton, and
(as we shall see) perform ocean-wide seasonal migrations. Reproduction
occurs in a few rather restricted parts of the ocean, south of Japan, in the
eastern Gulf of Mexico, in the Mediterranean,and (for the southern hemisphere population) off northwestern Australia. Figure 7.7 shows the extent of equatorially submerged populations of bluefin tuna through the
tropical Atlantic Ocean, usually at depths of at least 100 m, and sometimes twice that.

Swordfish (*Xiphias gladias*), sailfish (*Istiophorus platypterus*), and blue
marlin (*Makaira nigricans*) are, like yellowfin and skipjack, cosmopolitan
tropical surface water species; as in tuna, there are also differences between their specific distributions. Perhaps most importantly, sailfish have
a tendency to occur in greater abundance near land than in the open
ocean.

Besides these cosmopolitan species of oceanic scombroids, some species are restricted to one ocean basin, or to an even smaller region. The

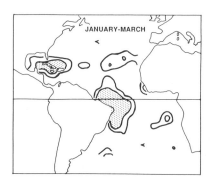
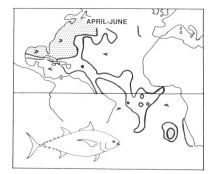

Fig. 7.7.   Seasonal distribution of bluefin tuna in the Atlantic Ocean showing summer
feeding and winter spawning regions. (From Wise and Davis, 1973.)

longtail tuna (*Thunnus tonggol*) and the kawakawa (*Euthynnus affinis*) are abundant in the western Indo-Pacific, rare in the eastern Pacific, and absent in the Atlantic; the black skipjack (*Euthynnus lineatus*) occurs only in the eastern Pacific, where it is very common. The striped marlin (*Tetrapterus audax*) and white marlin (*T. albidus*) occur only in the Indo-Pacific and Atlantic Oceans, respectively; the black marlin (*Makaira indica*) is widespread in the Indo-Pacific but is only occasional in the Atlantic. All species undertake seasonal and ontogenetic migrations, some of which are known to be ocean-wide, the fish traveling between spawning regions and feeding grounds appropriately for the size of the fish concerned. Tuna ecology and migrations are regulated by the oceanography of the surface waters of the ocean; temperature, transparency, and the existence of ocean fronts and other features are the principal physical determinants of their distribution, while the interaction between physical oceanography and the production of nektonic organisms suitable as food for large scombroids determines the existence of suitable feeding areas (Blackburn, 1965).

All species of tuna and billfish are migratory and their migrations can probably be simplified into a strategy for placing the adults during the reproductive periods in water masses most favorable for larval growth and survival, and for placing the juveniles and nonreproductive adults in water masses containing a high availability of suitable prey. Food items are usually nekton in the case of tuna, and both nekton and medium-sized pelagic fish for adult billfish. Migration is best developed in the temperate and subtropical species of tuna; albacore (*Thunnus alalunga*) annually cross the North Pacific Ocean in both directions, so that the same stock of adults appears off the coast of western North America (California–Oregon) in late summer, and off the east coast of Japan 6 months later; tagging demonstrates the continuity of the stock involved (Laurs and Lynn, 1977). There appears to be only a single reproductive region at about 20–25°N in the westerly half of the North Equatorial Current, from south of the Japanese islands to south of the Hawaiian islands. The North Atlantic albacore reproduce in the analogous part of the North Equatorial Current of that ocean, and migrate annually to the continental edge west of Europe, where albacore appear (as off California) in the late summer. Bluefin (*T. thynnus*) migrate annually from the reproductive areas of the large fish, and the overwintering areas of the prespawners, in the northern Caribbean and Canary Current (Rivas, 1978). These migrations carry the fish to the coast of western Europe and Scandinavia as well as to the Mediterranean and also along the eastern coast of North America as far as the Gulf of St. Lawrence. Though eastern and western Atlantic stocks are somewhat distinct, transoceanic migration does occur in May and June, when large, lean individuals cross from Newfoundland to Norway; smaller, very fat individuals probably cross from the southern Gulf Stream region to the Mediterranean for a subsequent spawning season, though this is not supported by tagging data. The smaller fish move in summer to feeding areas in the eastern Sargasso Sea. Towards the end of the summer, the O-group fish, spawned the previous winter, move from the Caribbean through the straits of Florida to the Cape Hatteras–Cape Cod section of the Gulf Stream and Slope Water; in the eastern Atlantic the O-group fish move to the Mediterranean, the Canary Current, and the Bay of Biscay. In early winter, the migrations are reversed, and the fish move back to their wintering areas, though there is not yet evidence that the large fish that crossed to Norway return to the western Atlantic. They may pass first to the Canaries wintering region. Individuals with body weights in excess of 500 kg participate in these spectacular transoceanic migrations, analogs of which occur also in the Pacific Ocean, but have not been so well worked out.

The two tropical surface water species are also migrators, but not on quite the same scale; genetic and some tagging evidence (Williams, 1972) suggests that the Pacific skipjack population comprises at least two distinct stocks: an eastcentral and a western, with a zone of seasonal replacement and interchange in the New Hebrides–Solomons–Marianas–Bonin–Japan arc. Each migrates within its own region of the Pacific, but apparently does not mix significantly with the other; within each stock, there is also evidence of northern and southern substocks with different reproductive seasons. The existence of migration by oceanic skipjack is clear from seasonal fishery data, but its exact pattern is not yet fully established, though it has been much studied in the eastcentral stock (e.g., Rothschild, 1965; Blackburn and Williams, 1975). The fundamental mechanism suggested by Rothschild to explain both the appearance of fish along the western coast of Mexico and Central America and also the distribution of larval skipjack (which are rare east of 130°W) is of a central reproductive region, varying in location and extent from year to year, from which a recruitment stock of fish less than 35 cm pass eastward toward the coast; these split into fish destined for the northern (Mexican) fishery and those of the southern (Galapagos–Ecuador) fishery. From these two adult feeding areas, maturing fish of 40–65 cm return eastward to the reproductive regions of the central Pacific. The extent to which these migrations are passive or active has been analyzed by Williams (1972) who examined various alternative hypotheses concerning how the eastward- and westward-flowing currents of the equatorial zone might be used during the migrations. It remains quite unclear to what extent east–west movement can be explained simply by passive transport within a current, or the directed use of the narrow equatorial gyres to assist migration. In this highly dynamic and variable region of the ocean, Rothschild's original comment (that a variety of mechanisms may be used) is probably correct; it is unlikely that the mechanism that places fish in the right location at the right time of year is simple. From the overall distribution of skipjack larvae in the three tropical oceans, it is clear that at least four reproductively isolated stocks occur, if indeed the Pacific Ocean is occupied by only two.

Like skipjack, yellowfin occur almost everywhere in the tropical oceans; a useful by-product of the Japanese global long-line fishery for high seas tuna of the late 1960s are extremely detailed plots of the distribution of yellowfin from which, with a little interpretation of the fishing data, a remarkable picture emerges (e.g., Fig. 7.8) of a tuna species occupying the whole extent of the tropical oceans. As the two seasonal plots of Fig. 7.9 show, yellowfin migrations do not cause it totally to forsake any region, but rather to shift its centers of major abundance between the two hemispheres; Figure 7.10 shows that larval yellowfin occur extremely widely through the same area of distribution as the adults, though there are three regions, east, central, and west, of apparently higher than average abundance. Data on length frequencies of adults from the whole Pa-

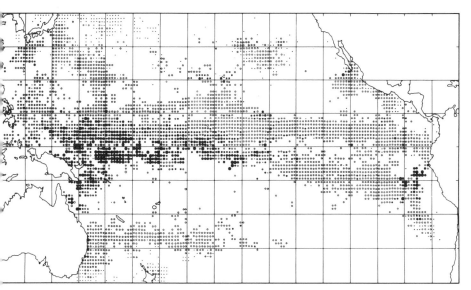

Fig. 7.8. Average catch of yellowfin tuna by Japanese long-line gear for the month of July, 1967–1972. The small dots indicate no catches, the other symbols increasing catch rates to a maximum of >3.0 fish per 100 hooks. (From Suzuki *et al.*, 1978.)

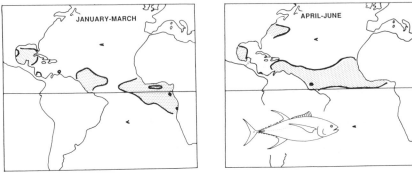

Fig. 7.9. Seasonal distribution of catches of yellowfin tuna in the Atlantic Ocean showing seasonal displacements. (From Wise and Davis, 1973.)

cific region covered by these two figures suggests that three semi-independent subpopulations (Kamimura and Honma, 1963) exist, corresponding with the three centers of high reproduction. Some data on the actual extent of migration have been provided by tagging and also from the dispersal of yellowfin contaminated by radioactive fall-out from the tests of U.S. nuclear weapons at Bikini atoll; these data suggest that some mixing takes place between subpopulations, but that it is not complete.

In the Atlantic, the Japanese long-line catches of the 1960s fishery show that yellowfin occur across the whole tropical ocean, though apparently

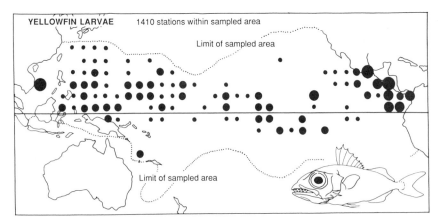

Fig. 7.10. General distribution of yellowfin tuna larvae in the Pacific Ocean, from many plankton surveys; relative abundance indicated by size of symbol. (From Wise and Davis, 1973.)

with greater seasonal migrations than skipjack (Fig. 7.9). For the eastern Atlantic subpopulation of yellowfin, it has been proposed from an analysis of the fishery catches and larval distribution (Richards, 1969) that reproduction occurs after large adults move into the Gulf of Guinea from the west, and small adults from the south, during the dry season when water temperatures are highest; larvae and juveniles then remain there for about 1 year. Some of the 1-year-old fish are taken by the surface fishery as they move southward to a feeding area off Angola, whence they return again to the Gulf of Guinea. This migration is repeated by 2-year-old fish, but subsequently the 3-year-old fish probably move westward to the central Atlantic rather than southward to Angolan waters.

The great mobility of tuna and billfish, and their high swimming speeds—50 kph sustained, 125 kph burst speeds for large species—are associated with a high degree of morphological and physiological adaptation. Speed is achieved by perfection of streamlining, by the management of boundary flow and turbulence, and by energy-effective propulsion mechanisms. Energy output is optimized by ram ventilation of the gill surfaces, by differential temperature control of organs, and by the partitioning of mechanical work between red and white muscle masses. These topics are reviewed well and comprehensively in a volume edited by Sharp and Dizon (1978).

The bodies of tuna and billfish are highly streamlined, with the pectoral fins fitting into shallow grooves along the flanks; the elongated bills of some species probably function both in reducing the frontal pressure wave and as a weapon to disable prey. The absence of hydrodynamic lift from the pectoral fins when sheathed during high-speed swimming is compensated for by bony grooves on the tongue that produce downwardly directed jets from the opercular region during ram ventilation of the gills. Boundary flow is managed by the anterior corselet of enlarged scales that prevents separation of the boundary layer and reduces form drag, as well as the rows of dorsal and anal finlets that serve as movable slots to control cross-flow in the tail area. Two converging lateral keels on the caudal peduncle produce a high velocity jet across the middle of the caudal fin, reducing separation of the boundary layer, and reducing turbulence at the tips of the tail fin. Some species also have a single, larger lateral keel on the caudal peduncle that increases the efficiency of the peduncle itself during high-speed oscillation in swimming. The aspect ratio of the caudal fin itself is highest in the faster-swimming species, and is much larger in billfish (built for high-speed dashes) than in the more generalized tunas.

Sustained high-speed swimming requires a relatively high level of food intake and its very efficient conversion into muscular activity. This is achieved by thermoregulation of individual organs, by an unusually large

heart, an unusually large gill surface area, and a large red muscle component. The high level of oxygen demand is satisfied by large relative gill
areas; in mackerel-like scombroids this ratio is about the same as for other
constantly swimming pelagic fish, but in the larger tuna the ratio is much
larger than in other fish, and approaches the relative lung surface area of
mammals (Muir, 1969). This is achieved both by long gill filaments, and by
the very close spacing of secondary lamellae. The large gills are associated with ram ventilation of the opercular cavities, a mechanism that both
conserves energy (compared with ventilation by branchial pumping), and
maximizes the efficiency of flow across the gill surfaces.

Metabolic heat is conserved in tuna but not billfish by vascular heat
exchangers, or *retia mirabilia,* in which arteries and veins are arranged
closely opposed so that heat, but not ions, may be exchanged. By this
mechanism, whose morphology differs in different genera, the viscera and
muscle masses are maintained at temperatures higher than ambient water
temperatures for efficient rapid digestion, and to speed up the biochemical
processes within the muscle masses. The red muscle masses are relatively
larger and more internal in the more highly evolved and larger species.

Tuna heart muscle closely resembles that of the mammalian heart, and
the masses of red body muscle are highly vascularized and rich in fat,
glycogen, and the enzymes of aerobic metabolism: the contribution of red
muscle to swimming is based on aerobic oxidation of lipid or carbohydrates (Hochachka *et al.,* 1978). The white muscle mass, on the other
hand, has unusually large amounts of glycogen and though aerobic metabolism occurs, high-speed, short dashes are driven by white muscle con-
traction that is based on exceedingly intense anaerobic glycolysis. If high-
speed swimming is sustained beyond an initial burst, then some aerobic
metabolism begins to occur also in the white muscle. As Hochachka *et al.*
point out, the great activity of tunas is supported not only by their highly
complex and variable muscle biochemistry, but also by the retention of
heat within the muscle masses; they point out that the change in physiological rates for both aerobic and anaerobic metabolism between basal and
maximum, burst-speed metabolism is doubled when accompanied by a
10°C rise in body temperature.

Billfish have a somewhat different way of life from tuna and this is
reflected in their morphology; over great ocean depths they perform regular diel migration from the surface at night to 400–600 m by day; they are
crepuscular fish having, as Carey and Robison (1981) put it "eyes as large
as oranges in a 150 kg fish which about touch in the mid-plane" and it is
known that they take prey at great depths. Surface basking in daytime of
larger individuals occurs regularly in some species when in cold water
masses, during their summer migrations beyond the tropics, and (most

curiously) resting on the bottom at continental shelf depths has been suggested. Where offshore banks occur in their habitat, individual billfish may have a home range over a bank to which they return each day (Carey and Robison, 1981).

Tuna exist in, and travel fast through, a medium in which visual cues are few and carry very little information; navigation for long-distance migration does not need to be very accurate, but it requires sufficient precision to maintain a general heading appropriate to the season and water-mass in which the tuna finds itself. It is possible that this is achieved by the use of a pineal window of transparency in the dorsal part of the cranium, below which there are tissues that possess the same characteristics as vertebrate retinal cells, and which are also suitably innervated. This curious photoreceptor organ is thought to function to orient the fish during migration, though there is apparently no experimental evidence to support the hypothesis.

In the near-absence of major visual stimuli in the open ocean, there is a universal attraction for fish to flotsam; even the smallest piece of wood, cuttle-fish bone, or even a small piece of plastic suffices to attract and retain small fish which remain in its vicinity for long periods. Tuna are no exception and occur regularly in association with large flotsam—dead whales, floating logs, and so on. The numbers of tuna found in the vicinity of large flotsam in the eastern tropical Pacific Ocean is so high that in the mid-1970s about one-fifth of all purse-seine sets reported to the Inter-American Tropical Tuna Commission were made around flotsam in order to catch the associated tuna. The relationship between large pelagic fish and flotsam has never been satisfactorily explained; clearly, for smaller species of fish, perhaps expatriates from their normal continental shelf environment, flotsam represents a refuge from predators, but this seems an unlikely explanation for large tunas. In some regions, artificial floating rafts have been used to aggregate tuna so that they can be caught with a purse-seine or hand-lined, notably in the Philippines where local devices known as "payaos" in fact provided the basis for the development of the regional tuna fishery.

A second association still lacking a complete explanation is that between tuna and porpoises. It was quickly discovered by the tuna purse-seiners, as their techniques evolved during the 1960s, that a school of porpoises in the open tropical ocean was frequently (or even usually) accompanied by a school of yellowfin tuna, the two species forming a two-layered school having porpoises above, tuna below. Though, once again, no experimental evidence is available the most likely explanation is a mutualism in the location of prey; tunas use visual hunting techniques, while porpoises also have the capability of echo-location of their prey organisms, which are approximately the same in any given region as those of the larger tuna.

The association was found profitable by tuna seiners, who began to circle porpoise schools with their seines for the sake of the tuna entrapped with them; though there is much variation by area, season, and ship size, the proportion of all sets that are now made on porpoise schools reaches 50% overall for seiners in the 1000-ton class, for instance. The global mean for number of porpoise taken per set is about 40 animals, and schools of several thousand have been recorded as being captured. The mortality of porpoises can be very high, since the encirclement apparently confuses them; this is perhaps not surprising for a mammalian species having no previous experience of any kind of solid interface. They may become catatonic when faced with a seine net over which they have the physical capability to jump very easily. One consequence of the high rate of mortality of porpoises was the U.S. Marine Mammal Protection Act, which required the industry rapidly to develop methods of releasing the porpoises from a seine before they drown.

Within a few years a technique for releasing porpoises had been developed and by 1984 all except 10 of 151 tuna-seiners surveyed by the Inter-American Tropical Tuna Commission were equipped with various forms of "safety panels" inserted into their seines. By "backing down" onto the net, it is possible to sink part of the headrope at the safety panel, and to herd porpoises out of the safety channel thus formed. Herding is accomplished by small outboard motor boats, with fishermen stationed at the exit from the safety channel to assist manually the porpoises to get out if needed, and to warn the skipper on the bridge by two-way radio if he risks losing his catch of tuna, too. Porpoise mortality is now much reduced by these techniques and probably also by the development of experience on the part of porpoises; mortality is highest in newly fished regions.

OCEANIC SHARKS

Large pelagic sharks, though Parin notes that their diet is not well investigated in tropical seas, are the apex predators of the pelagic ecosystem; as well as taking a great range of smaller fishes, the larger, more active sharks such as the mako (*Isurus* spp.) are capable of preying even on tuna and billfish (Talbot and Penrith, 1963).

However, of course, not all large sharks and rays are predators. Apart from the well-documented (and enormous, exceeding 17 m in total length) whale sharks that feed by filtering very small plankters from the zooplankton biomass maxima in the epipelagic zone, we now know of even more unusual tropical forms. In very recent years, the "megamouth" shark *Megachasma pelagios* has been captured below the epipelagic zone off Hawaii and California: 4.5 m long, black, soft-bodied, with a large gape surrounded by thick lips and having large numbers of tiny teeth and filtering cartilages of the gill rakers, these plankton-feeders are quite unlike the whale shark. The internal surface of the mouth is bright and silvery, and plankton may possibly be attracted to a stationary plankton pump, rather than gathered by forward swimming after the manner of a plankton net (Diamond, 1985).

RESOURCES OF SMALL MESOPELAGIC FISH OF THE OPEN TROPICAL OCEANS

Throughout all tropical oceans, and indeed in all oceans, there is a population of small fish that occur in the mesopelagic zone by day but may migrate at night to the epipelagic zone to feed. In tropical oceans, these mesopelagic fish are sometimes very abundant, and they form an important component of the acoustic deep-scattering zone along with diel migrant crustacea and siphonophores. There are reasons to believe that the biomass of mesopelagic fish is extremely large in tropical oceans, and trials have been made in recent years of the possibility of commercial exploitation of these fish, which are usually small (<10 cm) and often of bizarre form. In some regions, a few small fisheries exist; off South Africa, *Lampanyctodes* is taken with purse seines, and off West Africa the Soviet distant-water fleet attempted a fishery for myctophids. At Madeira, there is a deep hand-line fishery for large bathypelagic scabbard fish that themselves probably feed on many of the species discussed in this section. The study of mesopelagic resources has been much constrained by a lack of taxonomic and distributional information; mesopelagic fish can be studied only by the use of expensive high seas research vessels, and only in the last 10 or 15 years has our knowledge of them become sufficient to generalize either about their ecology or their potential as a fishery resource, and the extent to which their high wax ester content renders them unsuitable for human consumption.

The global mesopelagic fish resources have recently been reviewed for FAO (Gjøsaeter and Kawaguchi, 1980), and much of our material for this section is based on this report. These authors record that about 160 genera of 30 families occur as important components of this fauna globally, and that the majority occur in low latitudes. The families contain only a few highly evolved Perciform genera; 12 families are relatively unevolved salmoniform fish such as bathylagid and argentid smelts, gonostomatid hatchetfish, and others. The remainder include seven families of Myctophiformes, three families of elongate lampidiform ribbonfish, oarfish, and crestfish, two families of Beryciformes (relatives of the holocentrids of coral reefs discussed in Chapter 6), and the anguilliform snipe eels. Many species are black, many bear light organs, and many have telescopic or tubular eyes. Jaws are frequently highly specialized, and all have the morphology to be expected from small, active predatory fish. Gjøsaeter and Kawaguchi show that four families have the greatest diversity of genera: Gonostomatidae, Melanostomiatidae, Myctophidae, and Gempylidae account for about half of all genera in the fauna. Individuals of only two families, Myctophidae (lanternfishes) and Gempylidae (snake mackerels), account for between 50% and 90% of the global biomass of mesopelagic fish. These two families, therefore will bear the burden of any future serious exploitation of the mesopelagic fish fauna.

Species of mesopelagic fish may be regionally or widely distributed, and some are sufficiently cosmopolitan as to occur in both Atlantic and Pacific Oceans: *Gonostoma elongatum* occurs both in the central Pacific and in the subtropical Atlantic. Further, because they occur, at least in daytime, at depths where latitudinal gradients are slighter than in the epipelagic zone, some species have wide north–south distributions; *Maurolicus muelleri* occurs off Norway, in the Japan Sea, and off the coast of Morocco.

Backus and Craddock (1977) have analyzed the catches from more than 1000 samples of mesopelagic fish living from 800 m to the surface in the Atlantic Ocean, and have identified faunistic provinces whose boundaries coincide with species replacements among Myctophidae. The boundaries correspond with what would be expected from plankton zoogeography: according to Backus and Craddock, the tropical Atlantic can be divided into five subregions, and the subtropical into another five, with the Mauretanian upwelling region forming an independent faunistic region. In the warm-water zone of the Atlantic, some tropical species of Myctophidae, such as *Diaphus dumierli,* occupy all five tropical regions and extend into higher latitudes along the western boundary currents. Other species, such as *Diaphus vanhoeffeni,* occur only in one subregion, while yet others, such as *Ceratoscopelus warmingi,* occur throughout the tropical and subtropical regions (Fig. 7.11).

In fact, the geographical distribution of at least the Myctophidae follows the same general pattern as for oceanic zooplankton species and, like zooplankton, mesopelagic fish do not seem to perform directed horizontal migrations. However, advection and expatriation of adults and larvae in major ocean currents must occur, and has in fact been reported in some cases (O'Day and Nafpaktitis, 1967; Zurbrigg and Scott, 1972),

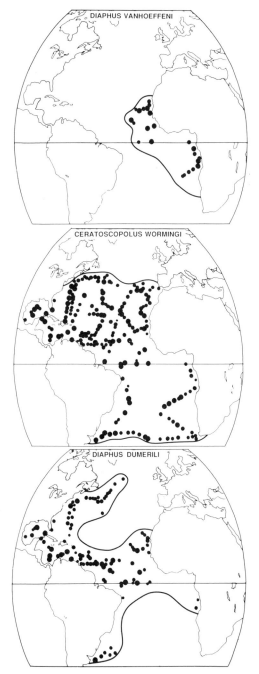

Fig. 7.11. Distribution of three species of small mesopelagic fish in the Atlantic Ocean indicating similarity to oceanic plankton distributions shown in Fig. 4.4. (Selected from Backus and Craddock, 1977.)

but it is unclear if advection, combined with seasonal vertical migration could not sometimes maintain a population in a suitable habitat, or whether expatriation and loss to the reproductive population is always the rule.

The migrations of mesopelagic fish, then, are dominated by the role that the individuals of many species play in the diel interzonal migration of pelagic organisms between the surface layers and the dark zone below. During the day, no species will normally be seen at the surface of the sea, but at night many can often be seen right at the surface under the lights of a research ship stopped on station in the tropical ocean. Active swimmers, they pass rapidly in and out of the lights together with swarms of red Cephalopoda that seem to be preying on them. The nocturnal activity of some tropical oceanic birds may well be related to the availability of these small fish at the surface at night as well as the availability of small migrant crustacea.

The depths of occurrence of individual species extend over about the upper 1 km of the water column, and many studies of the vertical distribution of the mesopelagic fish fauna, such as that of Roe and Babcock (1984) of *Benthosema glaciale,* show that smaller individuals lie above larger individuals of the same species (e.g., Fig. 7.12), in exactly the same manner as occurs between large and small individuals in species of invertebrate zooplankton. There are several different patterns of vertical migration. Many species (e.g., *Myctophum, Diaphus, Benthosema*) occur during daylight at some depth between 200 and 500 m, and migrate into the epipelagic zone at dusk, to descend again at dawn; others remain in a preferred depth range ·day and night (e.g., *Gonostoma, Cyclothone*), while a third group (e.g., *Argyropelecus*) performs a limited migration from farther down to some depth below the epipelagic layer, often migrating approximately over the range 600–200 m. Pearcy *et al.* (1977), and Roe (1983) give examples of these different patterns for a number of species, and Gjøsaeter and Kawaguchi (1980) noted that nonmigrating forms are distinguished from diel migrants by morphological adaptation to a nutrient-poor environment. The occurrence of dwarf males and sex reversal in nonmigrants, and their absence in migrants that have access to the food-rich epipelagic zone suggests a fundamental difference in the energetics of the two groups.

Because there is essentially no fishery for mesopelagic fish, all assessment of their stocks must be done with fishery-independent techniques, one of which is acoustic survey and echo-integration. Since a large proportion of the resource appears to be associated with the migrating deep scattering layer, a first-order question concerning acoustic surveys is: Does the gas-filled swim bladder (which will be an important component

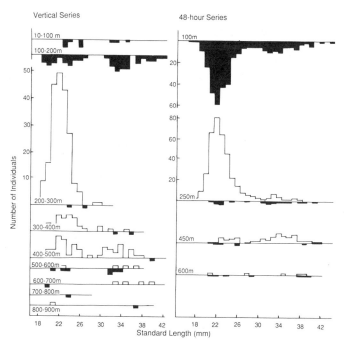

Fig. 7.12. Differential size distribution of small mesopelagic fish, by day (open histograms) and night (closed), showing smaller fish at shallower depths. (From Roe and Babcock, 1984. Reprinted with permission from *Progress in Oceanography* **13**, Copyright 1984 Pergamon Journals Ltd.)

of the acoustic reflectivity of the fish) of many species change in relative size during diel migration? This question appears to remain unresolved and is actually a difficult physiological problem; there may be an energetic constraint on vertical migration in small fish that does not exist for the many taxa of diel migrant crustacea. Many species of mesopelagic fish, including diel migrants, do have gas bladders to adjust their buoyancy; for vertical migration, there is a requirement for work to be performed to overcome the vertical hydrodynamic forces imparted by the buoyancy of the gas bladder, or else in adjusting the amount of gas in the bladder, by secretion or resorption by special tissues. D'Aust (1970) analyzed the advantages of gas resorption and secretion to a range of fish of different sizes and vertical migration ranges; he concluded that it was only the rate at which gas can be resorbed that limits the rate of downward migration, and the rate of secretion for upward migration. He calculated maximal rates of ascent and descent that are realistic, and suggested that some special spongy tissues found in the air bladders of migrating *Lampanyctus* may have the function of increasing secretion/resorption rates.

There is a great variety of feeding patterns and behavior associated with the different migration patterns exhibited by mesopelagic fish. The most usual pattern, such as that observed off West Africa for *Benthosema glaciale* by Kinzer (1977) or for *Lampanyctus alatus* in the eastern Gulf of Mexico by Hopkins and Baird (1985), is of a nocturnal feeding period near the surface, usually on a variety of small crustacean prey, followed by a quiescent period during the daylight hours at greater depths, possibly with a relatively low level of intake, especially late in the day. However, it is probably more usual than food items found in the stomachs during daytime are undigested remains from the previous night's feeding foray closer to the surface. Some species become quite torpid during their daytime residence at depth, hanging vertically in the water with their head downwards, apparently motionless for long periods (Barham, 1971); some species have also been observed to form daytime aggregations containing several thousands of such quiescent individuals, tightly packed together into a spherical shoal.

The vertical migration pattern of the diel migrants of the mesopelagic fish community is indirectly but well-described by the diel migrations of the deep scattering layer (DSL) that, as we have already seen, extends continuously in the tropics on an ocean-wide scale. The same, or apparently the same, features of the DSL can be traced for thousands of kilometers and such features appear to change their general pattern only when a change in oceanographic regimes, measured by other variables, also occurs. Figure 7.13 illustrates the details of the DSL in the tropical Indian Ocean; associated with this DSL were large numbers of mesopelagic fish, especially the stomiatoid *Vinciguerria nimbaria* and the myctophid *Notolychnus valdiviae,* while hatchetfish (*Argyropelecus*) were associated with it during the daytime, but did not complete a surfaceward migration at night. Other myctophids (*Diogenichthys panurgus, Diaphus regani,* and *D. lutkeni*) occurred at greater depths than the DSL during the daytime, but rose right into the surface layer at night.

Relying on data from a variety of survey techniques (micronekton net surveys, trawl surveys, acoustic surveys, and egg and larval surveys), Gjøsaeter and Kawaguchi (1980) have been able to review the global resources of mesopelagic fish: The available biomass appears to be very large. The total biomass for the tropical oceans is suggested by this survey to be of the order 650×10^6 tons, or about 10 times the total world catch of marine fish of all sorts; as we shall see in considering this estimate regionally, the actual indicated total is not to be taken too seriously—but it must nevertheless be a very large figure indeed.

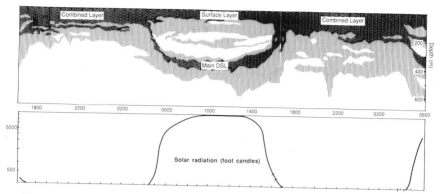

Fig. 7.13. Evolution of the "deep scattering layering" caused by acoustic reflection from planktonic and nektonic biota in the tropical Indian Ocean through day and night periods. Scale is hours. (From Bradbury, 1971.)

It is in the northern Arabian Sea that the biomass appears to be greatest, both regionally and per unit area, and we have already seen this region is anomalous both oceanographically and biologically. It is also unusual in the extent to which it has been surveyed in ways relevant to mesopelagic fish stocks; during the International Indian Ocean Expedition, egg and larval surveys were completed for the whole of the region, as well as for the remainder of the tropical Indian Ocean. In addition, acoustic and midwater trawl surveys have been performed on several occasions in the richest regions: the Norwegian ship *Fridtjof Nansen* surveyed the northern Arabian Sea five times from Mogadiscio to the Indus delta, and several FAO-led surveys of the same kind have covered the Gulf of Oman. Extremely high biomass (25–250 g/m²) was indicated by these surveys for the Gulf of Aden and the Gulf of Oman, as well as the western coastline of Pakistan; the remainder of the region had lower, but still high biomass (10–85 g/m²). In each region concentration of mesopelagic fish was highest just at, or beyond, the edge of the continental shelf where values of 50–500 g/m² were common. There are also preliminary results available for surveys of *Fridtjof Nansen* in the Mozambique Channel, around the Seychelles and off Sri Lanka which Gjøsaeter and Kawaguchi (1980) have put with the International Indian Ocean Expedition (IIOE) larval distribution data to suggest mean regional biomass of 60 g m⁻² for the northern Arabian Sea, 30 g m⁻² for the west coast of India, 0.5 g m⁻² for the main part of the open Indian Ocean between India and Africa, and 6 g m⁻² for the East African coastline. The main concentration of mesopelagic fish in the Arabian Sea occurs at 150–350 m by day, and in the upper 50 m at night, while in the open Indian Ocean the concentrations are deeper, from 250 to 500 m. The myctophid *Benthosema pterotum* dominates the Arabian Sea biomass, along with *B. fibulatum* (which is also important off Mozambique), *Hygophum proximum, Bolinichthys longipes,* and *Diaphus thiollierei.* In the open ocean to the south, where diversity is greater, *Notolychnus valdiviae* is important. Gjøsaeter and Kawaguchi (1980) suggest that a combined regional biomass for the western Indian Ocean might be in the region of 250 × 10⁶ tonnes.

For the eastern Indian Ocean, the survey data are much less comprehensive, and restricted to rather sparse micronekton net data, which suggest that highest biomass occurs along the eastern coasts, from the Bay of Bengal to western Australia, but with a mean regional biomass of only about 5 g m^{-2}; progressively toward the center of the eastern basin of the Indian Ocean, biomass drops to levels of about 2 g m^{-2}. As Gjøsaeter and Kawaguchi (1980) emphasize, there is much still to be learned about quantitative species composition of the mesopelagic fish fauna of the region. For the western Pacific, there are much better micronekton net data, almost 2000 samples being available from the surveys of Grandperrin and Rivaton (1966), Parin (1977), and Legand *et al.* (1972). However, these more extensive data appear to confirm the general abundances in the eastern Indian Ocean; over the whole of the shallow continental shelf region south of the Philippines, and down to Australia, biomass averages about 5 g m^{-2}, with progressively decreasing values towards the open Pacific Ocean, in the region of the equatorial currents, where biomass is about 2.5 g m^{-2}. The most abundant species appear to be *Vinciguerria nimbaria, Gonostoma elongatum, G. atlanticum, Cyclothone alba, C. pseudopallida,* and *C. acclidens* (all Gonostomatidae); *Diaphus fulgens* and six other species of this genus, *Lampanyctus niger* and three others, *Bolinichthys longipes,* and *Ceratoscopolus warmingi* (all Myctophidae); *Stenopteryx diaphana* and *Chauliodus sloani.*

Information on biomass is not satisfactory for much of the central Pacific, though Grandperrin and Rivaton (1966) show that there are significant changes in species composition from west to east over this enormous expanse of ocean. Myctophidae as a group occur in approximately equal numbers right across the ocean in the equatorial zone, though numbers of *Diaphus* decline steadily eastwards from at about 180°W. Sternoptychidae occur at a very low abundance in all longitudinal zones except in the eastern part of the ocean, east 110°W. Highest numbers of *Vinciguerria* occur near the American coast, with other regions of high abundance in the western Pacific. In the eastern tropical Pacific the survey data are once again based only upon micronekton net tows; the most extensive data sets being the CalCOFI surveys of the California Current, the

EASTROPAC surveys from 20°N to 20°S (Blackburn, 1968), and various surveys of the Peruvian region. Once again, the indicated biomass is in the region of 5 g m^{-2} in the more eutrophic coastal regions and about 1–2 g m^{-2} in the open ocean. In the open ocean regions, the following species dominate the biomass: *Ceratscopus warmingi, Lampanyctus steinbecki, Vinciguerria nimbaria,* and *Triphoturus nigrescens.*

For the western tropical Atlantic, there are a variety of surveys with micronekton nets, and some acoustic research on DSLs, though not using fishery stock assessment echointegration techniques like the *Fridtjof Nansen* surveys of the Indian Ocean; the western Atlantic data, according to Gjøstaeter and Kawaguchi (1980), indicate relatively low biomass regionally, nowhere above 0.2 g m^{-2}. This sparse fauna is dominated by *Notolychnus valdiviae, Diogenichthys atlanticus, Lampanyctus pusillus,* and *Vinciguerria attenuata.* Somewhat higher biomass (1–2 g m^{-2}) is indicated along the northeast coast of South America, where *Cyclothone microdon* appears to be the most abundant species, as indicated by micronekton net surveys.

The eastern tropical Atlantic Ocean seems to be somewhat richer than much of the remainder of the tropical oceans for which we have only micronekton net survey data; based on several extensive surveys, such as that by Voss (1969) of the Gulf of Guinea, the whole region appears to be about one order of magnitude richer than the Caribbean and Antillean region. The biomass in the Mauretanian upwelling region appears to be of the order 15 g m^{-2}, in the Gulf of Guinea about 6 g m^{-2}, and about the same in the north Equatorial Current.

Thus, it is clear that there is a very large discrepancy between the indicated abundance of this resource in the northern Indian Ocean and everywhere else in the tropics. This must result from one of three causes: Either micronekton nets underestimate the abundance of mesopelagic fish by a factor of at least 10, or the *Fridtjof Nansen* surveys were optimistic by about the same factor, or the northern Indian Ocean really is 10 times as rich, area for area, as the rest of the ocean. Intuitively, the last possibility seems unlikely to be correct. Thus, it is only with the greatest caution that we should interpret the apparent size of the potential resource base of mesopelagic fishes in the tropical oceans.

Chapter 8

Fish as Components
of Marine Ecosystems

In previous chapters we have examined benthic and planktonic ecosystems in tropical seas; in this chapter we will examine how fish are integrated into tropical marine ecosystems and some of the roles they play in planktonic and benthic ecosystems as herbivores, omnivores, and top predators. In this way we hope to approach the question of how tropical fisheries are maintained by their life-support systems, the benthic and planktonic invertebrates and marine plants.

We shall ask how the energy demands of tropical sea fish are met by the ecosystems of which they are a part, and how, in turn, these demands affect the energy flow and the mass budgets of organic material of tropical marine ecosystems. We shall examine what is known of the diet of some feeding guilds of fish, and how much flexibility they have to accomodate to regional variability of ecological conditions. For a very few cases, we shall be able to examine in quantitative terms the role of fish in the mass budgets of some exemplary tropical marine ecosystems.

To discuss how fish are able to structure, control, or contribute to the flow of energy through marine ecosystems, it is necessary to have at least a first-order description of their diet and, if possible, at what energy cost it is obtained, a subject we shall return to in Chapter 9 in our discussion of population biology of fish. Fish communities are usually modeled as including two components: demersal and pelagic, and a first approximation is that each component obtains its energy from food items captured from the benthic and planktonic ecosystems, respectively. However convenient, these categories are actually misleading, because many species that are normally (and conveniently) considered to be demersal fish obtain much of their food from the pelagic ecosystem, including from its pelagic fish component. In the opposite sense, some apparently pelagic species of fish include bottom-living invertebrates in their diets, at least at some seasons or parts of their life cycle, as large *Decapterus* do in the western Pacific.

Similarly, it might seem that a simple first step toward understanding the role of fish in ecosystem energy flow would be to categorize each fish species in a community according to its diet; in practice, this is a very difficult first step to take with any conviction. Not only are many species able to utilize a wide range of items according to availability (see Chapter 9), but each species of fish normally has at least three quite distinct dietary regimes during growth: the larval diet, the diet of juvenile fish, and the diet of adult fish. For species that undergo a metamorphosis (as in Pleuronectiformes) at the end of larval life, or a change in habitat from planktonic larvae to demersal juveniles, the switch between diets may be stepwise, but in other species the changes are gradual.

Apart from the primary requirement for co-occurrence, the food taken by larval fish is determined principally by the geometry of the prey and of the mouth of the larva; its minimum dimension (width rather than length) determines if a food item can be successfully ingested by a fish larva, but within this criterion the range of items actually ingested is very wide. Clupeoid larvae tend to be small mouthed compared with the larvae of more highly evolved teleosts such as scombroids, so that at the same larval length clupeoids utilize much smaller plankton: a 15-mm *Engraulis* larva, for instance, can only utilize particles of the same maximal size as a *Scomber* or *Trachurus* larva about 3 mm long. Though naupliar and larval stages of copepods are the typical prey of most fish larvae, their diet tends to be more varied in the very young stages, when tintinnids, ciliates, dinoflagellates, and various invertebrate larvae are important. An increase in prey size, and a concomitant change in prey species, during the growth of fish larvae is universal and occurs in every species studied (Hunter, 1981); the different range of prey sizes utilized during larval growth by different taxonomic groups of fish has important ecological consequences, and determines the larval survival strategy inherent in each species' reproductive pattern.

After metamorphosis, species that take active, swimming prey either near the bottom or in midwater, tend to feed heavily at first on natant crustacea, and to increase the percentage of fish in their diet progressively as they grow. This is especially prominent in many generalized demersal species, such as sciaenids and lutjanids, but perhaps less so in the case of highly specialized ichthyophagous species. A corollary is that there are many examples of the small species in a genus specializing in feeding on active crustacea though taking some fish, while the large species eat larger numbers of fish and fewer crustacea. In some clupeoids, a change in diet occurs between phytoplankton and zooplankton during growth. As we have already mentioned, both *Sardinella longiceps* and *Engraulis ringens* depend principally on zooplankton in their juvenile stages, but adults (as their specialized gill rakers develop) are increasingly able to utilize phytoplankton cells.

Thus, there can be no precise way of classifying even adult fish diets, once the principal ontogenetic changes have been accomplished; adult fish of the same species may have different diets in different habitats, and in the same habitat their diet may vary from year to year, or season to season, as the availability of items changes. Both these sources of variability have been recorded for tropical West Africa, as well as regular seasonal variation in the amount of food taken. For six of seven species of demersal fish, specimens taken in the Sierra Leone estuary have three to five times as many polychaetes in their stomach contents as similar-sized specimens taken only a few miles away on the open continental shelf (Longhurst, 1957b). Conversely, specimens from the continental shelf contain more molluscs and small crustacea. These differences clearly reflect differences in food availability from the estuarine and shelf benthic communities. For several species of demersal fish in the Sierra Leone estuary, an anomalous availability of *Lingula parva* during the wet season of 1 year (March–September) dominated the variability in stomach contents that was observed during a 3-year period of observations, occurring in more than 90% of stomachs of *Cynoglossus* examined during this period, but rather rarely during the rest of the sampling program (Fig. 8.1). A study of the food and feeding habits done in 1952 or 1954 would have concluded that the sole's diet was much more restricted than it actually is—just as *Balistes vetula* (see Chapter 10) was regarded as a specialized feeder on sea urchins until these disappeared in a pandemic. This is the opposite kind of case.

ECOSYSTEMS OF OPEN CONTINENTAL SHELVES AND ESTUARIES

We have already discussed (Chapter 6) the coupling of energy flow between pelagic and benthic components of continental shelf ecosystems in the tropics. We suggested that the evidence is strong that on continental shelves and in estuaries (though not in the open ocean) pelagic production/consumption ratios are high, and excess plant cells sink to the deposits to be utilized by the benthic system of the continental shelf, or of the deep ocean after slumping and turbidity flow from the continental

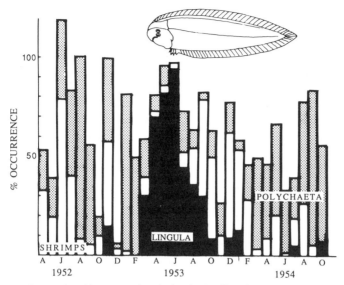

Fig. 8.1. Seasonal and interannual variation in the diet of the tongue-sole *Cynoglossus senegalensis* in the Sierra Leone estuary, showing temporary dominance of *Lingula* in stomach contents. (From Longhurst, 1957b.)

edge. Since this phenomenon is not restricted to the tropical seas, but on present evidence is apparently the normal balance in shallow seas at all latitudes (Joiris *et al.*, 1982), we might expect that the general balance between pelagic and demersal fish species, and between herbivores, omnivores, and predators to be similar in similar habitats (i.e., soft muddy bottoms) in both temperate and tropical seas. Yet, as we shall see, in tropical seas the pelagic component of fish faunas in each region appears to be relatively more numerous and varied than the demersal component, recalling some suggestions (e.g., Jones, 1982) that there are relatively more pelagic fish in warmer seas. We shall attempt below to examine to what extent this apparent difference is real, or is simply an aspect of the greater overall diversity of assemblages of tropical sea fish.

There are sufficient data on the diets of fish on the tropical West African coast to allow construction of an approximate model of their mutual interactions and their relations with the other components of their ecosystems, and for which first-order quantitative descriptions exist. For demersal fish, data from 27,800 individual stomach-content examinations are available from weekly sampling over a 3-year period of 26 species in the Sierra Leone estuary (Longhurst, 1957b) and for 71 species from many trawling stations on the continental shelf from Senegal to Nigeria (Longhurst,

1960b). Théodore Monod (1927), in one of his very first studies, reviewed the diet of many of the same species from Senegal and Cameroon, respectively, and, more recently, Cadenat (1954) and LeLoeuff and Intes (1973) have added more data from the Gulf of Guinea for the same group of demersal species. For pelagic species, Bainbridge (1963), Blay and Eyeson (1982), and Fagade and Olaniyan (1972) described the diet of *Ethmalosa dorsalis,* while Postel (1955, 1960), and Zei (1966) discussed that of *Sardinella aurita* along the West African coast. For many other species, diets can be deduced from morphology, as we shall discuss generally in Chapter 9.

From these data it is possible to assemble 160 species (which comprise most species except the smallest and rarest fish in the tropical region of the Gulf of Guinea) into eight feeding groups or guilds, as a first approximation to an analysis of the paths of energy flow through the fish communities in the continental shelf ecosystems. Table 8.1 shows how these may further be consolidated into only five groups: herbivores, 7 species (4%); zooplankton feeders, 8 (4%); benthos feeders, 36 (23%); benthos and fish, 59 (37%); and specialized fish feeders, 47 (30%). Such numbers and percentages are not very different, in fact, from a recent analysis of the trophic levels of the 200 species of fish inhabiting the North Sea (Yang, 1981); 22% planktophagic, 43% benthophagic, and 35% ichthyophagic.

The herbivorous component comprises three groups of species: *Ethmalosa dorsalis,* which specializes in the filtration of estuarine planktonic diatoms, five species of grey mullet (*Mugil, Lisa*), which browse on the benthic diatom mats on shallow-water estuarine mud banks, and the surgeon-fish *Acanthurus monroviae,* which browses on coralline algae on rocky reefs.

Zooplanktivorous species are, of course, dominated (part from larval forms of many species) by clupeoids (five species of *Sardinella, Cynothrissa, Engraulis,* and *Ilisha*) with one flying fish (*Cypsilurus*), one damsel fish (*Holocentrus*), and one small carangid (*Chloroscombrus*). As we have already noted *Sardinella aurita* has a somewhat restricted ecology in the upwelling regions, but the other species of zooplankton-feeding *Sardinella* are generalists and obtain their energy from available organisms in the plankton community of the coastal regions described in an earlier chapter. It is probable that the other zooplanktivores have an equally wide-spectrum diet.

Seventeen species satisfy their energy demands from the benthic infauna, principally the polychaetes, molluscs, echiuroids, brachiopods, and small interstitial crustacea of the *Macoma* and "ophiuroid" communities of estuarine and near-shore muddy deposits. These 17 species are

TABLE 8.1. **Allocation of Fish Species to Feeding Groups, for the Sierra Leone Continental Shelf, Based on the Diets of Adults**

Herbivores		Active benthos and fish	
Planktonic diatoms		Coastal species	
Shad—*Ethmalosa* spp.	1	Sharks—*Mustelis, Leptocharias*	2
Benthic diatoms		Rays—*Squatina, Raja*	2
Mullets—*Mugil* spp.	5	Flatfish—*Bothidae*	3
Reef algae		Scabbard-fish—*Trichiurus*	1
Surgeonfish—*Acanthurus* sp.	1	Sciaenidae—*Pseudotolithus*, etc.	4
	7		12
Zooplanktivores		Offshore species	
Sardinella spp.	5	Perches—*Epinephalus, Diagramma*	10
Exocoetes, Holocentrus, etc.	6	Boarfish—*Capros afer*	1
	11	Scorpaenidae—*Scorpaena*	2
Reef epifauna browsers		Triglidae—*Trigla, Lepidotrigla*	2
Pomacentrids	3	Others—*Dactylopertus, Uranoscopus*	2
Labrids	5		17
Tetraodontids	5	Omnivores	
	13	(Fish, crustacea, infauna)	
Benthic infauna		Shark—*Ginglistoma*	1
Rays—*Trygon, Pteromylaea*, etc.	6	Catfish—*Arius*	3
Flatfish—*Soleidae, Cynoglossidae*	4	Threadfins—*Galeoides, Pentanemus*	3
Sciaenidae (*Umbrina*)	2	Lizardfish—*Trachinocephalus*	1
Sheepshead—*Drepane, Ephippus*	3	Sciaenid—*Larimus*	1
Tetraodontidae (*Ephippion*)	2	Breams—*Dentex, Pagrus*, etc.	7
Gerres macrophthalmus	1	Grunts—*Pomadasys, Brachydeuterus*	4
	18	Snappers—*Lutjanus, Lethrinus*	7
Active benthos only		Others—*Upeneus, Balistes*	2
Rays—*Rhinobatis*	3		29
Ladyfish—*Albula vulpes*	1	Ichthyophages	
Snake-eel—*Ophithchus*	1	Pelagic predators	
Smaris	1	Sharks—*Hexanthus, Sphryna, Isurus*	11
	6	Bonefish—*Elops, Megalops*	2
		Bluefish—*Pomatomus*	1
		Coutas—*Sphyraena*	2
		Jacks—*Caranx, Trachinotus*	8
		Dolphinfish—*Coryphaena*	1
		Mackerel—*Auxis, Scomberomorus*	7
		Others—*Sciaena, Stromateus*, etc.	2
			34
		Benthic predators	
		Rays—*Squalus, Torpedo*	4
		Toadfish—*Batrachoides*	2
		Eels—*Muraena, Echidna*	3
			9

dominated by rays (*Trygon, Pteromylaea, Myliobatis*), flatfish (*Cynoglossus, Citharichthys, Synaptura*), and spadefish (*Drepane, Ephippus, Chaetodipterus*) and include two sciaenids (*Umbrina*) and two puffers (*Chaetodon, Ephippion*). The stingray (*Trygon margarita*) selects a range of large polychaetes that occur in almost 80% of all stomach contents examined (*Clymene, Pectinaria, Diopatra, Glycera*), small molluscs (*Tellina, Macoma*), and largely burrowing crustacea (*Squilla, Alpheus, Callianassa*) and small crabs that live largely below stones (e.g., *Porcellana*). A spadefish (*Drepane punctata*) selects a rather similar mix, though with more specialization on *Clymene* and *Pectinaria*, and with a substantial intake of benthic amphipods; the small brachiopod *Lingula*, a small pennatulid (*Virgularia*), and the lancelet (*Branchiostoma*) are all also important constituents of the diet. Another spadefish, *Chaetodipterus lippei*, has a quite different diet and its stomach contents are usually dominated by a mass of hydroid material. The tongue sole (*Cynoglossus senegalensis*) feeds largely on the polychaetes *Pectinaria* and *Glycera*, the crustacea *Squilla* and *Panope*, and the brachiopod *Lingula*, at least on the inshore grounds.

A few species of fish specialize in preying on the more active and mobile epifauna of the sea bed: three species of rays (*Rhinobatis*), *Albula vulpes*, the eel *Ophichthus semicinctus*, and *Smaris macrophthalmus*. These all seem to specialize in preying on small crabs and penaeid prawns, though undoubtedly at times their range of diet is widened to include some fish and some more sedentary epifaunal molluscs. The inshore sciaenid *Pinnacorvina epipercus* specializes in feeding on small crabs, which form more than 80% of its diet on a percentage occurrence basis.

The rather large group of omnivorous fish shown in Table 8.1 are placed together because their diet is highly varied and includes other fish, active epifauna, and some sedentary infauna. These 30 species fall naturally into two categories: (1) the threadfins (*Galeoides, Pentanemus, Polydactylus*) and catfish (*Arius*) of inshore and estuarine muddy deposits, and (2) many of the demersal fish of the deeper, sandier grounds of the outer continental shelf—the sparids (*Pagrus, Dentex, Pagellus, Boops*), grunts (*Pomadasys, Brachydeuterus*), snappers (*Lutjanus, Lethrinus*), the lizard fish *Trachinocephalus*, and the surmullet *Upeneus*. However, even among omnivores, there is some specialization. It is well known that some of these groups, such as the sparids, are able to excavate benthic infaunal organisms from the deposits, as well as being generalized benthic and pelagic predators. *Galeoides decadactylus* concentrates in the Sierra Leone estuary on small cumacea (*Uspelaspis, Cumopsis*) which it eats at a ratio of about 50:1 aginst all other food items, while *Polydactylus quadrifilis*, which is by far the largest polynemid, depends increasingly on fish as it grows in size itself, so that the largest individuals are entirely fish-eaters—in fact, from the stomach of a very large specimen found recently dead by one of us in the Sierra Leone estuary was recovered a new species of moray eel (in this case, the polynemid did not survive its meal). Taken in the same trawl hauls as *Galeoides decadactylus,* the grunt *Pomadasys jubelini* contain almost equal numbers of small cumaceans and *Pectinaria sourei,* which outnumbered other polychaetes by a factor of 9 : 2 in the stomachs.

A similar two-group situation occurs among those species which are somewhat more specialized than the true omnivores: those that obtain their diet from fish and from active epifaunal invertebrates, but take very little from the benthic infauna. Once again, it makes sense to recognize a coastal group, and an offshore, outer continental shelf group. The former comprises small sharks (*Mustelis, Leptocharis*), rays (*Squatina* and *Raja*), some flatfish, including *Psettodes belcheri,* which (like its Southeast Asia counterpart, *P. erumei*) is a fish and cephalopod specialist. Most importantly, this inshore group includes the principal sciaenids of fisheries interest (*Pseudotolithus*). The latter, offshore group is dominated by sea perches (*Epinephelus, Serranus, Diagramma,* etc), and a range of rather diverse benthic fish (*Scorpaena, Lepidotrigla, Uranoscopus,* etc.). The diet of the important sciaenids in this group is principally determined by their relative sizes: the relatively small *Pseudotolithus elongatus* in the estuaries has a diet dominated by mysids, with some cumaceans, penaeids, and small fish, while the larger *P. brachygnathus* from the same trawl stations has a percentage occurrence of fish in its stomach contents of about 75%.

Most of the specialized ichthyophagous genera listed in Table 8.1 are extremely fast swimmers, and most are generalized pelagic predators that take their prey in midwater by visual approach. We have noted those characteristics of the oceanic tunas, and coastal barracudas and jacks, that are essential for successful pelagic predators in Chapter 7. Some of these are somewhat specialized as "stalk and lunge" predators; barracudas, as we noted, approach their prey very slowly and lunge forward when within striking distance. Cornetfish (*Fistularia*) attack in the same way, except that their snake-like body is used in the same manner as a larval fish; an S-bend form is adopted prior to the final lunge and the capture of the prey with the highly extensible, tube-like mouth.

A few of the ichthyophagous (and some omnivorous) species are skulking predators that lie in wait camouflaged on the bottom; some of these, such as lizard fish (*Trachinocephalus,* closely related to the Indo-Pacific *Saurida*) dash upward from a camouflaged position and snatch their prey in their long, sharply toothed jaws, while others, such as large groupers (e.g., *Epinephelus aeneus*) rely on a very wide gape to engulf their prey with little forward movement from their hiding place. Finally, some predators, such as *Muraena,* obtain their prey from within the crevices of the rocky substrate they inhabit. For all ichthyophagous fish, selection of prey appears to be almost entirely determined by availability and size; the appropriate predator/prey size ratio is quite variable, jacks taking relatively small prey for their size, while barracuda take relatively large prey. Differences in relative predator/prey size is reflected in the different morphology of ichthyophagous groups.

The stomach content analyses from Sierra Leone, on which the above review of feeding groups was based, may be combined with information on benthic and pelagic ecosystems, and on the estimated biomass and energy demands of each group of fish, to evaluate the role of fish in the biological balance on this tropical continental shelf. What follows is based on data obtained off Sierra Leone in the mid-1950s by the staff of the West African Fishery Research Institute based in Freetown, as integrated and interpreted in a much later study (Longhurst, 1983). The first step in this task is to establish the relative biomass of elements of the marine ecosystems and of the feeding groups of fish communities which they support. Table 5.4 (which we introduced in an earlier chapter in our analysis on pelagic–benthic coupling) shows the results of this for the estuary and continental shelf of Sierra Leone. The biomass estimates (all in units of grams of carbon per square meter) for phytoplankton, zooplankton, and benthos are derived directly from ecological surveys from which such data were a planned output. In Table 5.5, the indicated phytoplankton standing stock is for large cells only, such as might be accessible to filtration by zooplanktonic metazoans, to filter-feeding benthos fauna, and to filter-feeding herbivorous fish.

For zooplankton and benthic macrofauna it was relatively simple to isolate the groups of organisms having similar feeding habits, as indicated from the species counts from the ecological surveys and to express them as biomass of planktonic predators and filter feeders, and benthic suspension and deposit feeders, and finally, to convert to organic carbon equivalents by appropriate conversion factors. However, in so doing, we must remember that any simple allocation to trophic groups conceals many interactions which act to recycle material within feeding groups, especially if these are assumed to be as broad as a Lindeman trophic level. As Committo and Ambrose (1985) have shown, by reviewing many hundreds of descriptions of predator/prey interactions in the benthos, many predators eat other predators. Thus, one must not assume that benthic predators (or, for that matter, any predators) consume only organisms that are herbivores or detritivores.

For fish, biomass estimates were available from trawling assessment surveys stratified by fish community—the estuarine and offshore sciaenid communities, and the offshore sparid community as described in Chapter 5. From the percentage biomass of each family in the trawl catches, and the allocation to feeding groups discussed above, it is simple to derive the suggested biomass for each feeding group indicated in Table 8.1 both for the estuary and the inner open continental shelf. For pelagic fish and specialized feeding groups, such as the estuarine mullets that browse on benthic diatoms, indirect (and less accurate) techniques were used, based on relative abundances of these and the demersal species in fishery landings at Sierra Leone ports. Using such methods, it was possible to assemble the comprehensive table of standing stocks for an ecosystem model at a rather high level of abstraction, in which 11 categories of organisms represented the entire marine ecosystem.

The next step was to attach to these biomass values a set of corresponding equivalent production rates, once again in units of organic carbon per square meter. For primary organic production by phytoplankton, the task was relatively straightforward, since values were available for water column chlorophyll, water transparency, and incident solar radiation. Production per unit area calculated from these variables, using reasonable assumptions for P_{max} and the simple equation for P_{eu} of Ryther and Yentsch (1957). Fortunately, there are proxy data from elsewhere in the Gulf of Guinea obtained with more modern ^{14}C tracer techniques (e.g., Nellen, 1966) suggesting that the production values of 835 and 73 mg C m^{-2} day^{-1} for the dry and wet season estuary are reasonable. Site-specific measurements of zooplankton production rates are lacking for Sierra Leone, but fortunately there are excellent estimates for other parts of the Gulf of Guinea, for communities comprising many of the same species of zooplankton; these have already been discussed in Chapter 6, and are based on P/B ratios obtained from C/N/P ratios of food and excretory products. Based on the actual composition of the plankton off Sierra Leone, daily P/B ratios between 0.408 and 0.605 appear to accomodate wet and dry season conditions in the estuary and offshore. For benthos, similar assumptions allowed us to set annual P/B ratios bewteen 1.12 and 2.56 for each community.

Production/biomass ratios for feeding groups of fish were calculated from data that are available from studies of the population dynamics of nine species of West African fish covering the major ecological groups; P/B can be calculated quite simply by the method of Allen (1971) based on a negative exponential model and the von Bertalanfy equation to describe mortality and growth, respectively. As Lévèque et al. (1977) have shown, the assumption that $P/B = Z$ (where Z is the total mortality of an equilibrium population) is valid for many species of vertebrates having life spans shorter than 5 years. As discussed in Chapter 9, there is a close relationship between the parameters of the von Bertalanfy growth function and M, natural population mortality. For an unfished stock, which was essentially the situation in Sierra Leone when these data were obtained in the early 1950s, we could assume that $M = Z$. Because the values for the von Bertalanfy growth function are quite robust for these West African fish, being based on modal class progression (see Chapter 9) and the measurement of more than 2×10^5 individual fish, we could obain what are probably accurate estimates of P/B for each year-class of the more important species in the communities, and integrate these (from the known age composition of the stocks at the time) to population P/B values ranging from 3.34 to 3.93 for demersal fish, clupeids, and predators.

The final step in integration was to estimate the requirement of each feeding group of fish and invertebrates for food to fuel their estimated production rates. When this can be done, it is then possible to partition the production at each level between consumers, and to begin to understand the role of each feeding group in the bionomics of the tropical continental shelf ecosystem. The demand estimates could be calculated from values of overall assimilation efficiencies, and for overall gross (K_1) and net (K_2) growth efficiencies. Knowing these, and with all diet reduced to the single unit of organic carbon, we could then estimate the intake of food required to fuel the calculated production of each feeding group. From published values of these parameters for demersal and pelagic fish, for plankton, and for benthos (e.g., Jones, 1976; Conover, 1975; Mann, 1982; Parsons et al., 1977), it was possible to establish reasonable values for K_1, which is the critical parameter we need: clupeids, 0.40; demersal fish, 0.35; plankton, 0.37; and benthos, 0.25. The inverse values of these estimates express the ratio (R/P) between ration and production for each feeding group.

From these calculations, we could summarize plankton, benthos, and fish production rates and required rations for two ecosystems, the estuary and inner shelf off Sierra Leone, for both wet and dry seasons, and these results are included in Table 5.5. It is immediately clear from this table that although the clupeid Ethmalosa dorsalis represents the largest component of herbivore biomass, its relatively very slow growth rate compared with zooplankton requires a significantly lower intake rate of phytoplankton. Herbivorous fish, therefore, only consume 20–25% as much phytoplankton in the estuary as the smaller biomass of filter-feeding zooplankton. Thus, although it is the high standing stocks of diatom cells inshore and especially in the estuary that appear to be essential to the feeding ecology of E. dorsalis, the population of this clupeid can have only a minor effect on phytoplankton standing stocks. As already noted, the fate of most phytoplankton cells is not to be eaten by a zooplankter or a fish (as classical marine ecology has it), but either to sink to the deposits or to be consumed by protozoan microplankton.

For fish dependent on benthic organisms for their energy source, the balance appears much closer both in the estuary and on the inner continental shelf. In both areas and seasons the apparent demand exceeds the apparent supply; however, our analysis of the diets of demersal fish, which in this simplest of all possible box models are shown to be dependent entirely on benthos, also show that these take a great deal of their food from mobile invertebrates (cephalopods, penaeid prawns) and from small fish, including many shown in the other fish components of the model. The model is not unreasonable, therefore, in suggesting that production of demersal fish (including at least three feeding groups discussed above) is fueled by the total output of benthic organisms plus the production of mobile invertebrates and some groups of smaller fish. We can expect, then, that demersal fish are a major controlling factor for benthic invertebrates and that the coupling of production and consumption may be fairly close—as indeed it seems to be, on the evidence of the variability of composition of food, apparently in response to availability.

Off Sierra Leone, the apparent abundance of zooplankton-feeding clupeids (e.g., *Sardinella*) is much lower than in regions of coastal upwelling where *Sardinella* rather than *Ethmalosa* is likely to dominate the clupeid fauna, so it is not surprising that there is apparently a surplus production of herbivorous and omnivorous zooplankton above the demands of planktivorous fish, and that this surplus is apparently consumed by predatory zooplankton. The balances in the zooplankton-fish system are sufficiently close as to fall within the anticipated uncertainties of a first-order budget of this kind.

The general conclusions of this budget are consistent with other tropical studies integrating fish with their ecosystems, such as the carbon budgets for the Peruvian upwelling region and Gulf of Mexico suggested by Walsh (1983). The Gulf of Mexico budget is outlined for two regions, of which one has high phytoplankton production rates and a large stock of the phytophagous clupeid *Brevoortia patronus* and the other where this fish was relatively much less abundant. Walsh suggests that the ratio between zooplankton and clupeid consumption of phytoplankton production in the Louisiana–Texas region was similar to what we have discussed for Sierra Leone: about 7 : 1 in favour of zooplankton. He also shows that even in the presence of a large biomass of phytophagous fish, about two-thirds of plant production passes directly to the detrital pool. Although his budget does not specify the production rates of different components of the ecosystem, his figures do suggest that demersal fish consumption is relatively close to benthic production.

However, the budget for the Peruvian upwelling system in the period prior to the collapse of the great anchovy stocks that existed until the early 1970s was apparently quite different. Here, Walsh suggests, the ecosystem may be divided into two regions: a northern, in which *Engraulis ringens* was abundant, and a southern, where it was much less abundant. In the northern region, only a small part of the phytoplankton production passed directly to the detrital pool by sinking to the benthic ecosystem, while the bulk of the phytoplankton production was consumed in about equal amounts by the herbivorous zooplankton and anchovies. In the southern region, where anchovies were relatively sparse, almost half of the total primary production passed directly to the detrital pool. The northern stock of anchovies was clearly imposing a demand on the ecosystem sufficiently strong to have profound effects on its structure. As Walsh suggests, the drastic reduction of anchovy grazing pressure following overfishing and the 1972 El Niño led to a situation in the northern region where the amount of plant production available for burial as organic detritus changed significantly, from about 8% to about 60%.

ECOSYSTEMS OF BOULDERS AND CORAL REEFS

The number of studies of the feeding relationships between fishes and the invertebrate ecosystems of coral and other reefs far exceeds what is available for the ecosystems of level sea bottoms on the continental shelf; though most approaches toward a comprehensive budget of energy flow between fish and the reef ecosystems that support them are still very tentative, we can say a great deal more in a descriptive sense of how the interactions occur. This is especially the case for coral reefs, where the feeding habits of fish have been observed extensively, and the effect of their activities on the coral reef community has been described exhaustively. A particularly useful compendium of the characteristic diets of the fish of reefs and their associated biotopes is that of Randall (1967), noted as a "Citation Classic" in Chapter 9.

Table 8.2 shows how the fish species on four coral reefs from the Caribbean to the Great Barrier Reef may be allocated to feeding groups. What is immediately obvious from these lists is the relatively high number of species classified as herbivores compared with any other fish communities: from 5 to 30% of all species, and as much as 25% of total fish biomass may have this mode of nutrition on coral reefs. Reef herbivores are all browsers on the epilithic algal community, which occurs either as algal

TABLE 8.2. Relative Numbers of Species of Reef Fish in Feeding Groups at Various Pacific Locations[a]

	Marshall Islands[b]	Virgin Islands and Puerto Rico[c]	Hawaiian Islands[d]
Herbivores	15	13	5
Planktivores	4	13	17
Benthic carnivores	56	42	52
Midwater carnivores	12	22	11
Coral feeders	4	0	6
Omnivores	7	10	8
Detritus feeders	2	0	1
Scavengers	0	0	0
Total number species	225	212	107

[a] From Sutton (1983).
[b] Hiatt and Strasburg (1960).
[c] Randall (1967).
[d] Hobson (1974).

filaments growing on the surface of dead coral rock, or as the turf of small, rapidly growing macrophytic and coralline algae that occurs on parts of the reef flat and the *Lithothamnium* ridge; filter-feeding herbivores are generally not considered to form part of the coral reef ecosystem. Many of the species listed in Table 8.2 as benthic carnivores have the browsing habit in common with herbivores; rasping attack on benthic epifauna is the normal mode of feeding of many of the species in this category. As we shall see, this method of obtaining food is so prevalent that it certainly plays a significant role in maintaining the characteristics of tropical reef and rocky ground communities, with their conspicuous lack of epifaunal invertebrates or of large algal fronds compared with temperate rocky coasts. There, the only important grazers are echinoids, which only rarely, and perhaps abnormally, lose balance with their substrate and denude a region of large algae. In fact, it is probable that energy flow in coral reefs differs from that in other marine benthic ecosystems in being dominated by grazing by herbivorous fish; only in the Caribbean, and there perhaps only because of modifications to the ecosystem by fishing, do echinoids play an important role in grazing. It is thought (Hatcher, 1983) that grazing not only controls the major fluxes of energy and material within the ecosystem but also controls the relative species composition of both plants and benthic animals, and even the supply of fragments of calcareous rock to the deposits of white sand surrounding coral reefs.

On Pacific and Caribbean coral reefs, respectively, Cloud (1959) and Bardach (1961) estimate that the production of fine sand by reef erosion by grazing fish can reach 200–600 t km^{-2} y^{-1}.

Hobson (1974) and Hiatt and Strasburg (1960) point out that grazing herbivores are still relatively few in number of species compared with their importance in energy flow, and suggest that herbivory of the kind performed by reef fish is an advanced evolutionary trait among teleost fish. Hobson further points out that like the benthic predators of reefs, herbivores tend to be diurnal, small-mouthed, colorful fish of families that include members that are themselves benthic predators. The dominant grazing herbivores are usually parrot fish (Scaridae), some damselfish (Pomacentridae), some comb-tooth blennies (Blennidae), and some Acanthuridae (surgeonfish). Most of the herbivores do not attack living corals, though Kaufman (1978) suggests that *Eupomacentrus* may suppress coral growth by establishing feeding territories of bare coral rock which can then support a mat of filamentous algae. A typical parrotfish, such as *Scarus,* is active by day, and continually grazes on coral rock bearing an algal coating, using the sides of its parrot-like jaws, and withdrawing after each rasping motion. This leaves the rock covered with tooth marks which typically come right up to the edge of living coral but do not encroach on it. At night, as Hobson and Chess (1978) have shown, at least some parrotfishes sleep under rocky overhangs, surrounded by a protective tent of mucous strands. Many surgeonfish (e.g., *Acanthurus*) browse differently from scarids, and specialize in the turf of small coralline and other macrophytic algae, from which they bite and tear fragments. Other species of Acanthuridae ingest calcareous sediment from around the coral heads, each species specializing in coarse or fine grade of fragments; their nutrition appears to be obtained from the associated interstitial diatoms and organic detritus. Much grazing occurs by similar species of fish also on the sea-grass meadows which often occur close by reefs.

The range of individual mechanisms and specializations by which the predators on benthic organisms obtain their prey is overwhelming; most are diurnal, probably, as Hobson (1974) suggests, because most mechanisms for finding prey are visual, and some require very delicate manipulative techniques impossible in the dark. The principal prey organisms of this group of fishes, in terms of overall biomass, are probably sessile invertebrates (sponges, tunicates, and many forms of coelenterates) as well as the larger motile epifauna (gasteropods, echinoderms). The fishes themselves comprise, principally wrasses (Labridae), angelfish, butterfly fish (Chaetodontidae), trigger fish (Balistidae), filefish (Monacanthidae), and many other families. All, as Hobson points out, are among the most highly evolved teleosts.

Sessile sponges are the speciality of some angelfish (*Holacanthus*), butterfly fish (*Chaetodon*), and Moorish Idol (*Zanclus*). These attack sponges simply by tearing fragments off them; most species that attack sponges and other soft epifauna, such as alcyonarians, tend to have somewhat elongated snouts. Rather few fish specialize in attacking corals directly, though some filefishes (e.g., *Oxymonacanthus*), blennies (e.g., *Exallius*), and butterfly fish (e.g., *Chaetodon*) do specialize in feeding on individual coral polyps and on nibbling off the tips of coral branches. Generally, predation on corals is restricted to a relatively small number of species, and coral fragments usually form only a small proportion of the material ingested even by these species. However, because nibbling may cause damage that allows algae to enter the coral skeleton and prevent regrowth of living coral tissue, more damage may result than simple tissue removal. Some boxfish, such as *Ostracion*, feed somewhat similarly on soft tunicates, polychaete tentacles, and coral polyps. There is, as Hobson observed, very fine specialization even within a feeding group such as this: *Holacanthus* feeds on the larger exposed sponges, while *Zanclus*, with its relatively more elongated snout, probes into crevices for smaller, cryptic sponges. Hiatt and Strasburg (1960) noted that monacanthids use their fine snouts to snip off individual coral polyps, while balistids have heavier dentition and shorter snouts, and so bite off larger pieces of coral.

Other common adaptions among this group of fish are for hovering, a few centimeters from the substrate, to search visually and minutely for the presence of small active crustaceans which are then carefully and individually removed. Hobson suggests that it is from such activity that the habit of cleaners evolved: Their ancestors expanded their substrate to include the skin of larger fish. Other species use very long snouts or barbels (in the case of Mullidae) to examine crevices and extract small organisms. Cryptic organisms are obtained by some labrids by overturning stones, and by some balistids by fanning away the overburden of sand with their fins. Finally, one should note the massive jaws and crushing teeth developed by some pufferfish (Tetraodontidae, Diodontidae) which enable them to crush the shells of very heavily armored organisms, perhaps especially echinoids and gastropods.

Many species obtain their energy from the zooplankton over the reef; some members of this group are very typical planktivorous fish (e.g., atherinids, holocentrids) but others appear quite anomalous, with only very minor specializations for this kind of feeding. The triggerfish *Xanthichthys* is specialized only by a slightly upturned mouth and a somewhat enlarged and strongly forked tail compared with its benthivorous relatives, yet its diet is almost entirely of calanoid copepods taken over the reef during daytime. Figure 8.2 shows the partitioning of space around a

Fig. 8.2. Partitioning of space by plankton-feeding fish at a coral reef off Hawaii, showing how the slimmer, faster-swimming species are able to occupy space further from potential refuges in the reef crevices. (From Hobson and Chess, 1978.)

reef-face by a group of five species of diurnal zooplanktivorous fish as a function of their degree of specialization as planktivores. It is clear that the most elongated, herring-like species are those whose behavior patterns take them furthest from the shelter of the coral heads, into whose interstices (Fig. 8.2) all of these species dive when threatened by a predator. Some specialization in prey selection occurs within this as other feeding groups; caesionids, for example, appear to specialize in gelatinous plankton to a remarkable degree. *Caesio* and *Pterocaesio* on the Great Barrier Reef feed heavily on tunicates and chaetognaths.

Finally, in this review of feeding habits of reef fishes, turning to predators we find the least divergence from the analogous feeding group in the open continental shelf ecosystem. Morays (Muraenidae), groupers (Serranidae), snappers (Lutjanidae), porgies (Sparidae), trumpet- and cornet-fish (Aulostomatidae, Fistularidae), lizardfish (Synodontidae), as well as the more pelagic predators are all active on the reef. Many of the more generalized predators have a crepuscular peak of activity as the diurnal schooling fish return to their nocturnal refuges and when light levels are such as to render their antipredator schooling mechanisms less effective than in the daytime. Obviously, many of the generalized predators (e.g., *Epinephelus*) use the hiding places offered by the crevices of a reef as lurking places from which to dash out and seize their prey.

Analysis of feeding groups of reef fish at Madagascar (Harmelin-Vivien, 1981) and the Great Barrier Reef (Williams and Hatcher, 1983) shows how individual species may be assembled into feeding guilds within each of the major trophic categories. Table 8.3 shows how these guilds are distributed on reef fronts of the Great Barrier Reef across the continental shelf of Queensland. From this table, we may infer that algal grazing is of greatest importance on the outer reefs, where algal mats are prominent and

TABLE 8.3. Relative Numbers of Species of Reef Fish in Feeding Groups at Various Locations on the Great Barrier Reef[a]

	Inshore	Midshelf	Outer shelf
Algal grazers	9.30	10.80	29.10
Suckers	0.00	1.30	11.70
Small croppers	4.60	2.80	7.70
Large croppers	2.70	1.60	3.50
Scrapers	2.00	5.50	6.20
Planktivores	45.40	70.90	43.00
Zooplankton	2.80	11.90	11.00
Omnivorous	15.40	8.40	2.90
Gelatinous	27.20	50.70	20.70
Algal	0.00	0.00	8.40
Benthivores	34.10	12.20	25.90
Mobile	21.00	8.10	21.40
Sessile	3.90	0.40	1.40
Omnivorous	3.60	1.20	0.70
Coral	5.60	2.50	2.30
Piscivores	11.20	6.10	2.00
Obligate	2.00	2.90	1.30
Facultative	9.30	3.20	0.70
Relative biomass	39	100	66

[a] From Williams and Hatcher (1983).

macroalgae (e.g., *Sargassum*) undeveloped, and that zooplanktivorous fish maintain their greatest biomass in the midcontinental shelf zone.

Some indication of the intensity of grazing pressure placed on both algal and benthic communities by reef fish can be judged from anecdotal evidence. It is a common experience of SCUBA divers that any disturbance of the sediments, or the overturning of a coral slab, immediately brings a crowd of fish of many species to snatch or graze down the exposed organisms. As the tide rises, fish crowd into the newly covered areas of reef, sometimes lying on their sides in water too shallow for them, in their eagerness to get at the algal and epifaunal food made accessible to them by the rising tide (Bakus, 1974). Exclusion experiments show how rapidly epifaunal and algal mats grow in the absence of fish but quite soon, in some cases, slough off the rock bearing them (Stephenson and Searle, 1960). It has been suggested, and perhaps demonstrated (Brock, 1979), that grazing increases the possibility of settlement and diversity of benthic organisms by decreasing the density of algal cover as well, of course, as increasing the overall rate of plant production by recycling nutrients and maintaining plant in their rapid growth phase.

Coral reef ecosystems are so complex that energy budgets encompassing all their components must be constructed at a very high level of abstraction, but models for energy balance have been attempted, and we shall describe one such below. Some early investigations (e.g., Bakus, 1967) suggested that only a small portion (<10%) of total production by Cyanophycae on a reef-flat was required to support the observed algal herbivores and their apparent intake, but this seems at odds with much of the anecdotal and observational data. Thus, Odum and Odum (1955), for example, found a biomass of herbivores of almost 20% of the standing stock of total plant material of all kinds of Eniwetok in the Marianas and suggested that the plant material itself was cycled at least 12 times annually.

The only comprehensive attempt to model energy flow through all components of a coral reef ecosystem appears to be the recent research (Polovina, 1984) on French Frigate Shoals in the Hawaiian Islands at 24°N. This is a crescent-shaped reef surrounding a lagoon about 36 km in diameter and quite isolated from other reefs. The study was designed to determine if the reef ecosystem was nutrient or predator limited, and to investigate the potential of reefs for sustained fishing; a simple box model (ECOPATH) was used to estimate biomass and production in each of 15 species groups, identified as the major components of the ecosystem down to 300 m depth (Fig. 8.3). ECOPATH comprises a system of simultaneous equations that solve the following expression for all 15 species groups: Production of biomass for species i − all predation on species i −

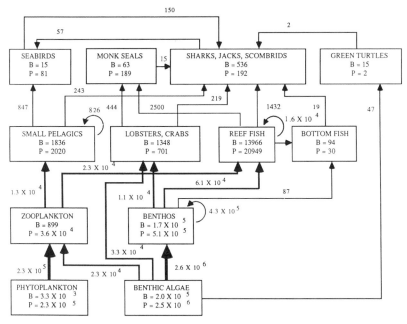

Fig. 8.3.   Analysis of the ecosystem at French Frigate Shoal, Hawaiian Islands, based on a production/biomass ratio model. Mean annual biomass (B) is in kg km^{-2} and production (P) is in kg km^{-2} y^{-1}. (From Polovina, 1984.)

nonpredatory biomass loss for species $i = 0$ for all i. The ecosystem, in the present version of the model, is simulated from the top down, driven by independent biomass estimates for the top predator groups. It is validated by comparison with field measurements of abundances and growth rates of lower trophic levels. The equations include parameters for P/B ratios (derived in a comparable fashion to those discussed above for the Sierra Leone estuary), a diet composition matrix, and food requirements to support maintenance and growth. Table 8.4 shows the values achieved when the model is initiated with field estimates of biomass for tiger sharks, seabirds, and monk seals.

Grigg *et al.* (1984) also made measurements of community respiration and primary production on French Frigate Shoals for validation of the model. Under very large transparent plastic tents covering coral knolls dominated by *Porites compressa* and *P. lobata,* and in the water passing across the reef, time-course measurements were made of oxygen, total dissolved CO_2, pH, and alkalinity. This respirometry technique established rates of benthic net primary production and net community production that could be compared with the output from the ECOPATH model.

TABLE 8.4. Quantitative Analysis of the Reef Ecosystem at
French Frigate Shoals, Northwest of the
Hawaiian Islands[a]

	Biomass (kg km^{-2})	Annual production (kg km^{-2})	P/B ratio (annual)
Tiger sharks	42	11	0.25
Birds	15	81	5.40
Monk seals	63	189	3.00
Reef sharks	38	7	0.18
Turtles	15	2	0.15
Small pelagic fish	1,836	2,020	1.10
Jacks	411	144	0.47
Reef fishes	23,941	35,912	1.50
Large crustaceans	2,311	1,202	0.52
Bottom fishes	377	121	0.32
Nearshore scombroids	60	40	0.66
Phytoplankton algae	3,295	230,679	70.00
Zooplankton	899	35,944	40.00
Benthos	289,181	867,543	3.00
Benthic algae	342,598	4,282,471	12.50

[a] From Polovina (1984).

Despite some obvious shortcomings in the model assumptions for lower trophic levels, agreement was remarkably close.

Field estimates for the standing stock of lower trophic groups were available for comparison with model output: for reef fishes the survey suggested 15,000 kg km^{-2}, against model output of 24,000 kg km^{-2}; for reef sharks 48 kg km^{-2} against 38 kg km^{-2}; the model suggests that benthic primary production of 4.3×10^6 is necessary to support the growth of higher trophic levels, and the field measurements gave a rate of 6.1×10^6 kg km^{-2} y^{-1}. There is some reason to believe, with the authors of these studies, that the model output does, in fact, represent the general outline of energy flow within the higher trophic levels of the French Frigate Shoals ecosystem. Grigg et al. (1984) then interpret their results in the following terms: (1) phytoplankton production over the reef is a very small fraction of total plant production; (2) about 5% of net benthic primary production passes up to nonbenthic consumers; (3) there is much internal predation within the reef benthos and reef fish species groups; and (4) an overall ecological efficiency of about 0.17 between trophic levels is indicated. From the behavior of the model, Grigg et al. conclude that the reef ecosystem at French Frigate Shoals is regulated primarily from the top down by predation pressure, while primary production of plant material is regulated by space constraints and by the limits imposed by metabolic rates, as well as ambient nutrient levels. Such a conclusion, they suggest, if it extends right down through this almost undisturbed ecosystem, also probably accounts for the lack of large fleshy algae on this and other tropical reefs; this is, as they point out, quite different from the situation in a terrestrial ecosystem where the top-down control of herbivores by predators does not extend down to the control of plant biomass by herbivores, since most plant biomass is consumed by microrganisms.

They conclude that the potential yield of top predators around reefs is limited also by the high levels of internal predation (and hence that there is a large number of trophic levels, six by their count) within the reef benthos and reef fish species groups. They suggest that fishery yield of selected species of reef fish might be increased by removal of top predators from the ecosystem, in this case by a tiger shark fishery.

One of the consequences of a complex ecosystem including a great diversity of symbiotic and free-living algae, and a very large total number of species of animals and macrophytic plants, is its chemical diversity. Many compounds occurring naturally in the organic mass that comprises the reef community are toxic to vertebrates, and some are capable of being passed from prey to predator, and of being accumulated up a food chain toward the top predators. It is not surprising, then, to find that some fish taken on reefs are toxic to humans. The exogenous toxins involved here appear to be more important than endogenous compounds, produced by the basic biochemistry of the fish. The exogenous nature of toxins is principally attested to by the highly variable way in which toxins are found in fish; the same species may be highly toxic or perfectly wholesome on different sides of the same island, and island folklore is full of anecdotal cautions to be observed in selecting fish for food. Some particular stretches of reef or coast are reputed to produce many dangerously toxic fish while in the Marshalls it was considered that only fish taken in the reef passes are toxic, and elsewhere the color, stomach contents, and time of capture are all considered to indicate safety or danger. Parallel development of folk remedies has resulted in a wide variety of potions, ranging from possibly useful infusions of the tissues of certain plants (the "itch-herb" *Messerschmidtia* in particular) to the suggestion to drink large tots of rum, which may enable the victim to die in style and with less pain, but is scarcely remedial.

Ichthyotoxins occur throughout the tropics, perhaps (and consistent with food chain intensification) especially in top predators: barracuda and jack are among the most commonly toxic species. Consistent with the exogenous origin of many toxins, is the wide range of species involved and symptoms specific to each: massive histidine–histamine poisoning from scombroids, hallucinogenic intoxication from mullets and surmullets, fatal neurotoxic effects from morays, and poisoning resembling *Gonyaulax* toxins from small inshore filter-feeding clupeids. However, the most dangerous and widespread ichthyotoxins fall under the general name of "ciguaterra," so named for marine invertebrate poisoning encountered by the first Spanish settlers of the Caribbean world. The term is now used for a range of neurotoxins occurring in many intermittently poisonous fish species in the tropics. Baslow (1977) lists about 450 tropical fish species, most from reef habitats, that have been implicated in this kind of poisoning. Prominent among these are the grazing species that ingest corals and algal turf materials.

Ichthyotoxins are sufficiently widely present in the reef fish of tropical oceans, and with a sufficiently unpredictable occurrence, as to render the large-scale commercialization of snappers, groupers, jack, and other large, marketable fish a chancy business; in some island groups, such as (it is claimed) Johnston and Phoenix, very high percentages of species are toxic. Where commercialization of large fish has occurred, such as for export from Midway and Christmas Island to Hawaii, the occurrence of the occasional ciguaterra-bearing batch is not only a public health tragedy, but may effectively also kill an industry, as occurred in the late 1940s in this particular instance. The Japanese armed forces operating in the South Pacific during WWII were acutely aware of the problem and commissioned a special guide to poisonous fish in the Pacific Islands prior to the outbreak of hostilities (van Campen, 1950).

PELAGIC ECOSYSTEM OF THE OPEN OCEAN

The pelagic ecosystem of the open ocean differs from that of coral reefs as fundamentally as the range of marine habitats permits; coral reefs are one of the two habitats of our planet that are considered to have the most, and the most structured, spatial heterogeneity. On the other hand, the pelagic zone of the open ocean is perhaps the most monotonous living space of our planet, with few visual cues to maintain spatial orientation and with its spatial heterogeneity effectively restricted to vertical gradients of light, temperature, and abundance of organisms. Only the air–sea interface at the surface of the ocean and the rapid thermal and density changes through the thermocline represent disontinuities in the spatial medium in which the ecosystem exists.

It should be no surprise therefore that the Tertiary radiation of teleost diversity has left little mark on this environment, where most fish are what Hobson describes as generalized predators because they take prey, smaller than themselves by an approximately constant ratio, by a simple "chase and grab" technique that requires none of the diverse form we have seen in coral reef fish, where most of the diversification in teleosts has occurred. It is not only the top predators that are generalized in the open ocean. For the most part, all predatory fish, of whatever size, function rather similarly and the only specializations that should be noted are among the mesopelagic fish, some of which have huge gapes and illuminated lures and perform in the dark below the euphotic zone or at night rather like the "ambush and grab" predators of continental shelves and reefs. Otherwise, perhaps only the billfishes may be specialized "sidle and lunge" predators functioning in the open ocean like barracudas near the coast. Thus, in the open ocean pelagic ecosystem, a chain of generalized predators leads, as we shall see, from planktonic crustacea through mesopelagic and some small surface fishes, up to the tunas and billfish.

As we have already emphasized for the planktonic ecosystem, and as LeGand et al. (1972) reemphasize for the oceanic predators, the functioning of the pelagic, open ocean realm can only be understood if it is treated as a layered ecosystem, each layer having its specialized ecological base and its specialized fish community. Analyzed thus, we can begin to see some of the order that must exist if the ecosystem is to remain stable and predictable.

Both the details of vertical structure and of trophic links in the food web differ regionally and seasonally; opportunism is the rule in prey selection and is driven by regional and seasonal differences in vertical physical structure of the upper 500 m of the water column. A good model is that of Legand et al. (1972), representative of the southern tropical and equatorial surface water masses of the central and western Pacific. Four components are required: top predatory fish, cephalopods, micronektonic fish, and euphausiids and other large crustacea. Each component has a layered microstructure, matching the layered ecosystem of the pelagic realm that we have discussed in Chapter 5.

The most abundant micronektonic fish in the this model fall into two groups: deep-water nonmigrant species of *Cyclothone* and *Sternoptyx,* which perform limited vertical migrations and have their greatest abundance at 500–800 m day and night, though they form up to 60% of all fish in the micronekton, and midwater migrants (mostly myctophids of the genera *Lampanyctus, Lepidophanes, Diaphus,* and *Vinciguerra),* which rise at night from various depths to near the surface, with migration ranges of 200–700 m in vertical extent.

The model recognizes four groups of euphausiids: small epipelagic species restricted to the upper 200 m (*Stylocheiron carinatum, Euphausia tenera*); midwater nonmigrant species always occurring from 50–500 m (*S. abbreviatum, Nematoscelis tenera*); deep nonmigrant species that occur always from 300–800 m day and night (*Nematobrachion boopis, Thysanopoda cristata*); and the mass of migrant interzonal species of *Euphausia, Thysanopoda, Nematoscelis,* and *Nematobrachion* rising to near the surface at night from their daytime residence depths of 400–800 m.

Other large crustacea are similarly layered: some large amphipods (*Phronima*) occur near the surface, others (*Platyscelus*) at 250–400 m. Carideans, sergestids, penaeids, and thecosomatous pteropods all occur deep in the daytime, but rise to the upper 200 m at night. Among cephalopods, juvenile cranchiids occupy the 50- to 450-m zone by day and night, while adults rise at night from 700 to 1250 m to about 500 m nearer the surface; enoploteuthids rise at night from a maximum at 400 m to closer to the surface with a maximum in the 0- to 100-m layer; onychoteuthids have a surface population, and a deeper one that does not rise surfacewards at night.

This, or something like it in other tropical regions of the open ocean, is the nutritional base for a population of large predatory fish, most of which occur in the mixed layer or upper thermocline: tunas, billfish, and smaller pelagic species (*Auxis, Sarda, Alepisaurus,* and other genera). There have been many studies of the stomach contents of yellowfin, bigeye, and skipjack, the principal tuna of the tropical zone and these have all shown that tuna are opportunistic feeders on fish, crustacea, and cephalopods. Parin (1970) reviews many of these studies and clearly demonstrates that though the general pattern remains the same, each species shows considerable variation regionally, so that differences between the diets of different species are not constant from region to region. Each species conforms to the anticipated pattern of dietary shift during growth. Fish dominate the diet of larger tuna, and between small and large species, crustacea are more important to skipjack than yellowfin. Tuna stomach contents represent the most available species of the right size in each region, not necessarily as indicated by micronekton net sampling; tuna include many species in their diet that are apparently very poorly sampled with nets. In the eastern Pacific, off Baja California and Mexico, one of the dominant items in the diet of yellowfin is the pelagic munid crab (*Pleuroncodes planipes*), which occurs in pelagic swarms only in this part of the tropical ocean; further south, off Panama to Peru, this item is almost totally absent (the Peruvian *Pleuroncodes* being principally benthic) but a portunid swimming crab (*Euphylax dovei*) replaces it in the ecosystem and in yellowfin stomachs. From examination of large numbers ($N = 6080$) of stomachs at tuna canneries in California, Alverson (1963) was able to describe in detail the diet of skipjack and yellowfin from the whole eastern tropical Pacific, stratified by season, region, and size of fish. Yellowfin were found to eat about 45% fish, 45% crustacea, and about 10% cephalopods globally, the crustacea mostly being pelagic crabs and the fish dominated by young skipjack and other small tunas, coastal fishes drifted out to sea, flying-fish, small scombrids and carangids, and a small proportion of mesopela-gic fish, mainly myctophids. Skipjack, on the other hand, have a diet dominated in this region by euphausiids and a few small swimming crabs, but with much higher proportions of mesopelagic fish than yellowfin.

Though the number of fish examined are much smaller ($N = 656$), the diet analysis for yellowfin and albacore in the central and western Pacific by LeGand and his associates, in connection with the model for the nu-trional base described above, is probably typical of other open ocean areas. The large tuna eat mainly cephalopods and fish, the latter including very few of the diel migrant species of mesopelagic fish which rise to the surface at night; tuna are diurnal feeders and prey only on epipelagic species, such as individuals of suitable size of *Mola, Exocoetus, Alepi-saurus, Gempylus, Auxis,* and small Thunnidae, and on the more shallow-living mesopelagic species of *Sternoptyx, Argyropelecus, Vinciguerra,* and *Cubiceps,* for example. The most numerous interzonal, diel migrant species of mesopelagic fish of the families Myctophidae and Gonostomati-dae do not figure largely in the diet of tuna. In short, the diet is what can be located within the lighted zone in daytime from the surface down to about 400 m. Only the bigeye (*T. obesus*) among the truly tropical tuna, which lives deeper than yellowfin and skipjack, appears to consume inter-zonal diel migrant mesopelagic fish importantly.

Other large fish compete with tuna, the species of greatest fishery inter-est, in the same ecosystem, some having more specialized diets. *Cor-yphaena,* the dolphin fish, for example, has a very special trophic rela-tionship with flying fish, curiously established very early in life, though over continental shelves, in the presence of a more varied set of potential prey for *Coryphaena,* the relationship is not so close. As Parin describes, snake mackerel (*Gempylus*) rise to close under the surface at night, where they also prey on flying fish, as well as myctophids, young tuna, young *Coryphaena,* and other epipelagic species. Billfish and swordfish are gen-erally more ichthyophagous than tuna, and take prey up to and including tuna of moderate size, using their spears to disable larger prey. Billfish, as we have seen, dive deep and feed deeper than most tuna, and are very active and aggressive fish: the case of the marlin that attacked the re-search submersible ALVIN from the Woods Hole Oceanographic Insti-tute at a depth of more than 600 m has become classic. This, and cases of attack on whales and small boats, is probably the result of confusion between the large mass attacked and a tightly gathered school of small fish. The community of small mesopelagic fish that occurs in the upper kilometer of the water column depends principally for its food on plankton crustacea (ostracods, copepods, euphausiids), molluscs (pteropods), and juvenile and larval fish; there is both specialization and selectivity in their diet, as well (most probably) as regional differences that are yet to be described. LeGand and Rivaton (1969) found that 10 of 11 species of myctophids investigated by them south of Java fed predominantly on copepods, and Hopkins and Baird (1977) found that small crustacea comprised 70% of the diet of myctophids, gonostomatids, and sternoptichyids, with interzonal diel migration bringing deep species to feed near the surface at night. In the eastern Gulf of Mexico the same authors, in a 1985 paper, described very complex differentiation between the feeding strategies of the four abundant mesopelagic fish: *Argyropelecus aculeatus* fed in the epipelagic zone in the early evening (small fish feeding on ostracods and copepods, larger fish on pteropods, euphausiids, and fish). *A. hemigymnus* feeds in the late afternoon at 300–500 m, and fish of all size classes fed only on copepods and ostracods. *Sternoptyx* spp. feed around the clock, and at deeper levels, one species at 500–800 m and another at >800 m, both on a range of pelagic organisms appropriate to their size.

Merrett and Roe (1974) analyzed in detail the stomach contents of the most numerous species at a station in the eastern subtropical Atlantic Ocean. They found that *Valencienellus tripunctulatus, Argyropelecus hemigymnus,* and several other species specialized in copepods, especially *Pleuromamma,* while *A. aculeatus* preyed almost exclusively on ostracoda (especially *Conchoecia curta*). The larger mesopelagic fish *Chauliodus danae* principally ate the euphausiids *Thysanopoda, Stylocheiron,* and *Nematoscelis.* Parin notes that *Gonichthys tenuiculus* feeds almost exclusively on neuston and pleuston: pontellids, *Janthina,* and the sea-skater *Halobates.*

In the central and western Pacific, as we have noted, the model of Roger and Grandperrin suggests that tuna do not utilize the interzonal, diel migrant species of fish, nor the interzonal migrant crustacea but rely mostly on epipelagic and shallow fish and cephalopods; however, of the 667 stomach analyses made on small fish themselves removed from tuna stomachs, about 25% had been eating euphausiids, and these authors suggest that about 7% of all yellowfin and albacore tuna diet in this part of the Pacific is indirectly based on euphausiids. Figure 8.4 shows the general network of feeding relationships between fish and planktonic invertebrates in the epipelagic realm of the tropical ocean, using a trophic level concept in its analysis.

It is not easy to synthesize the available observations and measure-

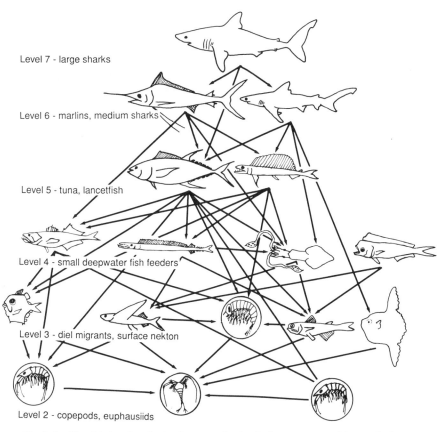

Level 7 - large sharks

Level 6 - marlins, medium sharks

Level 5 - tuna, lancetfish

Level 4 - small deepwater fish feeders

Level 3 - diel migrants, surface nekton

Level 2 - copepods, euphausiids

Fig. 8.4. Trophic level analysis of a generalized pelagic ecosystem of the tropical ocean. (From Parin, 1970.)

ments into a numerical, balancing budget for the ecosystem of the oceanic pelagic ecosystem, but a start in this direction has been made by Blackburn (1977), who attempted to produce a consistent set of biomass estimates for all trophic levels, by Sharp and Francis (1976) who sketched out an energy flow budget for the eastern tropical Pacific yellowfin population, and by Mann (1984), who assembled all the information available to him into a model synthesis. Several estimates of the relative biomass of diel migrant mesopelagic fish and the resident epipelagic small fish agree that there is 5–10 times more migrant biomass, and that an average value is 1–2 g m^{-2}; Clarke (1973) suggested that with an annual P/B ratio of 2, production of 0.2 gC m^{-2} y^{-1} of mesopelagic fish in a region where a reasonable mean value for primary production from phytoplankton would be about 50 gC m^{-2} y^{-1}. This, Clarke suggests, implies no more than two to three trophic steps between phytoplankton and mesopelagic fish. Mann's model, based on these suggestions, but using the biomass data of Gjøsaeter and Kawaguchi (1980), suggests the following mean values, as gC m^{-2} and gC m^{-2} y^{-1}: tuna = biomass 0.03, production 0.02–0.03; small epipelagic fish = biomass 0.5, production 0.5–0.13; mesopelagic fish = biomass 1.75–3.5, production 0.9–1.8.

This generalized model is consistent with the special model for energetics of yellowfin tuna and their energy base in the eastern tropical Pacific formulated by Sharp and Francis (1976). This model, based on tuna population analysis from catch statistics maintained by the Inter-American Tropical Tuna Commission, is based on estimates of the energy requirements of the exploited and a theoretical nonexploited population; the individual catch statistics were made to yield estimates of the age structure of the population for each of three subareas, based on two semestral cohorts annually. Specific dynamic action (loss of energy due to the inefficiency of the digestive process) was set at 30–40% of total consumed calories. These parameters and related assumptions were incorporated into the simulation model ENSIM to estimate the caloric requirement of each semestral cohort, in each quarter of the year, for exploited and nonexploited populations, for the years 1964–1972. Based on a biomass of yellowfin of 35.4 mg m^{-2}, or 42.5 cal m^{-2}, the energy demands of yellowfin are 1.5–2.5 cal m^{-2} d^{-1}.

This can be set against similar estimates for regional primary production and production of micronektonic fish. The EASTROPAC data (Owen and Zeitzschel, 1970) show that primary production in this region averages 205 mgC m^{-2} d^{-1}, or 2.34 kcal m^{-2} d^{-1}. The production of forage organisms in the range 1–10 cm can also be calculated from the EASTRO-PAC data for a standing stock equivalent to 1.25 kcal m^{-2} as being approximately 125 cal m^{-2} d^{-1}. These calculations suggest that there is at least one order of magnitude "oversupply" of small micronekton, suitable as tuna food, compared with the actual population demand. This requires an assumption that yellowfin tuna represent less than 10% of that part of the top predator biomass in the eastern tropical Pacific depending on the 1- to 10-cm range micronekton for food. This is an unlikely (though perhaps not impossible) suggestion. That a sophisticated model of the energy balance and food requirements of a single species can still leave us with such a high level of uncertainty as to its role in an ecosystem, which has been the subject of a great deal of modern, quantitative ecology, is an excellent example of how difficult it is to anwer what seem to be very simple questions about the dynamic balances between fish and the ecosystems that support their populations.

Chapter 9

Dynamics of Tropical Fish Populations

OVERCOMING A MYTHOLOGY

Progress in understanding the dynamics of tropical fish populations has been hampered by the limited research capability of many tropical countries, which have relatively few marine scientists as well a general absence of long time series of catch, effort, and age composition data for their fisheries. Such data are available for many northern stocks, and have enabled analysis of the populations of paradigmatic species such as cod, plaice, and herring to have been carried to a high level of sophistication.

Thus, it is still difficult and sometimes impossible in the tropics to apply the principal concepts and methods that have emerged from the study and management of the great high-latitude fisheries. This problem is compounded because concepts and methods that are really perfectly applicable to tropical fish are often perceived to be irrelevant in the tropical context, and because of perceived fundamental differences between temperate and tropical fishes that are actually reflections of underlying quantitative differences in, for instance, rates of growth or mortality.

The lack of scientific capability in tropical countries, and how this constrains their access to the "Common Heritage of Mankind" in the seas, has been widely discussed in relation to the general development of the Third World (Goodwin and Nacht, 1986; Marr, 1982) as well as in works dealing with transfer of science and technology (Ziman, 1976) and we shall not enlarge on this problem here, except to note in passing that it seems to us that progress is now being made rather rapidly, and that the situation in the marine science programs of many tropical countries has much improved over the last 10 years or so.

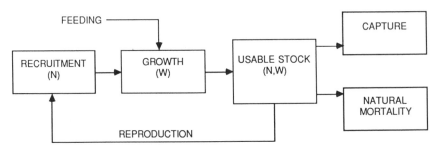

Fig. 9.1. Main factors investigated in fish population dynamics: : recruitment, growth (positive factors), capture and natural mortality (negative) and stock size. N, Numbers; W, weight. (From Pauly, 1982a, modified from Ricker, 1975.)

Because we believe that only a conscious attempt to deal with the multilayered mythology shrouding the dynamics of tropical fish populations can provide a base upon which progress will occur, dispelling this mythology will be the central theme of this chapter. So, in what follows, we shall attempt to separate what we believe are characteristic and basic features of tropical marine fish populations from features that have been attributed to them after only cursory analysis, based on hasty comparison with much better-known temperate fish populations. We shall review the nutrition, growth, mortality, reproduction, and recruitment of tropical fish populations, that is, those attributes that determine their potential yield to fisheries (Fig. 9.1).

GROWTH OF FISHES

One of the earliest scientific contributions on the age and growth of tropical fishes (Mohr, 1921) rationalized such studies in the following words:

> In modern fisheries research, ageing of fishes is very important because . . . through age determination we have the means to identify the age composition of our fish population, and so it can be determined to what degree the various age classes are utilized by the fishery. Only an exact knowledge of the age of the fish allows inferences on the appropriateness or the need for management measures such as closed seasons or minimum sizes . . . .''

Mohr then went on to describe what she perceived (already then!) as the major problem of biological studies on tropical fish, namely ''the suggestion, and later the dogma, that the scales of tropical fish should have no annual rings.'' This myth persisted widely among fishery biologists (including one of us) until the 1960s, and although, as we shall discuss below, conditions do occur where tropical fishes do not seem to generate annual rings on their hard parts, such rings are now known to occur in scales, otoliths, and vertebrae of tropical fish under far too wide a wide range of conditions to be dismissed as exceptions that prove the rule.

Another related myth states that spawning in the tropics is continuous, and therefore analysis of the growth of tropical fish based on the study of length-frequency data must fail. Once again, although there have been some puzzling failures of length-frequency analysis to generate growth data, such as the case of *Ethmalosa dorsalis* in West Africa (Salzen, 1958), this technique is now widely and effectively applied to tropical fish populations.

In fact, as we shall discuss below, while spawning of tropical fish is often more protracted than that of temperate fish, it is usually concentrated in one or two periods each year. Also we must remember that it is not necessarily the fluctuations of spawning themselves that generate the peaks and troughs of length-frequency data. The same effect can be caused by the occurrence during larval and juvenile life of recruitment windows (Bakun *et al.*, 1982), which can vary seasonally even when egg production is constant. Later, we shall present several length-frequency data sets from typical tropical fish, all displaying well-structured peaks and troughs, and which can be rather simply analyzed using Petersen's method (Petersen, 1891), Modal Progression Analysis (George and Banerji, 1964), or computer-based modifications of earlier techniques (Pauly and David, 1981).

The final myth we shall attempt to dispel is the frequently repeated statement that tropical marine fish are not subject to growth oscillations, or seasonal variations in their rate of growth. Of course, growth checks in tropical freshwater fish are well known, and known to be caused by wet-dry season alternation in riverine or lacustrine environments. Models could therefore be derived to account for growth fluctuations in yield computations for tropical freshwater fisheries (e.g., Daget and Ecoutin, 1976). We shall emphasize in this chapter that the growth of tropical marine fish can also be significantly checked by seasonal changes in their environment, and that faulty concepts and analyses probably explain occasional failures to identify seasonal changes in growth rate in tropical marine fishes.

A Simple Model of Fish Growth

Many general models exist to describe and explain the growth of fish. Most of these have been formulated for cold-water fishes, and treat tropical forms as an exception (see, for example, Ursin, 1967). In this section

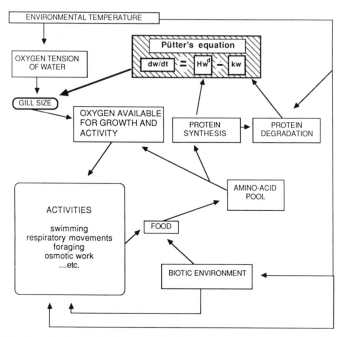

Fig. 9.2. Graphical model of fish growth with excretion process omitted, with emphasis on the limiting role for oxygen availability for protein synthesis. (Modified from Pauly, 1981.)

we propose a simple, general growth model which is applicable to fish of all latitudes.

Figure 9.2 presents a simple metabolic model of fish growth from which various forms of the well-known von Bertalanffy equation can be derived (von Bertalanffy, 1938; Pauly, 1981) and which accomodates, among other things, the tendency for tropical fish or warm water biota in general to remain smaller (i.e., to reach smaller asymptotic sizes) than their temperate counterparts.

This model is based on the following elementary observations (Pauly, 1981):

1. Fish are aerobic heterotrophs that need oxygen both for maintenance and for the synthesis of new body substances.

2. Oxygen cannot be stored by fish in more than insignificant quantities.

3. Oxygen enters the body of fish through a surface (gills, or gills + body surface) which on geometric grounds cannot grow as fast as body volume (weight) and hence oxygen demand.

From these three observations it follows that factors such as elevated temperature or stress (Selye, 1980), which increase routine metabolism, will decrease the amount of oxygen available per unit time for growth (Fig. 9.2). Such factors must lead, given species-specific anatomical constraints, to a reduction of the largest size reached by the oldest animals of a given population (Fig. 9.3).

The version of the von Bertalanffy growth model used here can be derived also from the formulation

Oxygen	Oxygen supplied	Oxygen needed by	
available =	through the	− entire fish	
for	gills and other	for maintenance and	(9.1)
growth	body surfaces	other activities	

It should be noted in interpreting this formulation that the oxygen consumption of fish, as measured experimentally (e.g., Scholander *et al.*, 1953), is the oxygen available to the tissues of the fish and not its whole-organism oxygen demand, as is frequently assumed.

Equation (9.1) leads, since synthesis of body substance is oxygen dependent, to the better-known equation of Pütter (1920)

$$dw/dt = Hw^d - kw \qquad (9.2)$$

where dw/dt is the growth rate, H the rate of synthesis of body substance, k a constant related to body oxygen requirements, w the body weight, and

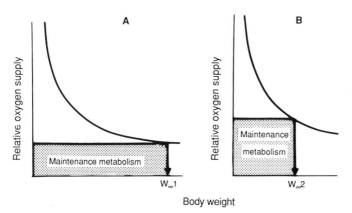

Fig. 9.3. Relationship between oxygen supply per unit of body weight (i.e., gill surface area per unit body weight) and the body weight of growing fish. Note that in A growth in weight must cease when, at W_∞, O_2 supply is equal to maintenance requirements and that for B any factor increasing maintenance requirements (elevated ambient temperature, reduced food availability, stress, etc.) will tend to reduce the weight at which O_2 supply becomes limiting to further growth and hence reduces W_∞. (From Pauly, 1984.)

d an exponent smaller than unity and often set equal to 2/3 (Bertalanffy, 1938; Pauly, 1981). Equation (9.2) can be integrated to the von Bertalanffy growth formula (VBGF), often used to model the growth of tropical fishes

$$L_t = L_\infty (1 - e^{-K(t-t_0)}) \tag{9.3}$$

where L_t is the length of fish at age t, L_∞ is the average length of fish of the stock in question would reach if they were to live indefinitely, K is a constant equal to 1/3 of k in Eq. (9.2) and t_0 is the (usually negative) age the fish would have at zero length in the unlikely case that adult growth patterns also applied to newly hatched fish, so that t_0 is usually only a convenience parameter. The corresponding equation for growth in weight is

$$W_t = L_\infty (1 - e^{-K(t-t_0)})^3 \tag{9.4}$$

where W_t and W_∞ are the weights corresponding to L_t and L_∞, respectively. Other versions of the VBGF exist, but the two versions presented here are sufficient to support the following discussion.

The model in Eqs. (9.1, 9.2) and Fig. 9.2 implies that factors which increase oxygen demand will lead to an increase in the parameter K of the VGBF, and a reduction of its parameters L_∞ or W_∞. For a given species, this leads to a near constancy of the parameter ϕ', defined by

$$\phi' = \log_{10} K + 2 \log_{10} L_\infty \tag{9.5}$$

and of the parameter ϕ, defined by

$$\phi = \log_{10} K + 2/3 \log_{10} W_\infty \tag{9.6}$$

To illustrate this, Fig. 9.4 shows the nearly normal, and rather narrow, frequency distributions of ϕ' values in various stocks of skipjack tuna (*Katsuwonus pelamis*).

Thus, while fitting of growth curves to empirical size-at-age data may lead in any given species to a wide range of growth parameter values (see Table 9.1), the underlying growth performance of the species in question can be captured by a single parameter based on the two growth parameters L_∞ and K and which, moreover, can be shown to be related to gill size, as required by a theory that links growth performance and oxygen supply (Fig. 9.5).

Seasonal growth oscillation of tropical fish can be expressed by a modified version of the VBGF of the form

$$L_t = L_\infty[(1 - \exp[-K(t - t_0)] + [(CK/2\pi) \sin 2\pi (t - t_s)])] \tag{9.7}$$

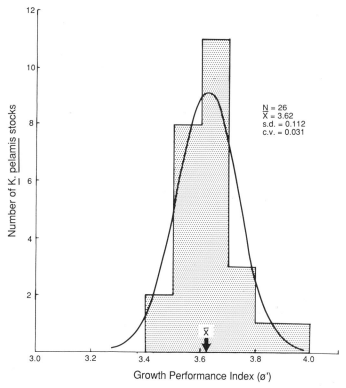

Fig. 9.4. Frequency distribution of ϕ' ($\log_{10} K + 2 \log_{10} L_\infty$) in 26 stocks of skipjack tuna (*K. pelamis*).

where L_t, t, L_∞, and K are defined as above, and where C and t_s are new parameters. The first of these parameters (C) expresses the amplitude of the seasonal growth oscillations (Fig. 9.6), and usually takes values ranging between 0 (no growth oscillations) and 1 (strong growth oscillations, generating one complete growth stop per year). The second parameter (t_s) fixes the start of a sinusoidal growth oscillation in relation to $t = 0$ and need not concern us further here.

The model and the seven equations presented in this section, along with Figs. 9.2 and 9.3, suffice to explain and illustrate the fundamental aspects of the growth of tropical fishes (and invertebrates, too, as we shall discuss in Chapter 10). We can therefore proceed to an examination of how this model can be applied in practice to analyzing the population biology of tropical sea fish.

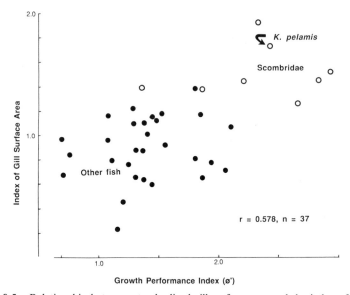

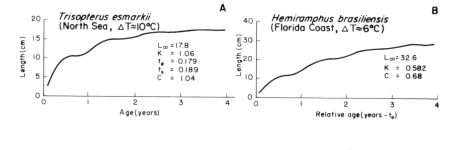

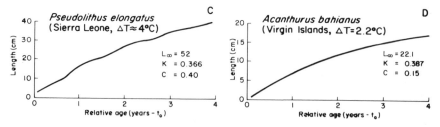

Fig. 9.5. Relationship between standardised gill surface area and the index of growth performance ϕ' ($\log_{10} K + 2 \log_{10} L_\infty$) in various marine fishes. (Source of data is Pauly, 1981, 1985a.)

Fig. 9.6. Seasonally oscillating growth curves in four species of fishes. Note that variations are not visible in the *Acanthurus* curve although they significantly affected the growth rates of the tagged fish upon which the parameters of this curve are based (Pauly and Gaschütz, 1979; Gordon, 1977; Berkeley and Houde, 1978; Pauly and Ingles, 1981; Randall, 1962).

TABLE 9.1. Estimates of the Growth Parameter (*K*) in Skipjack Tuna of a Range of Fork Length (*FL*), with Computed Values of the Growth Performance Parameter ϕ' [a]

Area	FL (cm)	K (y⁻¹)	ϕ'	Source
Hawaii	84.60	1.16	3.92	Uchiyama and Struhsaker, 1981
Hawaii	85.10	0.95	3.84	Brock, 1954
Eastern tropical Pacific	85.10	0.44	3.50	Schaefer, 1961
Hawaii	82.30	0.77	3.72	Rothschild, 1967
Pacific Ocean 10°N 100°W	80.50	0.63	3.61	Joseph and Calkins, 1969
North of Madagascar	62.30	0.98	3.58	Marcille and Stequert, 1976
Central Pacific	102.00	0.55	3.76	Uchiyama and Struhsaker, 1981
Hawaii	92.40	0.47	3.60	Uchiyama and Struhsaker, 1981
Eastern Pacific	142.50	0.29	3.77	Uchiyama and Struhsaker, 1981
Eastern Pacific	72.90	0.83	3.64	Joseph and Calkins, 1969
Eastern Pacific	107.50	0.41	3.68	Joseph and Calkins, 1969
Eastern Pacific	88.10	0.43	3.52	Joseph and Calkins, 1969
Papua New Guinea	65.47	0.95	3.61	Josse et al., 1979
Papua New Guinea	74.80	0.52	3.46	Wankowski, 1981
Papua New Guinea	65.00	0.92	3.59	Kerney, 1974
Philippines (southern)	84.50	0.51	3.56	White, 1982
Philippines (southern)	82.20	0.48	3.51	Tandog, 1984
Hawaii	101.10	0.39	3.60	Skillman, 1981
Taiwan	103.60	0.30	3.51	Chi and Yang, 1973
Taiwan	103.80	0.43	3.67	Chi and Yang, 1973
Eastern Pacific	79.06	0.64	3.60	Josse et al., 1979
Western Pacific	61.30	1.25	3.67	Sibert et al., 1983
Eastern Pacific	75.50	0.77	3.64	Sibert et al., 1983
East tropical Atlantic Ocean	80.00	0.60	3.58	Bard and Antoine, 1983
Vanuatu, western Pacific (1)	60.00	0.75	3.43	Brouard et al., 1984
Vanuatu, western Pacific (2)	62.00	1.10	3.63	Brouard et al., 1984

[a] Sources of data as indicated.

Diel Growth Variation and Daily Otolith Rings

It was realized during the 1960s and 1970s that the main outlines of fish growth could be determined for many tropical stocks by analysis of seasonal growth rings on otoliths (Bayagbona et al., 1963; Poinsard and Troadec, 1966) as well as by size-frequency analysis, thus finally dispelling the myth that Mohr (1921) had written about nearly half a century earlier. Further progress in the study of fish growth then awaited the observation of Panella (1971) that the very fine striations within otoliths, previously described but not correctly interpreted by Hickling (1931), were often laid down daily. Figure 9.7 illustrates daily and other rings in the otoliths of a number of tropical fish species. Diel growth striations,

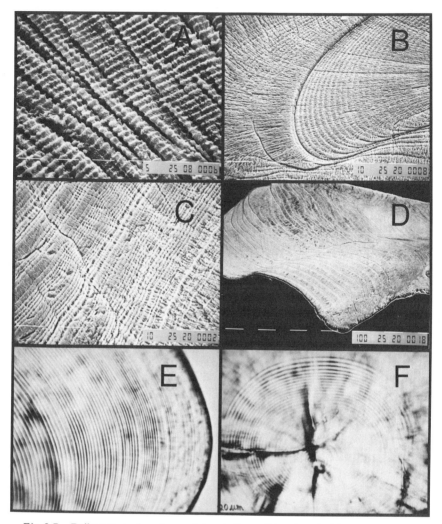

Fig. 9.7.   Daily, seasonal, and annual rings in the otoliths of tropical fish. (A) *Pomadasys argenteus* (Kuwait) 56 cm SL, very regular daily growth increments on sagitta; (B) *Scarus sexvittatus* (Australia) 24.5 cm SL, transition at end of larval stage; (C) *Pomadasys argenteus,* shows irregular, possibly tidal, spacing of increments; (D) *Stegastes* sp. (Australia) 13 cm SL, annual zones in sectioned sagitta; (E) *Scarus iserti* (Australia) 6 cm SL, daily rings of early larval period; (F) *Hyporhamphus unifasciatus* (Virgin Islands) 10 cm SL, extremely regular daily rings. A–D are scanning electron micrographs, E–F are light micrographs. Size of scale bars in μm indicated by left-hand number. (Photos by E. B. Brothers, EFS Consultants.)

presumably representing physiological changes driven by the solar or tidal cycles, had previously been observed in the skeletons of both living and fossil corals and molluscs. With their recognition in fish otoliths by Panella, it finally became possible not only to settle long-standing controversies about the growth rates of a number of important commercial species (Shubnikov, 1976), but also to probe more deeply into the fundamental factors that affect fish growth in all latitudes.

For example, it became possible to estimate the duration of the larval stage of coral reef fishes, to assess the relationships between repeated spawning or tidal cycles and growth, and even to age long-lived tropical fish which do not display annual rings (Ralston, 1985). One topic which has received little attention, however, is the cause for the occurrence of daily, or primary rings in virtually all fish species so far investigated (Gjøsaeter *et al.*, 1984). Attempts have been made to argue that temperature (Brothers, 1981) or endogenous rhythms cause such rings, but the explanations so far advanced fail to explain their ubiquitous nature, as is apparent from the recent review of diel rings by Jones (1986).

Proceeding from the growth model proposed in the previous section it seems probable that daily growth rings in the otolith of fish (and in the analogous organs of invertebrates) are simply the consequence of diurnal changes in activity levels, and concomitant changes in the oxygen budget of the tissues. Such a model for the origin of diel striations in invertebrates has been proposed by Lutz and Rhoads (1974) and there is abundant circumstantial evidence to support the concept (e.g., Panella, 1974; Brothers and MacFarland, 1981). Thus, in diurnal fish, high levels of activity during feeding, sustained high-speed swimming to escape from predators, social displays, etc., will induce anaerobiosis and high levels of lactic acid in body tissues during daytime (Burggren and Cameron, 1980; Wardle, 1978). These conditions must occur also in the fluids surrounding the otoliths, and must result in alterations of the chemistry of the outermost layer of the otolith. At night, when diurnal fish are relatively quiescent, they will compensate for the residual oxygen debt they have incurred during the day, and resume growth based on amino acid stores built up during the daytime. Such a model provides a sufficient explanation for Brothers' (1981) observation that "the protein-rich portion of simple daily growth units is formed at night." This proposition could be confirmed if it were found that the pattern of protein-rich/protein-poor striations was inverted between day and night in nocturnal fish such as the Holocentridae.

This model represents a special case of the more general growth model presented in the previous section, which itself appears to explain a wide variety of otherwise unrelated phenomena, as will be discussed below. It should be mentioned, however, that ageing by means of daily rings is a rather time-consuming and error-prone process, and also that the various counting techniques are still being refined (see, for example, Gjøsaeter *et al.*, 1984). It is therefore at present of more use as a technique for special research projects, rather than in routine aging for fish stock management. The technique is also, of course, susceptible to the biasing of results that can be induced by improper sampling from the catch in exactly the same manner as for routine techniques of growth parameter estimation based on annual rings or length-frequency data.

Seasonal Growth Variations

Figure 9.6 shows seasonally varying growth curves in four fish populations in which estimated values of the parameter C (Eq. 9.7) varies significantly. This figure, as well as Fig. 9.8, shows that in fishes and invertebrates the value of C, the amplitude of seasonal changes in growth rate, is

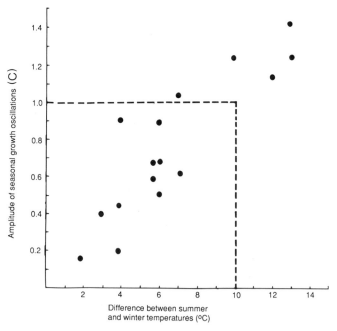

Fig. 9.8. Relations between the amplitude (C) of seasonal growth oscillations in fish, penaeid shrimp, and squid and the summer–winter temperature difference of their habitat. Values of $C = >1$ imply a period of no growth, not discussed here because it occurs only in tropical freshwater and high latitude habitats. (Modified from Pauly, 1985b.)

related to the range of change in water temperature between the warmer and cooler seasons. This relationship is not necessarily causal, and obviously it might be due to another factor, such as the availability of food, itself driven by temperature change. However, the fact that the value of C varies with temperature throughout the tropics as well as in temperate seas, for invertebrates as well as fish, and despite a diversity of specific life-history strategies, strongly suggests that temperature is the causative agent. This is strengthened by the fact that fish seem able to continue growing (i.e., have values of $C < 1$) within a range of $\Delta T = 10°C$, which is approximately the range over which enzyme systems will usually function, although at temperature-dependent rates (Rose, 1967).

Given appropriate models and fitting methods, seasonal variation in growth rate can be detected in several different kinds of data sets, such as sizes-at-age, length frequencies, and size increments from tagging and recapture (Fig. 9.6). Indeed, sensitive analysis can show unequivocal seasonal growth variations in data sets previously used to demonstrate the absence of seasonal growth variability in tropical fish (Ingles and Pauly, 1981). Such findings reveal the interesting paradox that while growth of a species within a region is accelerated by higher seasonal temperatures, the ultimate size of adult fish is reduced in regions having higher sea water temperatures integrated over the whole year, a paradox fully compatible with the metabolic growth model presented here.

The discovery that significant seasonal growth oscillations are frequently encountered in tropical fish should lead to improved growth parameter estimates, as well as to improvement of some of the yield models used for managing tropical fisheries. This is probably another example of how the mythology surrounding tropical fish population studies has in the past prevented progress. The myth of continuous, rapid growth of tropical fish clearly prevented seasonal growth oscillations from being observed, even where they now appear obvious.

Seasonal Rings in the Hard Parts of Tropical Fishes

Given that growth varies seasonally in so many tropical fish species, it is not surprising that daily growth rings in otoliths themselves show seasonal changes that integrate into identifiable zones representing seasonal changes in growth rate. These zones usually represent a winter growth check in high-latitude fish, while in tropical fish they probably reflect growth checks that are caused by whatever are the principal seasonal changes in each habitat.

The occurrence of seasonal marks from which fish may be aged by routine techniques developed for high-latitude fisheries is of more significance to practical fishery management than the occurrence of daily rings which are more time-consuming to count. However, most of the studies of growth marks in tropical fish over the last 15 years have concentrated on daily rings (see the references and lists of studies in Gjøsaeter *et al.*, 1984); the very simple techniques introduced by Bayagbona *et al.* (1963), Bayagbona (1966), Poinsard and Troadec (1966), and LeGuen (1970, 1976) have been little exploited in tropical fishery management. These authors showed that the "burnt otolith" technique of Christensen (1964), which was developed to darken rings on otherwise hard-to-read otoliths from North Sea fish, gave extremely easily-read, brown, hyaline rings on the otoliths of several species of West African sciaenids. These protein-rich rings are laid down during periods of warmer water, and alternate with white opaque rings laid down during cooler seasons. Four rings are laid down each year in sciaenids on the coast of Zaire, corresponding to the two maxima and two minima that occur annually in the temperature regime in the eastern Gulf of Guinea. The physiological basis for these rings has not been directly established, though the influence of environmental temperature, feeding regime, and reproductive cycle—and hence the cycle of lipid sequestration in body fat—have been considered. A similar discussion of the nature of seasonal rings also made visible by the burnt otolith technique in the flatfish *Psettodes erumei* from the Gulf of Thailand has been given by Kühlmorgen-Hille (1976); in this fish, it is suggested that narrow, hyaline rings represent spawning checks.

Although usually found to be less reliable than otolith rings especially in older fish (O'Gorman *et al.*, 1987), scales have been found to bear seasonal rings in at least some groups of tropical–subtropical fish, and their presence or absence can give us clues with regards to the summer–winter temperature differences needed to induce readable annual rings in tropical fish (Fig. 9.9). The evidence presently available suggests that a seasonal temperature difference of $\Delta T = 4$–$5°C$ may be necessary to induce readable annual rings in tropical fish, and that such rings may be therefore expected everywhere such differences occur. However, the existence of clear otolith rings on sciaenids in the Gulf of Guinea in habitats where ΔT can be less than this value suggests that other factors than temperature can also induce growth rings (and growth oscillations, see above) in tropical fish. It is to be expected that once the exploration of daily rings, and what they can tell us about growth in larval and juvenile fish, has run its course, examination of seasonal rings on otoliths and perhaps other hard parts will become a routine tool in tropical fishery management.

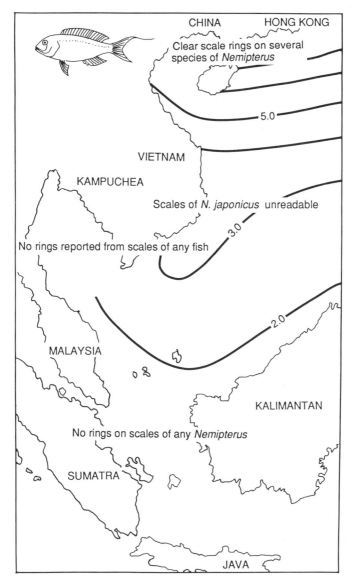

Fig. 9.9. Relationships between readability of scale rings in threadfin bream (*Nemipterus* spp.) and summer–winter temperature differences in °C (Eggleston, 1972; Chevey, 1934; Pauly, field obs.).

Length–Frequency Data and Fish Growth

Prior to the discovery of daily and seasonal rings in otoliths and scales, the analysis of length–frequency data was the only method that could be applied routinely to draw inferences on the growth of tropical fishes. Marking-recapture studies, and direct observation of captive fish, have been used in tropical fish research but have not yielded extensive and useful data. Because there is always some residual uncertainty in growth analysis based on a single technique, the use of length–frequency data will probably always be required alongside the reading of seasonal and diel growth checks on skeletal tissue and scales.

A great deal of confused terminology surrounds the use of length–frequency data in growth analysis and before discussing the available techniques it is necessary to clarify the terms we shall use. We shall restrict the term "Petersen's method," which is usually very loosely used, to the analysis of one length–frequency sample at a time. Such analysis consists of identification of class modes or means in the sample in question, using either a purely visual approach (Petersen, 1892), graphical techniques (Harding, 1949; Cassie, 1954; Bhattacharya, 1967), or computer-based methods of class separation (Abramson, 1970; Yong and Skillman, 1975). These techniques are reviewed in Everitt and Hand (1981). After the identification of class modes, the next step in Petersen's technique is the subjective attribution of relative ages to the modal or mean lengths thus identified. This then enables the derivation of growth parameters from the identified mean lengths, and the relative ages attributed to them.

Petersen's method for length frequency analysis is illustrated in Fig. 9.10a, which also illustrates what happens if the subjective attribution of relative ages goes wrong (which happens quite frequently, and is behind many of the statements that the VBGF is not suitable as a model for growth of this or that fish).

We shall use the term "Modal Progression Analysis" (MPA) for the method in which several length-frequency samples are plotted sequentially, and where the apparent shift of modes is used to infer growth. In MPA the identification of class modal or mean lengths is done as in Petersen's method, after which subjective identification of modes perceived to belong to the same cohort of fish enables the tracing of growth curves, or computation of growth increments between modes. Growth parameters are then estimated from the growth curve traced by eye using a Ford–Walford plot, or from the computed growth increments using the method of Gulland and Holt (1959). Thus, in MPA, the critical issue is not the attribution of "ages" to the various group, as in the Petersen method,

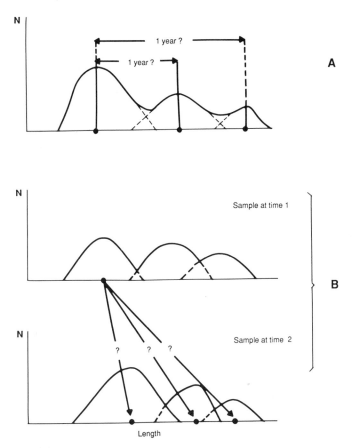

Fig. 9.10. Basic principles involved in traditional methods for length–frequency analysis. (A) Application of the Petersen method to a length–frequency sample; note that the time separating peaks must be assumed, a difficult task in species that may spawn several times annually. (B) Application of the modal class progression analysis to a set of two samples obtained at known times; note that the problem here is the proper identification of peaks to be interconnected (Pauly *et al.*, 1984b).

but the linking of peaks perceived to belong to the same cohort (Fig. 9.10b).

The subjective judgment required in the second steps of both of these methods and the large potential errors thus induced have prompted a search for ways of reformulating these two traditional methods so that improved solutions are obtained. Two major lines of inquiry have been followed:

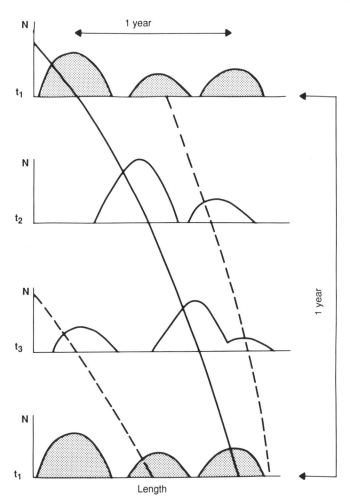

Fig. 9.11. An application of the "integrated method" to a hypothetical set of length–frequency samples. The attribution of a relative age to the third peak of sample t_1 is confirmed by the modal class progression which suggests a growth curve passing through the major peaks of samples t_1, t_2, and t_3 and through the third peak of t_1 repeated after 1 year. A smooth curve is thus achieved with more reliability than by the application of either of the two earlier methods (from Pauly *et al.*, 1984b).

1. Constraining possible solutions through the use of preselected growth functions, usually the VBGF and especially its seasonally oscillating formulation.

2. Combining the length–frequency data with other information, such as growth increment data from tagging-recapture experiments, or length-at-age data obtained from otolith analysis (Morgan, 1987).

Petersen's method and MPA can be integrated into a technique (Fig. 9.11) that can be implemented either using one single large sample, collected during a short period within a year, or a number of samples representing different periods of the year. However, in the latter case, seasonal growth oscillations (as defined and discussed above) must be considered explicitly lest some of the peaks in a set of data should be missed by the growth curve.

Moreover, whether one sample or several are analyzed, account must be taken of the possibility that two cohorts of unequal strength may be in the data, yet prior knowledge of this is usually not available. Finally, as in all analyses using length–frequency data, it must be recognized that such data tend to be biased by the fishing gear used to obtain them, so that they usually contain fewer small fish than really occur in the sampled population.

These various problems, which occur simultaneously in real-life situations, make paper-and-pencil methods, integrated or not, for the analysis of length–frequency data rather unreliable, and several computer-based methods have been developed that can objectively extract growth parameters and related information from length–frequency data, while accounting for seasonal growth, variable cohort strength, and gear selection.

The ELEFAN I and II software described by Pauly and David (1981) and Pauly (1982b) belong to this class of model, and have been implemented on microcomputers that are now widely available in tropical countries (Brey and Pauly, 1986; Saeger and Gayanilo, 1986). Alternative methods (Sparre, 1987) based on parametric statistics usually require access to a library of routines on mainframe computers, and are thus not so readily available to isolated researchers. Parametric techniques are also usually applicable only to length–frequency data that are weighted by a population abundance index, such as catch per unit of effort.

The following two examples of length–frequency analysis have been worked out with ELEFAN software to illustrate some of the practical problems in two very different tropical fisheries. In each example, the growth curves were fitted after taking account of incomplete recruitment of small fish through gear selection (Pauly, 1986a; Hampton and Majkowski, 1987).

Our first example is the analysis of growth during the period 1963–1965 of the Peruvian anchoveta (*Engraulis ringens*), whose biology was briefly discussed in Chapter 7. Though the temperature range off the Peruvian coast is relatively small ($\Delta T = 3$–$5°C$), it is notable from Fig. 9.12 that lowest size-adjusted growth rates for *Engraulis ringens* occur during the period of lowest mixed-layer temperatures off Central Peru. The best growth curve for the main 1963 cohort was found to be one that oscillates seasonally, though not strongly ($C = 0.31$), with minimum growth rate in

September (Fig. 9.12). This cohort entered the fishery in October 1963 at a mean length of 7–8 cm and disappeared about 20 months later, in mid-1965. The growth of some cohorts preceding or succeeding the main cohort is also shown in Fig. 9.12. It will be noted that the assumption of annually repeated growth curves used in Fig. 9.11 is reasonable as far as this data set is concerned.

The second example provided here is the small West African croaker *Pseudotolithus elongatus* found in estuarine and inshore habitats from Senegal to northern Angola. Our analysis is based on length–frequency data for 1953–1954 in the Sierra Leone estuary. These data had been previously been analyzed using MPA, and had yielded a growth curve with parameters very similar to that shown in Fig. 9.6 (Longhurst, 1963, 1964b). Figure 9.13 shows an ELEFAN growth curve for the same data,

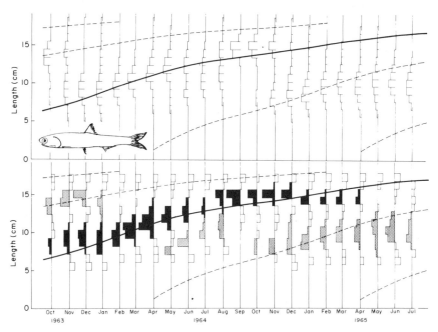

Fig. 9.12. Analysis of length–frequency data on the Peruvian anchovy (*Engraulis ringens*, northern/central stock) using the ELEFAN I program (Pauly and David, 1981). The upper graph shows the original length frequency data (courtesy, I. Tsukuyama, IMARPE) with superimposed growth curve. The lower graph shows the restructured length–frequency data, with peaks in black or grey and the troughs separating peaks as open histograms. The best curve ($L_\infty = 19.75$ cm, $K = 0.71$, $C = 0.31$, and Winter Point = 0.7, i.e., September) was found by varying the growth parameters until a maximum number of positive points were accumulated, and troughs avoided as much as consistent with a continuous curve.

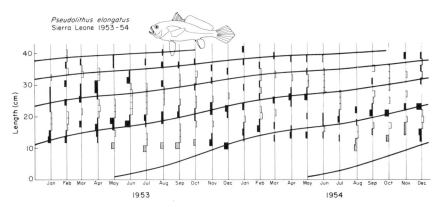

Fig. 9.13. Growth of "gwangwa" (*Pseudotolithus elongatus*) in the Sierra Leone estuary as estimated by the ELEFAN I program (L_∞ = 52 cm, K = 0.366, C = 0.4, WP = 0.37). (Based on data in Longhurst, 1964a.)

superimposed on the restructured length–frequency data to which the curve was actually fitted. These restructured data, a characteristic output of ELEFAN software, express the positions of the peaks and troughs in a length–frequency data set as positive and negative histograms, while the absolute values express how distinct a peak (or a trough) is. The curve itself is fitted by varying the parameters of an equation such as Eq. (9.7) until a maximum of positive points is accumulated, so that the curve passes through as many peaks as possible, while avoiding troughs.

It will be noted that the succession of size classes for *Pseudotolithus elongatus* is rather clear for the first 2–3 years, then becomes relatively difficult to follow, as is to be expected in a species in which the very oldest known individuals are less than 10 years (LeGuen, 1970) and in which the numbers of fish at age 4 are already only about 1% of those taken in commercial trawl samples (Longhurst, 1963). For this reason the fitting of the data to a given growth function (in this case, a seasonally oscillating VGBF) is helpful in identifying which sequence of peaks most probably belong to the same cohort.

These two examples will serve to illustrate the utility of length–frequency-based growth analysis in tropical fish, but special caution is appropriate concerning quality and quantity of the data needed for such analyses. In particular, users of these methods must be aware of their limitations, especially the possibility that because of behavioral features of the fish sampled, or bias of the sampling gear, they should not be applied (see contributions in Pauly and Morgan, 1987). A sufficient number of measurements, well distributed over time, is also obviously necessary for more than cursory analyses (see Table 9.2).

TABLE 9.2. Criteria for Assessing the Reliability of
Length–Frequency Samples for Studying
the Population Dynamics of Fish.
Reliability, on a Scale of 0–5[a]

	Months over which sample is accumulated				
Total sample size	1	2	4	6	12
1–99	0	0	0	0	0
100–499	0	0	1	2	2
500–999	1	1	2	3	4
1000–1499	1	2	3	4	5
1500–∞	2	3	4	5	5+

[a] From Pauly (1984b), based on Munro (1980).

The analysis of sciaenid growth discussed above was based on measurements of 9847 individual fish, and additional special samples were obtained with small-meshed trawls properly to sample small fish that would have escaped through the standard survey gear; a similar study of *P. senegalensis* and *P. typus* off Nigeria was based on about 147,000 fish measured over a 3-year period. Such numbers are quite easy and cheap to obtain with a proper catch-sampling program and are self-regulating, in the sense that the data themselves, as they are accumulated, clearly demonstrate when a sufficient sample of fish has been measured to identify its major size classes.

Environmental Factors Affecting Growth

In a Darwinian world, not all members of a given population of fish or marine invertebrates are equal. Thus, given a scarcity of resources, or adverse environmental factors, environmental pressure will be experienced differently by different members of the population. Both specialized predators and more catholic carnivores take a proportionately higher number of less fit and hence vulnerable animals of a given prey species (Jones, 1982). In the marine environment, which offers relatively open connection between different subsystems, and especially so in the tropics where interactions may be more intense than in temperate systems (Robinson, 1978), such removal of vulnerable individuals should be rather intense, resulting in the evidence for environmental effects on fish growth being systematically and quite literally eaten away.

We have already discussed some of the ways in which environmental temperature may determine growth rates in marine fish, and in the last section of this chapter we shall discuss the energetics of fish nutrition, and the manner in which energy deficits are avoided during feeding activities. However, a number of other factors may cause changes in growth rate, and we shall now discuss two of these—social interaction and water depth.

It is in the defense of a home territory, or in the maintenance of a stable pair-bond, that vertebrate social interactions are strongest, yet for marine fish the defense of space is important only in some families of reef fish, and it is among these that we should look for effects of social interactions on growth rates. In reef-dwelling species of *Amphiprion,* dominant females continually chase all males present in their habitat except one, with which a pair-bond is formed. This stress is sufficient to keep all unmated males in the habitat smaller than the large mated male. In the event that the pair-bond is broken by removal of the mated male, the female ceases to chase the unmated males, and these begin rapidly to grow until one of them can replace the missing large male (Allen, 1975; Fricke and Fricke, 1977). This is in line with Eq. (9.1), which implies that release from stress should indeed result in increased growth rate. Other examples of social interactions affecting growth rates of tropical fish are given in Keenleyside (1979).

Water depth as a factor is usually not discussed in connection to growth. Johannes (1981) noted that "an increase in mean size of fish with depth has often been noted by biologists, but no one as yet has come up with a generally accepted explanation for this trend." The model of growth outlined in this chapter allows us to accept Johannes' challenge and to comment on what is also known as Heincke's Law, which states that in the North Sea the size of individual plaice depends on stock abundance and also on distance from the coast and depth (Heincke, 1913; Harden-Jones, 1968). In fact, all that is needed to explain the tendency for larger fish to occur in deeper water is to postulate that routine metabolism decreases in deeper water because of decreased water temperature, as is predicted by our simple growth model (Figs. 9.14a and 9.15).

MORTALITY OF TROPICAL MARINE FISHES

Some Definitions of Mortality

The rate of mortality is an important number in the management of fish populations and it can be quantified by several quite different and complementary techniques. One approach to quantifying natural mortality is to measure the abundance and food demand of major classes of predators

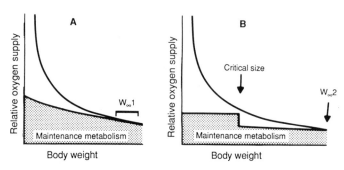

Fig. 9.14. Schematic representation of the effect on tissue oxygen supply available for growth of (A) gradual decline of maintenance metabolism achieved by migration to colder (deeper) water and (B) when a given abundant food item becomes available once a critical body size has been reached which renders the prey obtainable. A will lead to a growth curve with a long phase of steady growth, and B to a diphasic growth curve.

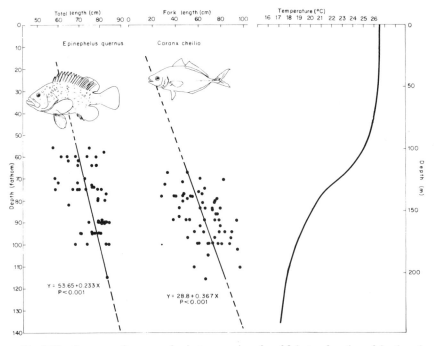

Fig. 9.15. Increase of average size in two species of reef fish as a function of depth and temperature at French Frigate Shoals, Hawaian Islands (S. Ralston, 1981; J. J. Polovina, personal communication).

(Potts *et al.*, 1986) so as to draw inferences about their impact on the prey population, and we have discussed several examples of this approach in previous chapters. Provided satisfactory tagging techniques can be developed for the tropics (which is not always the case) classical release-recapture methods are available for transfer from temperate fisheries biology. The principal biological problems to be overcome before this method gives good results are to understand the size of the area over which individuals of the tagged stock are distributed and to use tags that cannot be shed and are readable after some months in the sea. This technique is perhaps most easily applied in lakes, lagoons, and other enclosed bodies of water where the landed catch is easily monitored.

Finally, one can infer mortalities, if certain assumptions about recruitment and growth can be made, from the size or age frequency distribution of fish sampled over a given period. We shall not dwell here on the methodological aspect of these approaches which are covered in general population dynamics texts (Gulland, 1983; Ricker, 1975; Pauly, 1984a). Rather we shall discuss the results of some studies that have led to the identification of specific features of the natural mortality of tropical marine fishes. We shall not discuss eggs and larvae, which are dealt with in another section of this chapter, nor fishing mortality except where fishing and natural mortalities interact. We shall express all mortality as instantaneous rates, defined by

$$N_t = N_0{}^{exp} - Mt \qquad (9.8)$$

where N_0 is the number of fish at a given time, and N_t the number of fish still alive after time t, given a natural mortality rate M. In this equation, M can be replaced by total mortality (Z), defined as the sum of natural mortality (M) and fishing mortality (F). All mortality rates that we shall quote (and which have the dimension 1/time) are on the same annual basis as the parameter K of the VGBF.

In this general book, which is not devoted specifically to issues of fish stock assessment and fisheries management, we cannot deal in detail with the interactions between natural and fishing mortality. However, it is appropriate to mention here some implications of Fig. 9.16, illustrating Munro's concept of the interaction of F and M and based on studies of Jamaican reefs (Munro, 1983). This concept implies a high natural mortality in unexploited stocks and a partial replacement of natural by fishing mortality under exploitation, resulting from a higher rate of removal of predators than prey species. Evidence for the ubiquity of this phenomenon is beginning to emerge and would probably be accepted as conclusive, were it not for all the inherent difficulties in estimating M reliably in exploited stocks.

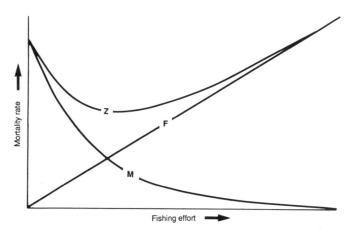

Fig. 9.16. Theoretical relationship between natural mortality rate (*M*), fishing mortality rate (*F*), and total mortality rate (*Z*) which will exist if natural mortality rates in an exploited community decline as a result of concurrent exploitation of predatory species. (From Munro, 1980.)

Straightforward methods for the estimation of *Z* (or *M* in unexploited stocks) use the mean length of fish in catch samples (Beverton and Holt, 1956) or length-converted catch curves (Pauly, 1984b) of which an example is provided in Chapter 10. The generally straight catch curves that are obtained when suitable data are analyzed suggest that the assumption of constant mortality rate implied in Eq. (9.8) is, in most cases, a reasonable one.

Relations between Growth Parameters, Temperature, and Natural Mortality

Beverton and Holt (1959) were the first to point out that in fishes there is a strong relationship between the parameter *K* of the VGBF, and *M* as defined above. Their inference was based on a data compilation which covered a narrow range of sizes, and very few tropical fish. For this reason, they failed to identify the equally strong relationship between *M* and temperature and the parameters L_∞ or W_∞ of the VGBF (Pauly, 1980a).

When we examine taxa that are relatively similar in size, and are drawn from habitats with similar temperatures, the ratio *M/K* holds almost constant, however. Thus, in the case of Serranidae and Lutjanidae, Ralston (1986) found that *M/K* = 2 (Fig. 9.17). We present in Chapter 10 another such relationship for sea urchins. Similar ratios could be derived for a

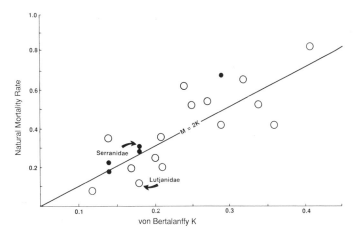

Fig. 9.17. Relationship between natural mortality (M) and the growth constant K in Serranidae and Lutjanidae; K is on an annual basis. (From Ralston, 1986.)

number of fishes covering different ranges of sizes and occupying different ranges of mean environmental temperature, using the compilation of data points in Pauly (1980a, 1985a), for instance.

Another generalization relevant to tropical fish is

$$\log_{10} M = 0.065 - 0.287 \log_{10} L_\infty + 0.604 \log_{10} K + 0.513 \log_{10} T \quad (9.9)$$

which expresses the relationship between M, L_∞ (in cm), K, and mean habitat temperature (°C). Equation (9.9) was derived from 175 specific data sets for fish ranging from 2 to 1226 cm, and for temperatures ranging from 3°C to 30°C (Pauly, 1980a, 1985a). This data set can also be used to establish a direct, significant relationship between M/K and mean environmental temperature, which results in

$$\log_e (M/K) = -0.22 + 0.30 \log_e T \quad (9.10)$$

Equations (9.9) and (9.10), which are both based on a very large data set, show unequivocally that the natural mortality of tropical fishes is markedly higher than that of cold water fishes, other things being equal. This finding has considerable theoretical implications because the species-specific mechanisms compensating for higher adult mortality, such as egg production and recruitment, must compensate these elevated mortalities in tropical species, while predation, which probably accounts for most of the natural mortality of fish, should bear more heavily on tropical than temperate species.

Additionally, hypotheses such as the utilization of r versus K reproductive strategies, placing emphasis on life-history strategies of individual fish species, may be incomplete since biological rates such as growth and reproductive output appear to be constrained by the basic geometry which limits relative growth of gills and gonads, and temperature which regulates metabolic and mortality rates. This point is discussed further below in relation to fish, and in Chapter 10 in relation to sea urchins and squids, but this book is not the place for a comprehensive reinterpretation of the published evidence. It is simply appropriate for us to discuss a few points to illustrate the implications of relatively constant ratios between VGBF parameters, mortality, and temperature.

We have suggested that tropical fishes should produce more eggs per unit time, other things being equal, than their temperate counterparts, and Fig. 9.18 provides evidence that this does in fact occur. Moreover, recent discoveries related to the spawning dynamics of clupeoids, notably *Engraulis mordax,* strongly suggest that previous approaches to estimating egg production in teleosteans have seriously underestimated this process, particularly in serial spawners which are abundant among tropical fishes (Lasker, 1985; Breder and Rosen, 1966).

It is easy to document, at least for fish held under experimental conditions, that there is a positive relationship between food intake and temperature (Menzel, 1960; Kraljevic, 1984), and hence we infer a relationship between temperature and predation pressure. To demonstrate a positive effect of mean habitat temperature on mean food consumption of fish under natural conditions is more difficult, however, given the generally low precision of all estimates of food intake for fish swimming free in the ocean.

Our suggestion concerning the limited utility of analysis of life history strategies will probably be seen as contentious, given how a strong a role such analysis plays in the articulation of biological thought at the present time. However, while simulations of life history stategies may confirm that individuals in a given population maximize their fitness or reproductive output by, for instance, opting to remain small, the question must be asked, what is the cue for the adjustment of their growth rates, and their size at first reproduction so that maximum benefit accrues at the population level. Thorpe (1986) is one of the few authors to have perceived this problem. He writes,

> the concept of a critical size at which a major developmental conversion . . . may occur begs the question of how the organism recognizes how large it is. It needs some reference standard.

The solution proposed below, in our section on the reproduction of tropical fishes, is based on Pauly (1985b) and is essentially the same as

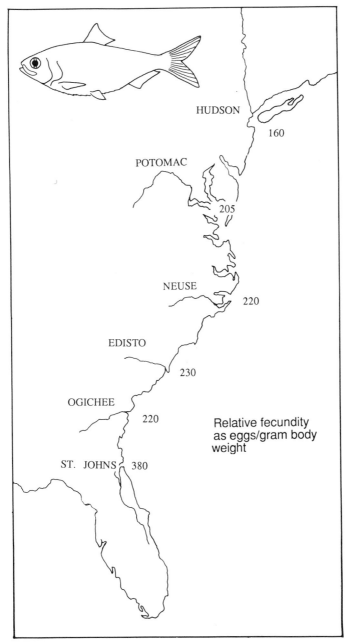

Fig. 9.18. Latitudinal trend of relative fecundity in anadromous shad *Alosa sapidissima* along the eastern coast of the United States. (Adapted from Davis, 1957.)

that proposed by Thorpe (1986). It contradicts much of the conventional view of the interrelationships between growth and reproduction, while being fully compatible, however, with the generalized growth model proposed earlier, and which provides the backbone of this chapter.

REPRODUCTION AND RECRUITMENT
OF TROPICAL MARINE FISHES

Relationship between Growth and Reproduction

Instead of reviewing the very many observations that have accumulated of the patterns of recruitment in tropical marine fishes and presenting them as a compendium of tabular statements similar to those collected by Breder and Rosen (1969), we shall review one recently proposed hypothesis which seems to provide an explanation for a large class of observations concerning tropical fish reproduction.

Based on the metabolic growth model presented above, this hypothesis explains why fish should first spawn at a predictable size, or more precisely, at a predictable fraction of their asymptotic size (L_∞ or W_∞). Conceptually, three time scales can be defined for spawning and related processes.

1. A short time scale (seconds to days) for processes that occur just before shedding gametes (e.g., reproductive behavior, final maturation, etc).

2. A medium time scale (weeks to months) during which gonad maturation and spawning migrations take place, as triggered and influenced by environmental stimuli.

3. A long time scale (months to years) during which the age, and particularly the size (L_m or W_m), is determined at which a fish becomes sexually mature for the first time.

It is only scale (3) to which the hypothesis applies, the elements involved in (1) and (2) being adequately covered in the literature (e.g., Breder and Rosen, 1966) for fish inhabiting all latitudes.

The hypothesis itself is simple. To explain it, we return to Fig. 9.14 to observe that fishes at W_∞ have no surplus energy (expressed as no surplus oxygen) that could be devoted to the elaboration of gonads, and hence they must first spawn at some smaller size. Indeed, it is a trivial observation that the ratio L_m/L_∞ usually ranges between 0.4 and 0.9, is usually smaller in large fishes and higher in small fishes, and is remarkably constant within families comprising fish of approximately similar dimension (Beverton and Holt, 1959; Cushing, 1981).

The question might be posed: How do the physiological processes of individual fish "know" the value of their future L_∞ so that L_m can occur at the appropriate time? However, absurd questions lead to absurd answers, and we should probably rather ask if the relative constancy of the L_m/L_∞ ratio within fish taxa is due to the fact that both L_m and L_∞ are determined by the same process. It was suggested above that asymptotic size in fishes is determined mainly be the interactions between oxygen supply and demand, as mediated by species-specific gill surface area. It is suggested here that the basic mechanism outlined above determines size and hence also age at first maturity. All that is needed now is to postulate a genetically fixed threshold (i.e., a multiple of the routine metabolic level) below which one or more substances are produced that enable the brain of a fish to respond to environmental stimuli for reproduction.

Figure 9.19 illustrates this concept by contrasting two fish stocks of the same species, one living in a cool, food-rich water mass leading to high values of L_∞ and hence also high values of L_m, the other living in a food-limited, marginal, or warm-water habitat leading to low values of L_∞ and L_m. Figure 9.20 illustrates the method by which a preliminary estimate of the threshold level alluded to above was obtained. This figure shows that fishes, whether guppies, cod, or tuna, reach first maturity when their overall metabolic level (that is, active + maintenance metabolism) is about 1.4 times the level of their maintenance metabolism.

This hypothesis, which has a minimum of *ad hoc* assumptions and derived hypotheses, complements the explanation given above as to why

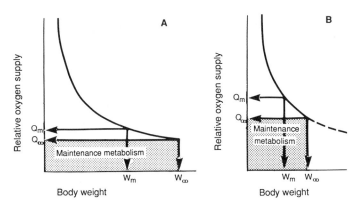

Fig. 9.19. Interrelationships between oxygen supply at maintenance level (Q_∞), oxygen supply near first maturity (Q_m), and the weight of fish. Note that the Q_m/Q_∞ ratio is constant for two different habitats one of which (A) is linked with large asymptotic size and a large size at first maturity, the other (B) being associated with the opposite situation. Note also that the constancy of Q_m/Q_∞ can only be imperfectly mirrored in the ratio L_∞/L_m, which can be derived from W_∞ and W_m. (Modified from Pauly, 1984b.)

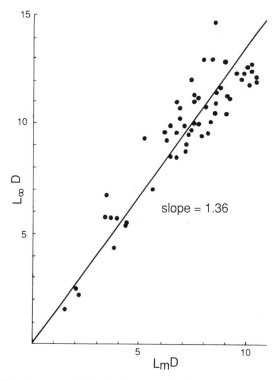

Fig. 9.20. Plot of asymptotic length (L_∞, cm) raised to the power D on length at first maturity (L_m, cm) also raised to the power D in 56 species of fishes, ranging from guppies to tuna. The exponent D is equal to $b\,(1 - d)$, where b is the exponent of the length–weight relationship and d is the exponent of the relationship linking the body weight and the gill surface area of fishes. Note linearity of this plot, which suggests that fish reach first maturity when their total metabolism is equal to about 1.4 times their maintenance metabolism (Pauly, 1984b).

warm-water fish stay smaller than their cold-water counterparts with a mechanism explaining how size at maturity is adjusted to these smaller asymptotic sizes, without the need to postulate the operation of an absolute biological clock for the long-term time scale mentioned above. It also explains the abortive, or "trial run" maturations discussed by Hickling (1930) and Iles (1974) and is equally applicable to invertebrates as to fish.

Obviously, the relationship between L_m and L_∞ does not cover all aspects of the relationships between growth and reproduction, which can be rather complex and involve a variety of other relations, such as those between body weight and apparent fecundity, itself mediated by the relationship between body weight and gonad weight (Fig. 9.21). Other rela-

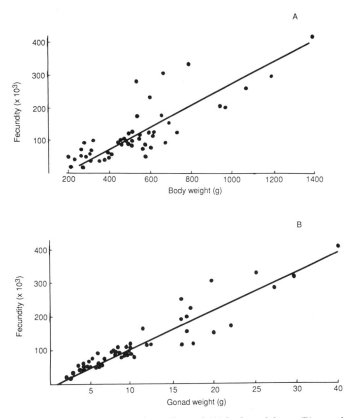

Fig. 9.21. Relationship between fecundity and (A) body weight or (B) gonad weight in *Pseudotolithus elongatus* off West Africa (Fontana and LeGuen, 1969).

tions between growth and reproduction are even more intricate, and may involve high fecundities that are mediated via reduction of egg size, which itself may be related to temperature-induced reduction of body size within and between different species of fish.

Egg size and fecundity also varies seasonally, as does the reproductive status of fish populations (Fig. 9.22), and knowledge of these seasonal changes is invaluable in understanding population biology. For this reason, great care must be given to the development and standardization of gonad maturity scales that are truly reflective of the peculiar features of tropical fishes, especially the rapid rematuration of spawned fish (Fig. 9.23).

Few of the generalizations thought to be relevant to tropical fishes have been as conceptually barren as the commonly held notion that tropical

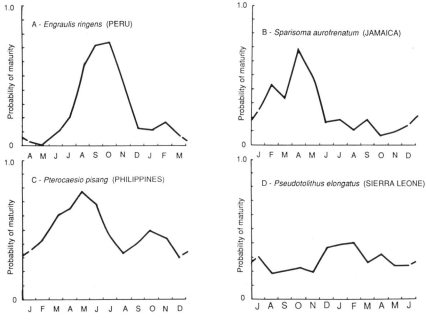

Fig. 9.22. Approximate probabilities for the adults of four tropical fishes to be ripe, or mature, by month, showing some reproductive activity throughout the year, but with seasonal peaks. (A, Average for 1966–1970, Jordan, 1980; B, Reeson, 1983; C, data adjusted from Cabanban, 1984; data from Longhurst, 1963).

fish reproduce throughout the year, or continuously. This view was probably prompted by the perennial luxuriance of some terrestrial habitats in the tropics, but is clearly incorrect because most tropical fishes do display a marked reproductive seasonality (Fig. 9.22), while many temperate fishes also have long spawning seasons, with at least some spawning through most of the year (Murphy, 1982). Thus, the difference between seasonality of spawning (and recruitment) between temperate and tropical fishes is really only a quantitative difference in the relative spread of the recruitment seasons.

The development time of marine fish eggs follows well-known physiological laws and is a negative function of water temperature. This can be expressed by the equation

$$\log_{10} D = 7.1 + 0.608 \log_{10} E_d - 4.09 \log_{10} (T + 26) \qquad (9.11)$$

where D is the egg development time in days, E_d is egg diameter in mm, and T is the water temperature in °C. This equation is based on 140 sets of observations of spherical eggs of marine fish, ranging from 0.6 to 3.4 mm

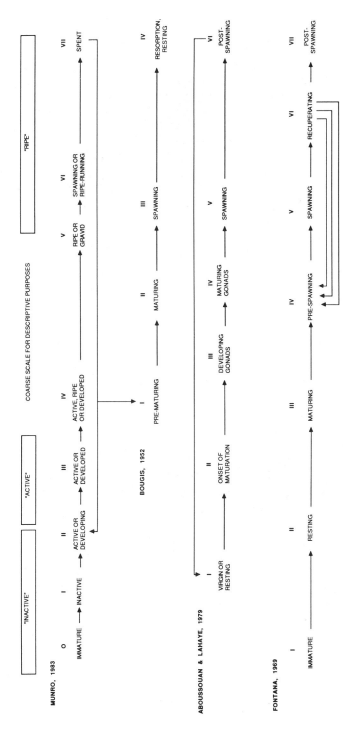

Fig. 9.23. Maturation scales proposed by various authors for tropical and subtropical fishes.

in diameter, at temperatures ranging from 3 to 30°C (Pauly and Pullin, 1987). Note that the slope (0.608) linking $\log_{10} D$ and $\log_{10} \phi$ shows that, for a given temperature, egg development time is not proportional to egg volume (which would require a slope of 3), nor to its surface (a slope of 2), nor yet to its radius (a slope of 1), but to a "physiologically effective radius," determined by an active transport system within the egg.

Equation (9.11) has, besides theoretical and anatomical implications, also some eminently practical consequences; biomass of temperate fish stocks are routinely determined by egg and larval surveys (Saville, 1977; Lasker, 1985), but these may be difficult or impossible to implement in tropical waters, since the eggs of tropical fish may occur too ephemerally in the plankton to be readily quantifiable. Off Sierra Leone, a search for the spawning grounds of the shad *Ethmalosa dorsalis* during normal working days proved quite fruitless, until it was discovered that the species spawned in the early evening, and that hatching was completed between 15 and 18 h later: thus, no eggs were available to survey during the afternoon (Bainbridge, 1961). In contrast, the eggs of the Peruvian anchoveta *Engraulis ringens* remain in the plankton for about 50 h before hatching and are available for semi-synoptic surveys to measure adult spawning stock size. This is one consequence of the differential effect of temperature of 27–28°C off Sierra Leone and 15–17°C off Peru at rather similar latitudes.

The Recruitment Problem

Aristotle was probably the first naturalist to write about recruitment, when he recorded that the fishermen of ancient Greece distinguished three size and age groups of tuna, which they called "auxids," "pelamides," and "thynni" or full-grown tuna (Hist. Anim. B. vi. c. 17). The fishermen were also reported to have observed that a scarcity of pelamides 1 year was followed by a failure of the fishery for thynni in the following year, which makes Aristotle's account the first ever on the still unresolved "recruitment problem."

Similar observations are available from medieval herring fisheries of Northern Europe, and stock availability (and hence recruitment) was one of the earliest preoccupations of nascent fishery biology in the late nineteenth century, but it is only since Ricker's seminal paper of 1954 that a clear-cut statement of this problem has been available: What is the relationship linking a parent stock to the number of its progeny entering the fishery as a result of each years reproduction? For many years after its formulation by Ricker, environmental variability was neglected and research was almost entirely directed at attempts to derive an equation, with parameters including coefficients for prerecruit mortality, to compile long time series of indices of parent stock size and of recruitment and force the former onto the latter.

Even if the neglect of environmental forcing used in this approach had been valid, much of the basic research done on recruitment in high latitudes could not have been relevant to tropical fisheries because for most tropical fish stocks sufficiently long time series may not be available, given the difficulties inherent in collecting reliable single-species statistics in multispecies, multigear fisheries (Bakun *et al.*, 1982). Moreover, the very notion of assuming that a single variable, parent stock, should be sufficient to explain between-year variability of recruitment is probably misleading (Fig. 9.24), especially since the demonstration that multivariate approaches (Csirke, 1980; Fig. 9.25) structured around key environmental variables (Fig. 9.26) can extract much more information from available data. The classical approach is in any case inappropriate for short-lived tropical species, which tend to have two, rather than only one, pulse of recruitment per year (Figs. 9.22 and 9.27) so that an approach using one data point per year would underutilize available information.

In fact, a whole new methodology for the utilization of within-year information on recruitment has recently been developed in conjunction with research on the recruitment of *Engraulis mordax* (Methot, 1984;

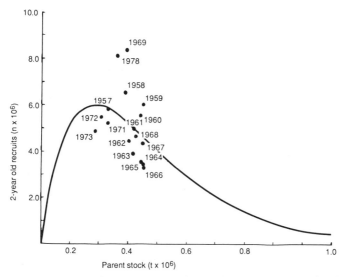

Fig. 9.24. Ricker curve fitted to the spawning stock biomass and recruits of southern bluefin tuna. Stock and recruit estimates by cohort analysis, showing how little of the variance is actually explained by this relationship (Murphy, 1982).

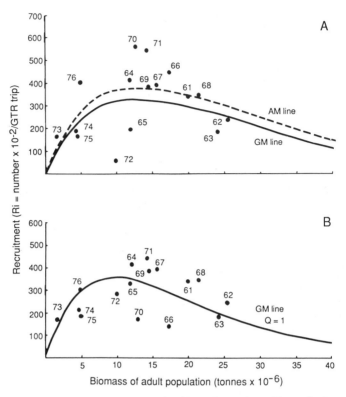

Fig. 9.25. Stock/recruitment relationship of Peruvian anchovy *Engraulis ringens* showing effect of incorporating an additional variable into the relationship. (A) Simple Ricker plot, showing rather bad fit and correspondingly low correlation of observed to expected recruitment (GM line, $r = 0.494$). (B) Plot of the residuals of a multivariate relationship involving recruitment, parent stock and concentration index Q related to occurrence of ENSO's, onto stock recruitment relationship, drawn for an average value of Q. This shows an improved fit with a correlation of observed to expected recruitment of $r = 0.893$. (Based on Csirke, 1980.)

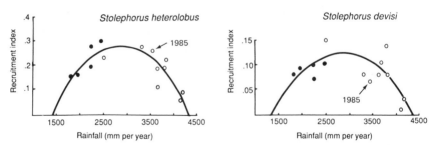

Fig. 9.26. Relationship between rainfall and recruitment in two tropical anchovy species. Recruitment is expressed as catch per effort for unit area of area fished in two regions off northern Papua New Guinea. Ysabel Passage shown as open circles, Cape Lambert as closed circles. The points for 1985 were added after parabolas were fitted (Dalzell, 1984a and personal communication).

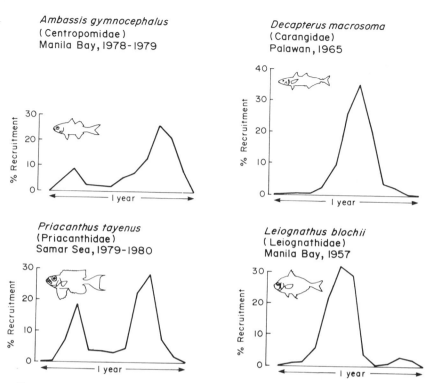

Ambassis gymnocephalus
(Centropomidae)
Manila Bay, 1978-1979

Decapterus macrosoma
(Carangidae)
Palawan, 1965

Priacanthus tayenus
(Priacanthidae)
Samar Sea, 1979-1980

Leiognathus blochii
(Leiognathidae)
Manila Bay, 1957

Fig. 9.27. Examples of recruitment patterns, based on backward projection of length-frequency data onto the time axis, of four species of Philippine fish. (From Ingles and Pauly, 1984.)

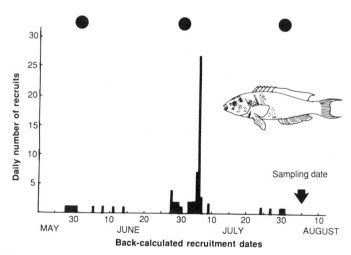

Fig. 9.28. Back-calculated settlement dates obtained by daily rings in the wrasse *Thallassoma bifasciatum* at San Blas Island, Panama (Victor, 1982.)

295

Lasker, 1985) which is potentially applicable to tropical fishes and which should lead to a whole array of new insights, as demonstrated by the successful application of aspects of this methodology to reef fishes such as *Thalassoma bifasciatum* and *Halichoeres bivittatus* by Victor (1982). Fine temporal resolution from the analysis of daily rings is a powerful tool in testing hypotheses on how recruitment is determined (Lasker, 1978; Fig. 9.28). On the other hand, detailed analysis of length–frequency data can only be performed at a coarser temporal scale. Thus, recruitment patterns obtained by projection of length–frequency data onto a time axis obtained with ELEFAN II software (Pauly, 1982b) can then be related to events on a similar temporal scale, such as the monsoon seasons (Pauly and Navaluna, 1983; see also Fig. 9.27).

In fact, the mechanisms beyond simplistic stock-recruitment relationships that determine recruitment in continental shelf fish remain enigmatic, are obviously diverse, and are now widely accepted as perhaps the principal determinant of abundance variability in most stocks. The environmental control of recruitment is perhaps the most important problem facing fisheries scientists today, and how it functions in tropical seas is very little known.

Any recruitment mechanism must, by definition, satisfy each of three simple criteria: larvae must (1) encounter adequate food supplies and (2) a tolerable level of predation to ensure survival of sufficient numbers to maintain the adult population, while (3) they must complete their larval life in a suitable location to enter the next stage in their life history. These criteria are usually found to be satisfied by a reproductive strategy that ensures that larvae are injected into the marine ecosystem at a suitable location and time. Since both spatial and temporal cues are subject to environmental variability, as are the conditions governing successful recruitment, a variety of reproductive strategies are utilized in all latitudes.

Reproductive strategies of fish form a continuous series whose end-members have been called (Lambert and Ware, 1984) the "big bang" and the "hedged bet" strategies, respectively; in the former, all individuals spawn more or less synchronously while, in the latter, serial spawning extends throughout a longer or shorter season. Generally, in temperate latitudes, pelagic spawners with small, fast-growing larvae adopt the big bang strategy while benthic spawners, with relatively slow-growing larvae, hedge their bets by producing many individual cohorts, spaced usually about 3 weeks apart. Synchronous spawners in temperate regions appear to use cues that enable them to spawn when suitable food occurs in maximum abundance, but predator swamping is an alternative and in some ways more attractive hypothesis to explain this reproductive behavior. Serial spawning, on the other hand, is usually interpreted as a means of ensuring that at least one cohort shall encounter conditions that ensure good survival.

Can we transfer this analysis of reproductive strategies to the tropics? As we have already noted, most species of fish of tropical continental shelves have rather extended spawning seasons (Weber, 1976). Reviews of the reproductive cycles of several hundreds of fish species of many families from the coasts of India and western Pacific reveal spawning seasons extending over several months, and some tendency for reproductive quiescence during the seasons of strong monsoon winds (Table 9.3). This kind of reproductive cycle occurs also in tropical Atlantic continental shelf fish.

Detailed studies (Johannes, 1978) of the reproductive behavior patterns of demersal fish along tropical coasts with fringing reefs have revealed a variety of mechanisms clearly related to predator avoidance both by the spawning fish themselves and for their young larvae. The corals, hydroids, crustaceans, and small fish of the reef fauna are efficient at trapping zooplankton (and hence fish larvae) by a myriad of different techniques, and spawning reef fish tend to release their eggs in locations that ensure they will be carried offshore and held there long enough for local dissemination to occur at the end of the larval period.

Paradoxically, as we have already discussed, the larvae of many species of fish of open continental shelves have an opposite tendency—to seek enclosed coastal lagoons that will serve (in the same manner as for the penaeid prawns of the continental shelf) as nursery grounds. Though this is an essential factor in the life histories of many of the important commercial species on subtropical coasts, it is not yet clear to what extent it has the same importance on tropical coasts unencumbered with fringing reefs (see Chapter 7).

It is not unreasonable to generalize concerning the significance of these observations in terms of how environmental variability might affect recruitment success in the tropical seas. Two explanations suggest themselves for the general observation that spawning of continental shalf species occurs at times of low wind stress. First, this is when the water column is most likely to be well stratified and when layers of abundant planktonic food organisms will be available for the nourishment of fish larvae which, at least under some circumstances, are thought to be unable to obtain a sufficient energy return from searching for prey in a well-mixed water column where plankton is unstratified (Lasker, 1975). Second, during periods of quieter winds it is less likely that massive off-shelf advection of fish larvae would occur. At least under some circumstances this is thought to lead to poor recruitment in some species of continental shelf fish. Some organisms, such as larval rock lobsters and penaeid

TABLE 9.3. Spawning Periods of Some Fish Species Along the West Coast of India[a]

Area	Species	J	F	M	A	M	J	J	A	S	O	N	D
Bombay	Psettodes erumei	●								●	●		
Bombay	Otolithus ruber		●							●	●		
Bombay	Johnius dussumieri						●				●	●	
Bombay	Muraenesox talabonoides			●	●	●							
Bombay	Polynemus heptadactylus			●	●	●	●		●	●	●	●	●
Bombay	Pseudosciaena diacanthus				●	●	●		●	●	●	●	●
Bombay	Polydactylus indicus				●	●							
Bombay	Polydactylus indicus												●
Gulf of Kutch	Several spp.	●						●	●	●	●	●	●
Karwar	Rastrelliger kanagurta		●				●	●	●	●	●	●	●
Karwar	Sardinella longiceps			●			●	●	●	●	●	●	●
Karwar	Opistopterus tardoore	●	●	●	●		●		●				
Karwar	Sardinella fimbriata	●	●	●									
Karwar	Sardinella longiceps	●					●	●	●				●
Mangalore	Nematalosa nasus												
Mangalore	Anodontostoma chacunda	●	●	●		●	●	●	●		●	●	●
Mangalore	Otolithus argenteus	●											●
Mangalore	Saurida tumbil	●											
Mangalore	Sardinella longiceps						●	●	●	●	●	●	●
Mangalore	Rastrelliger kanagurta		●				●	●	●	●			●
Cannanore	Cynoglossus semifasciatus										●		
Calicut	Trichiurus haumela		●	●	●	●							
Calicut	Sardinella longiceps						●	●	●	●			
Calicut	Sardinella longiceps			●	●	●	●		●	●			
Calicut	Leiognathus bindus	●	●										
Calicut	Caranx kalla	●	●										
Calicut	Cynoglossus semifasciatus	●	●	●	●	●							
Calicut	Thrissocles mystax	●	●	●	●	●							
Calicut	Sardinella longiceps						●	●	●	●	●	●	●
Cochin	Rastrelliger kanagurta		●				●	●	●	●	●	●	●
Cochin	Sardinella fimbriata							●	●	●	●		●
	Total per month	11	9	9	10	10	14	15	16	18	14	11	11

prawns, have developed ontogenetic or diel vertical migration patterns that interact with differential current vectors at different depths in the water column to achieve transport that will maintain them in region where survival is likely to be high (Rothlisberg *et al.*, 1983). In the open ocean, Taylor columns and leeward gyres may act to maintain populations of pelagic larvae near islands of sea-mounts so that sufficient numbers are able to recruit demersal or coastal pelagic populations (Bakun, 1986).

Because the relevant factors do not lend themselves to simple quantification, it is not yet possible statistically to isolate the causal relationships between timing of reproductive processes and the factors leading to recruitment level. In many cases, it seems as if more than one purpose is served by a reproductive strategy and only in some extreme cases is it obvious which one is the principal: The extreme cases of synchronous spawning are more likely to be related to predator swamping than to achieving simultaneity with a food source, or avoiding transport away from juvenile nursery areas. At the other extreme, the serial spawning that occurs in so many species of continental shelf species throughout a relatively long season is more likely to be a mechanism that maximizes the utilization of planktonic food for larvae (which itself has a relatively long season of abundance on tropical continental shelves) and enhances the probability that at least some of the many larval cohorts will have a high rate of survival.

What level of variability in recruitment in tropical species is driven by the action of environmental variability on these different reproductive strategies? There is far less information on recruitment-driven changes in stock size in any tropical fisheries than in temperate fisheries where year–class structure of populations is a routine measurement in stock management. It is usually not clear whether the variability which is observed in relative abundance is caused by differences in year–class (or cohort) strength or if it is an effect of different settlement or migration patterns. The apparent abundance of pomacentrids at several locations on the Great Barrier, between years and over a 4-year period, varies by a factor of ×2–9, but this is probably an effect of settlement patchiness, not overall recruitment variability (Williams, 1983). In the same region, statistical analysis of relative settlement of labrid larvae on seven reefs showed that abundance did not covary between reefs or between years, nor yet between closely related species (Eckert, 1984).

We should also note here that analysis of age or length-based catch curves, especially for demersal stocks, normally provide estimates of total mortality that are compatible with expectations for natural and fishing mortality, and therefore that recruitment to the stocks represented in these catch curves must have been fairly constant, since constant recruitment is one of the necessary conditions for catch curves to be straight, and for their slope to provide reasonable estimates of total mortality (see Ingles and Pauly, 1984, for a large number of catch curves from typical tropical stocks). Thus, at least as far as demersal stocks are concerned, it does seem that recruitment fluctuations are either relatively small or damped by density-dependent mortality at the juvenile stage, or both.

FOOD AND FEEDING OF TROPICAL MARINE FISHES

Food provides energy, but there is also an energy cost in capturing, processing, and defending it, and these are the topics analyzed by optimum foraging theory (Mittelbach, 1981; Tricas, 1986). If fish, as they grow larger, gradually get access to food with a higher relative energy content, or if they become more efficient at capturing their food, the cost line in Fig. 9.14 will gradually decline, and enable them to grow to a larger size than might have been originally predicted. If these same fish, however, also move into deeper water as they grow, this effect will not be distinguishable from the effects on their metabolism of lower temperatures. If, on the other hand, such a transition from energy-poorer to energy-richer food should be very rapid (Fig. 9.14), this may result in the second growth phase noted by various authors, a phenomenon one might call "second wind" or "deuxième souffle" growth.

Two lines of inquiry are open for the study of the food and feeding habits of tropical fish. One is to accept complexity, and to try to track it in ever increasing details. The other is to attempt to relate to food and feeding of fish as an energy flow to a quantity that can be parameterized, such as biomass per unit time (Longhurst, 1984). It is obvious that these two lines of inquiry have between them a middle ground in which the complexity of the trophic relationships of fish can be somehow reduced by intermediate models, and thus made accessible to holistic modeling.

Analyzing food intake from stomach contents or underwater observation is usually the first step that to be performed when studying the food of fish. As we did in the previous chapter, species can then be assigned to broad classes (planktivore, piscivore, omnivores, etc.) or to finer groupings based on the detailed taxonomics of the ingested prey and cluster analysis of the intake data (Table 9.4).

Such studies usually provide few deep insights, beyond confirming that fish have catholic tastes and illustrating the now well-known rule that larger vertebrate predators habitually take larger prey (Fig. 9.29). Most such studies fail to report absolute prey size related to predator size or

TABLE 9.4. Some Examples of Major Studies on the Food and Feeding Habits of Tropical Marine Fish

Topic	# Stomachs examined	Method	Remarks	Source
Tropical flatfish	N.A.	Various	Relations between food anatomy, and behavior	de Groot, 1971, 1973
Indian sea fish	N.A.	Various	Energetics of feeding, review of Indian studies	Qasim, 1972
Gulf of Guinea: 71 conshelf species, 26 estuarine species	28,320	% Occurrence	3 Years standard data allows study of dynamics of feeding	Longhurst, 1957b, 1960b
Caribbean reef fish (212 species)	5,226	% Volume of contents	Description of each species ("Citation Classic")	Randall, 1967, 1985
Jamaica reef fish	N.A.	Various	Descriptive, reviews of relevant literature	Munro, 1983
Saurida spp., Indian coasts	2,940	Various	Compares items in stomachs with their abundance in sea	Rao, 1981
Coastal fishes, south Gulf of Mexico	N.A.	Various	Describes linkages between different habitats	Yáñez-Arancibia, 1985
Shelf fishes, on Venezuelan shelf	>10,000	Various	Descriptive, identifies primary prey species	Penchaszadeh and Salaya, 1984
Reef fishes, on the Hawaiian coast	1,547	% Volume, occurrence, and much visual observation	Descriptive, emphasizes precise trophic role of each studied species on the reef	Hobson, 1974
Yellowfin, skipjack tuna, in eastern Pacific	6,080	% Volume % Occurrence	Descriptive account, stratified by 14 areas, 3 size classes	Alverson, 1963

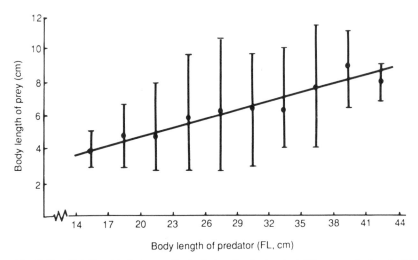

Fig. 9.29. Relationship between the body length of a bar jack (*Caranx ruber*) and of its prey, which is mainly labrids, scarids, and shrimps, on the Cuban continental shelf. (Modified from Siebert and Popova, 1982.)

prey weights, and do not enable absolute energy flow be computed. Most studies, instead, report ingested prey in the form of individual counts, or as percentage occurrence of stomach contents. MacDonald and Green (1983) have shown conclusively that these various available (and hotly debated) methods all give essentially the same results, as was also noted by Longhurst much earlier (1957b).

Even more important is the fact that stomach contents do not by themselves necessarily rank prey according to the fundamental feeding preferences of the fish, nor of the absolute or long-term relative abundances of their prey in the environment. Rather, stomach contents are a simple function of local prey availability and suitability, this latter often simply being a function of size and the predator's previous experience with that prey (Ursin, 1973; Fig. 9.29).

A complementary approach that has led to considerable insights on the food and feeding habits of tropical marine fishes is analysis of their comparative anatomy and physiology, combined with stomach content studies. Thus, for example, the mouths, teeth, eyes, and other body structures of different species of tropical flatfish (Pleuronectoidea) have been related to details of their behavior and their feeding habits by de Groot (1971, 1973), while the prey spectrum of different reef fishes have been related to their degree of evolutionary development by Hobson (1974).

The specialized feeding habits of the cichlid *Sarotherodon me-*

TABLE 9.5. Comparison of Feeding-Related Features of *Sarotherodon melanotheron* (Cichlidae) and *Mugil cephalus* (Mugilidae)[a]

Feature	S. melanotheron	M. cephalus
Feeding behavior	Picks up sediment, and swallows after some sorting in mouth	Skims off surface layer of deposits, and filters ingested material
Gill rakers	Absent	Present
Gut/body length ratio at 8 cm length	About 7 : 1	About 3 : 1
Gut clearance rate	3 per day	5 per day
Food item/sediment ratio in stomach contents	2 : 1	100 : 1

[a] From Pauly (1976) and Odum (1970).

lanotheron and the mullet *Mugil cephalus,* both of which utilize the organic fraction of the soft mud deposited in coastal lagoons is made possible by a number of characteristic anatomical and behavioral adaptations (Table 9.5) of the kind we refer to, and these enable them to concentrate this fraction by a factor of 2 for *S. melanotheron* and 100 for *M. cephalus.*

An interesting approach to holistic analysis of the nutrition of fish, illustrated by the work of Elliott and Person (1978), involves estimation of only two parameters, mean weight of stomach contents over time and gastric evacuation rate. Given certain assumptions, the product of these two parameters provides the rate of food consumption of the fish for which the parameters have been estimated. This approach was applied to tropical fish (Table 9.6) by Olson and Boggs (1986).

TABLE 9.6. Estimated Food Consumption per Unit Biomass in Some Selected Fish Populations

Species	Daily	Annual
Epinephalus tauvina[a]	0.64	2.34
Epinephalus guttatus[a]	0.76	2.78
Sparus auratus[a]	0.88	3.22
Acanthopagrus cuvieri[a]	0.64	2.33
Gadus morhua (60–100 cm)[b]	0.80	2.92
Thunnus albacares[c]	3.90	14.24

[a] Pauly and Palomares (1985).
[b] Daan (1973).
[c] From Olson and Boggs (1986) based on Elliott and Person (1978).

Finally, we propose a model for food consumption of natural fish population that is intended to be consistent with the key equations presented in this chapter. It uses gross food conversion efficiency, (K_1) as its starting point. This quantity, defined as

$$K_1 = (\text{growth increment/food ingested}) \qquad (9.12)$$

and which can easily be determined by feeding experimentally (Menzel 1960), can be expressed as a function of body weight

$$K_1 = 1 - (W/W_\infty)^\beta \qquad (9.13)$$

where W_∞, is the weight of a fish after cessation of growth, and whose conversion efficiency is therefore zero (Figs. 9.3 and 9.30). Note that this model can be derived from Eq. (9.1) and that it explains the decrease of K_1 with size as the result of a lack of oxygen for growth (Silvert and Pauly, 1987; also see Figs. 9.2 and 9.3).

Various manipulations involving Eqs. (9.4) and (9.8), and incorporation of an equation for biomass per recruit (Beverton and Holt, 1957) lead finally to a compact model for estimating the food of a fish population per unit of its biomass, i.e.,

$$\frac{Q}{B} = \frac{3K \int_{t_r}^{\infty} \dfrac{(1 - e^{-Kr_1})^2 \cdot e^{-(Kr_1 + Zr_3)}}{1 - (1 - e^{-Kr_1})^{3\beta}}}{\left(\dfrac{1}{Z} - \dfrac{3e^{Kr_2}}{Z + K} + \dfrac{3e^{-2Kr_2}}{Z + 2K} - \dfrac{e^{-3Kr_2}}{Z + 3K}\right)} \qquad (9.14)$$

where $r_1 = t - t_0$; $r_2 = t - t_r$ and $r_3 = t - t_r$, t_r is the age of the youngest fish considered, K, t_0, and Z are as defined above, and β is the exponent in Eq. (9.13) and the slope in graphs such as Fig. 9.30.

Table 9.6 presents some results obtained with the application of variants of this model, to three warm-water demersal species, as compared with results obtained for cod using a more convenitional approach.

This chapter differs from the rest of our book. In it we have tried to present a comprehensive model of fish population biology that has largely been assembled from the studies of the second author. Because some new concepts are introduced, a brief overview of the whole model is perhaps in order in closing this chapter.

What are the key elements of the theory proposed here? They can be summarized as five statements in simple language, as follows:

1. Fish morphology and the physics of oxygen solution and diffusion work together to limit the oxygen available to the tissues of any growing fish, whatever its environment, and whatever evolutionary or behavioral measures it may take to maximize its oxygen availability.

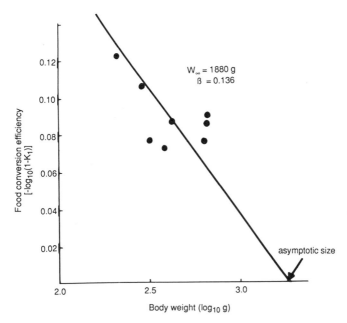

Fig. 9.30. Relationship of food conversion efficiency to body weight in red hind *Epinephalus guttatus*. (From Pauly, 1986b, based on Menzel, 1960, and Thompson and Munro, 1983.)

2. Thus, factors affecting O_2 consumption must limit not only the growth rate of fishes, but also their ultimate size, and any model of the dynamics of fish growth, and the distribution of their maximum sizes in space and time that cannot accomodate this effect must be fundamentally unsatisfactory.

3. Because of the limits thus placed on growth rates, natural mortality rates of fishes are constrained, for any combination of growth parameters, to a rather narrow and broadly predictable range of values.

4. Fishes do not stop growing because they start elaborating gonads; rather they initiate gonad maturation when their growth rate and metabolic rate drops below a predictable threshold.

5. Because the food conversion efficiency of a fish species diminishes with size, models that include growth as a linear function of food consumption will usually be unsatisfactory, and will suggest variations of growth rate of a magnitude that does not occur in nature.

Each of these five statements is open to refutation so that the model is "scientific" in the (trivial) Popperian sense. More interesting is to ask why this model is offered only in the mid-1980s, rather than 20 years earlier when all of its constituent elements had become available (von Bertalanffy, 1951; Beverton and Holt, 1959; Beverton, 1963; Taylor, 1958, 1962; Gray, 1954). Possibly, the large amount of detailed information on the paradigmatic, high-latitude fishes alluded to above (cod, herring, salmon, etc.) is precisely what prevented workers in high-latitude areas from deriving rules applicable to the whole range of fish species that actually occurs in the oceans. Working in the tropics, on the other hand, forces one to seek much wider patterns and their possible causes, empirical relationships of the type advocated by authors such as Parker and Larkin (1959), or Roff (1980) being simply impossible to establish for more than a handful of species. Whether the model presented here is accepted or not will not depend only on whether propositions 1–5 pass independent experimental tests, however.

An occupational hazard of tropical biologists has been to hear their models described as "rubbish, may apply in the tropics, but not here," as one thankfully anonymous reviewer described some of the research described in this chapter. We suggest that what may apply in the tropics may also come to be found to be very important "here" because, as we should not forget, most basic life forms of fish and aquatic invertebrates evolved in environments resembling the present-day tropics, and radiated from there to more marginal habitats in higher latitudes.

Chapter 10

Population Biology of Large Marine Invertebrates

In this chapter we have chosen to illustrate, using three representative groups as examples, how techniques developed for fish population dynamics can be applied to the population biology of invertebrates. Our reason for making this point explicitly is that, somewhat like the mythology surrounding tropical fish growth rates, it is easy to believe that the basic biology of invertebrates is so different from fish that population analysis techniques relevant to fish must be inappropriate for invertebrates (e.g., Juanico, 1983). It is our intention to illustrate the extent to which models of population growth and decay are independent of the phylum to which the modeled population belongs. Echinoids, penaeid shrimps, and cephalopods are the products of separate lines of evolutionary development and each possess fundamentally different morphology and physiology, yet (as we shall show) their populations are amenable to analysis by population models formulated for fish stock management purposes. Tropical invertebrate populations are increasingly coming under heavy exploitation and their rational management will require the careful application of such models. Further, the dynamics of natural population interactions within ecosystems may be studied by techniques specifically developed to analyze the interaction between fishermen and fish and, fortunately, this is increasingly recognized to be the case. The growth of organisms as diverse as giant mud-clams in Sierra Leone estuaries (Okera, 1976), prawns off Australia (Rothlisberg *et al.*, 1985a), and rock lobsters off Cuba (Cruz *et al.*, 1981) have been analyzed by fish population models. In this chapter, we shall also try to point out some attributes of tropical invertebrates that cannot be accomodated in unmodified fish population models.

LIFE-HISTORY STRATEGIES OF SEA URCHINS

Although collected for direct human consumption of their gonads in a number of countries, sea urchins (Echinoidea, Echinodermata) are economically important mainly because of their impact as grazers of algae that are cultivated or harvested commercially (Doty, 1973; Gomez et al., 1983). For the same reason, their ecological importance in structuring macrophyte-based littoral ecosystems is also considerable, and grazing by sea urchins may affect the whole structure of shallow-water tropical ecosystems (Hay, 1984a,b) as well as temperate ones (e.g., Mann, 1977). Sea urchin population biology has thus received considerable attention in recent years (Jangoux and Lawrence, 1983), especially with regard to what controls their recruitment (Ebert, 1983; Bacolod and Dy, 1986) and the rapid growth and decline of their populations.

One approach to understanding the dynamics of sea urchin populations has been through comparative studies of the parameters of their growth (Fig. 10.1) and mortality, such as the work of Ebert (1975) on postlarval echinoids of several species. Figure 10.2 shows a plot of values for M on their corresponding values of K, based on the sea urchin data in Table 10.1. This shows clearly that sea urchins exhibit a close relationship between K and M, and that this relationship is similar to the one discussed for fish in Chapter 9.

This suggests that other relationships reported for fish, such as the inverse correlation between K and asymptotic size, should also hold for sea urchins. But Ebert (1975) concluded that no correlation (or else a positive correlation) exists for sea urchins between asymptotic size and K, whereas in fish this correlation is negative. However, the sea urchins used by Ebert ranged only from 1.7 to 11.0 cm, or less than one order of magnitude in dimension. Where inverse relationship between K and asymptotic size have been found for fish populations, this has involved either analyzing a very wide range of sizes in different taxa or comparing several stocks of the same species (Beverton and Holt, 1959; Pauly, 1980a). And, as expected, we find that analysis of growth parameters in different stocks of a single species of tropical sea urchin (*Diadema antillarum*) does indeed lead to the anticipated inverse relationship between K and asymptotic size having a slope almost exactly equal to the mean slope reported for fishes (Fig. 10.3), and used to derive the growth performance index ϕ' (Eq. 9.5).

Ebert also suggested that tropical species of sea urchins have a significant inverse relationship between asymptotic size K or M and the upper limits of inshore occurrence (Table 10.1), so that tropical sea urchins increase in size, growth rate, and mortality rate downward from the surf

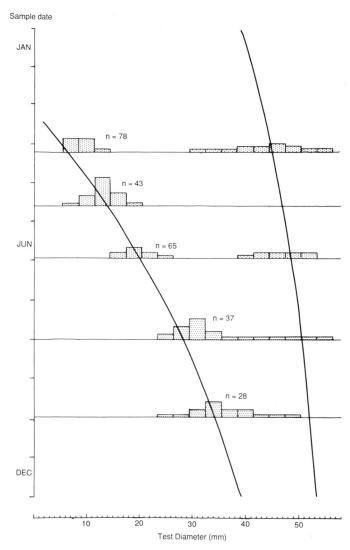

Fig. 10.1. Growth of the sea urchin *Diadema antillarum*, based on the size–frequency data in Bauer (1976) analyzed with the ELEFAN I program (see also Table 10.9).

zone. He suggested that individual urchins living in the surf zone survive longer, on the average, because subtidal individuals, which live in a less stressful environment, are able to expend fewer resources on maintainance and more on growth in size and for reproduction.

The metabolic growth model presented in Chapter 9 explains why sea

TABLE 10.1. Population Parameters of Tropical Sea Urchins Ranked by Upper Limit of Preferred Depths

Depth rank[b]		Species	$L_\infty{}^c$	$K(1/y)$	$M(1/y)$	$\phi'{}^d$
1[a]	1	*Colobocentrus atratus*	3.00	0.45	0.33	0.61
2	4.5	*Eucidarus tribuloides*	5.27	0.67	0.43	1.27
3	6.5	*Tripneustes ventricosus*	9.00	1.24	0.95	2.00
4	6.5	*Echinotrix diadema*	9.00	0.82	0.64	1.82
5	8	*Diadema antillarum*	7.56	1.18	1.30	1.83
6	8	*Diadema antillarum*	5.80	1.22	1.90	1.61
7	8	*Diadema antillarum*	5.85[e]	1.35[e]	2.07[f]	1.66
8[g]	8	*Diadema antillarum*	3.87	4.14	—	1.79
9[g]	8	*Diadema antillarum*	3.91	4.31	—	1.82

[a] 1–6 from Ebert (1975), 8–9 computed by method of Gulland and Holt (1959) and growth curves of Bauer (1982).
[b] Depth ranks: surf zone = 1, subtidal = 8.
[c] L_∞ is test diameter, cm.
[d] $\phi' = \log_{10} K + 2 \log_{10} L_\infty$.
[e] Estimated by ELEFAN I (based on data in Bauer, 1976).
[f] Length-converted catch curve based on Fig. 10.1.
[g] 8–9 are possibly stressed laboratory animals.

urchins living in the surf zone should remain smaller than those living in the less stressful subtidal. However, that their K (and hence M) should be the lowest on record is indeed puzzling but most probably will prove to be a consequence of some aspect of the detailed ecology of a sessile, grazing organism in the intertidal ecosystem. This may be a good example of the occasional need to modify fish population models because of some distinctly unfishlike aspect of the ecology or physiology of a group of invertebrates.

A further example of how a theoretical population model may be incapable of tracking actual population numbers is the case of a major population crash, due to some cause not normally accomodated in population models, such as a pollutant, or a pandemic pathogen. The sea urchin *Diadema antillarum,* which ranges from Florida to Surinam (Bauer, 1980), has been studied by several authors, notably Scoffin et al. (1980) and Hay (1984a,b), and shown to be a major factor affecting community structure in tropical neritic ecosystems. *Diadema* reaches densities of >50 individuals per square meter and erodes calcium from reef rock while grazing algal turf, eats live coral, and is a highly competitive herbivore able to cause large areas of barren rock, in the same way as kelp beds are turned into "urchin barrens" by the North Atlantic urchin *Strongylocentrotus.*

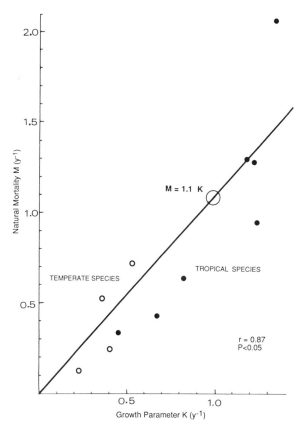

Fig. 10.2. Relationship between the growth parameter K and natural mortality estimates (M) (see also Table 10.1) in sea urchins. (Based on data in Ebert, 1975.)

In the 12 months following January 1983, populations of *D. antillarum* throughout the Caribbean were reduced to about 1% of their former abundance by the spread of a pathogen from an initial infection near Panama (Lessios *et al.*, 1984a). The timing of the infection is entirely consistent with the spread of a pathogenic organism by the current systems of the Caribbean (Gordon, 1969), and it is safe to assume that this was a phenomenon similar to the population crashes of *Strongylocentrotus* through infection with an amoeboid pathogen (*Paramoeba*) that occur in years only when water temperatures are anomalously high (Scheibling and Stevenson, 1984). One cannot help noting that the spread of the pathogen through the Caribbean occurred in the months of maximum warm SST anomalies in the eastern Pacific related to the 1982–1983 El Niño–South-

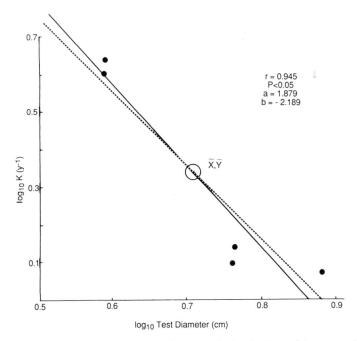

Fig. 10.3. Relationship between K and asymptotic size for S set of these growth parameters in the sea urchin *Diadema antillarum*, based on Table 10.1. Note how these data (solid line) match with the relationship expected on theoretical grounds (dashed line), as explained in the text.

ern Oscillation (ENSO) event, and when there was some evidence of warm anomalies also in the tropical western Atlantic.

It can be predicted that communities previously structured by grazing *D. antillarum* will, for several years, show different abundance and distribution of algae. Lessios *et al.* (1984b) also expected that the sudden decline in the availability of *Diadema* probably increased the predation pressure on the few individuals that survived the disease, and on other species of predators.

This prediction was confirmed by the switch during 1983 of the queen triggerfish *Balistes vetula* from a diet consisting predominantly of *D. antillarum* to one mainly of crustaceans (Fig. 10.4). Thus, what previously appeared to be a specialized predator (Reinthal *et al.*, 1984) is in fact a generalist, with a wide ethological repertoire (Fricke, 1975), which enables it to handle and utilize a large number of preys, and which concentrates on a difficult prey (*D. antillarum* in this case) only if its high abundance is sufficient to offset (in terms of net energy yields) the relatively high cost of handling it.

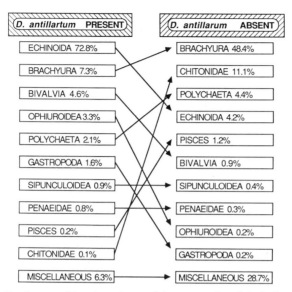

Fig. 10.4. Food composition, by volume, of the queen triggerfish, *Balistes vetula* in the presence of *Diadema antillarum* based on 95 specimens from the Virgin Islands, and absence based on 23 specimens from Belize (Randall, 1967; Reinthal *et al.*, 1984).

There is no known case of pathogenic fish mortality of this scope, and practical fish population models usually accomodate fish disease only as an unexceptional and stochastic element within *M*, the natural mortality parameter. This is an example of how some detailed aspects of invertebrate biology may require some revision to fish population models prior to their application in the management of invertebrates.

RECRUITMENT OF PENAEID SHRIMPS

Importance of Tropical Resources

Because of their economic importance to some tropical nations (Table 10.2), shrimps of the family Penaeidae have been much studied, and many aspects of their biology and numerical population dynamics have been clarified in recent years (Gulland, 1971; Holthuis, 1980; Garcia and LeReste, 1981; International Development & Research Council, 1982; Rothlisberg *et al.*, 1985a; Yáñez-Arancibia, 1984). The application of fish models to shrimp population dynamics has now become routine, and the FAO manual published by Garcia and LeReste (1981) is an excellent entry

TABLE 10.2. Nominal Global Catches of Shrimps

"Top twenty" countries, total shrimp catches

Country	t y^{-1}	How reported	Country	t y^{-1}	How reported
India	214,980	All as "Natant decapods"	Brazil	50,660	All as "Natant decapods"
China	185,790	90% are *Acetes*	Vietnam	49,100	All as "Natant decapods"
Thailand	173,967	11% *Sergestes*, rest *Penaeus*	Greenland	41,243	*Pandalus borealis*
Indonesia	129,610	71% Reported as penaeids	Ecuador	36,600	All as "Natant decapods"
USA	119,906	89% Penaeids, 9% *Pandalus*	Korea	36,424	47% are *Acetes*, 9% penaeids
Malaysia	76,475	16% Sergestids, rest penaeids	USSR	29,394	*Crangon* spp.
Norway	75,035	*Pandalus jordani*	Pakistan	27,502	All reported as penaeids
Mexico	65,586	All as "Natant decapods"	Australia	20,814	All as "Natant decapods"
Japan	61,943	8% Penaeids	Argentina	19,289	*Artemisia, Pleoticus*
Philippines	55,748	47% Sergestids, 53% penaeids	Panama	13,491	91% as penaeids

314

Tropical shrimp catches by species, as reported

P. aztecus	64,034	Metapenaeus spp.	46,240
P. merguiensis	59,454	M. joinieri	1,386
P. californiensis	598	P. longisrostris	6,424
P. duorarum	14,007	P. atlantica	122
P. japonicus	6,576	Atlantic sea-bobs	1,954
P. monodon	20,755	Artemisia spp.	314
P. chinensis	18,346	Pacific sea-bobs	10,677
P. kerathurus	10,666		
P. semisulcatus	1,806	P. edwardsianus	510
P. setiferus	30,064	Pleoticus, Sicyonia	20,603
P. brevirostris	2,311		
P. latisulcatus	1,208	Sergestes, Acetes	57,776
P. notialis	5,950		
Penaeus spp.	265,963	"Natantia"	1,753,339
(includes above species)			
Total			2,354,843

Temperate shrimp catches

Palaemon spp.	2,977
Crangon spp.	56,999
Pandalus borealis	147,944
Other pandalids	32,580
"Natantia"	24,875
Total	265,375
All shrimps	2,620,218

[a] Data rearranged from FAO landing statistics for 1983. Data as tonnes.

into this field. Measurement of effort, the use of tagging techniques, the calculation of growth, fecundity, and mortality, the relationship between parameters of the von Bertalanffy growth equation (VGBF), the analysis of the stock-recruitment relationship, and consideration of optimal exploitation levels can now all be handled for penaeids by techniques developed for fish models.

Many tropical countries are expected to increase their shrimp exports through intensification of the present fisheries and the development of aquaculture (New and Rabanal, 1985; Lawrence, 1985). However, besides market saturation effects (Palomares, 1985), the feasibility of these plans will depend both on the availability of wild stocks of shrimp to support inceased fishing pressure, and on whether shrimp farming will modify coastal wetland habitat so as to destroy more wild stock potential than is gained by output from the farms (as is already the case in Ecuador) to the extent that almost half of the mangrove area was converted to shrimp ponds by 1985, and it is projected that almost all will be converted by 1990 (Terchunian *et al.*, 1986). Conversion at a comparable rate is also occurring in Bangladesh (Quader *et al.*, 1986) where the Sunderbans, the largest mangrove area (59×10^4 ha) remaining anywhere, is undergoing rapid conversion.

This problem therefore makes shrimp recruitment models of peculiar importance today. We examine recruitment in detail, because it is here that the special biology of penaeids may cause some aspects of fish recruitment models to become irrelevant.

Parental Stock Biomass as a Factor Affecting Recruitment

For young shrimp to be recruited to a stock, there must be some parents to produce viable eggs so that at least one point (the origin) of the stock–recruitment relationship will always be known. Sharp and Csirke (1983) have criticized both the fish stock–recruitment models involving computation of one point per year, with subsequent fitting of a single curve of the Ricker (1954) or Beverton and Holt (1959) type, and also the more flexible model of Shepherd (1982).

Garcia (1983) has made the same comments for the application of such models to shrimp. Figure 10.5 (left side of figure) gives several apparent stock–recruitment relationships in shrimps from various areas that seem to imply a relatively linear relationship between parent stock size and subsequent recruitment. However, Garcia questions whether the apparent relationship parallels the stock–recruit relationship in fish, citing the strong serial correlation between stock and recruitment inherent in a fishery based on very short-lived species, and the impossibility that the sim-

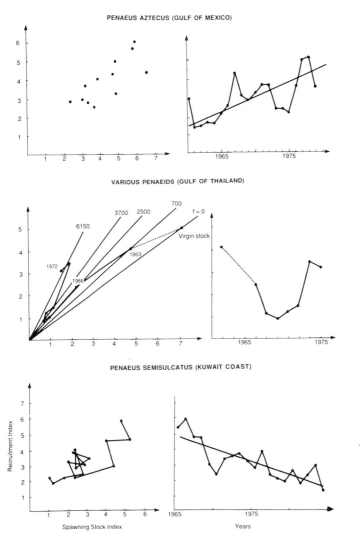

Fig. 10.5. Stock-recruitment relationships and temporal trends of recruitment in some penaeid shrimp stocks (Garcia, 1983; Mathews, 1985).

ple relationship should be linear over such a wide range of fishing effort. The temporal trends (right side of Fig. 10.5) on the other hand, are real, and they describe three different processes in the shrimp stocks in question: (1) A trend toward increased recruitment in *P. aztecus,* for which an explanation is wanting; (2) an initial decrease, subsequently followed by an increase in the stock-size of Gulf of Thailand shrimp, attributed to decreased predation on pre-recruit shrimp as the biomass of small fish species decreased (Pauly, 1982c, 1984c); (3) a steadily decreasing stock-size of Kuwait shrimps, attributed to a steady loss of intertidal area for the juveniles in the Gulf region (see below). It is most important to know if these observed changes in stock size are, in fact, due to recruitment being modified in some way by environmental modification, or by a feedback mechanism inherent in a linear stock–recruitment relationship.

Seasonal Fluctuations in Recruitment

Penaeid shrimps are short-lived, recruitment occurring to the coastal populations as little as 4 months after spawning. Thus, within-year events that are often neglected in population models for fish with a greater lon-gevity must be considered explicitly when formulating shrimp population models. It is essential, for instance, to accomodate seasonally fluctuating growth rates in such models, because totally erroneous inferences may be drawn on life cycles and relative cohort strength if this is not explicitly done (Garcia and LeReste, 1981; Pauly, 1985b; Garcia, 1985).

We have already described in Chapter 4 the essentials of penaeid life cycles. Figure 10.6 illustrates, as a special case, the life cycle of *Metape-naeus affinis* in Kuwait waters, along with some features of its environ-ment. The figure illustrates that growth oscillates seasonally, that spawn-ing occurs in two pulses per year, one being of shorter duration, and that recruitment consequently occurs as two cohorts, one smaller than the other.

Detailed investigation of the interaction between oceanographic, physi-

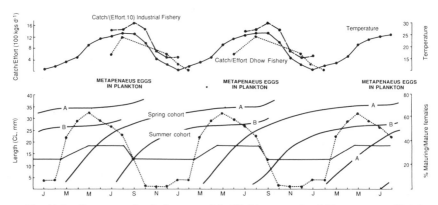

Fig. 10.6. Summary of main features of the life-history cycle of *Metapenaeus affinis* in Kuwait waters (Mathews *et al.,* 1987b).

ological, and behavioral processes has shown why *P. merguiensis* has two recruitment events per year in the Gulf of Carpentaria (Rothlisberg *et al.,* 1985b). These studies suggest that *P. merguiensis* in the Gulf of Carpentaria may be maladapted to its environment, which is not surprising since the Gulf is a Quaternary feature, about 5000 years old, while penaeid shrimp species are, in some cases, unchanged morphologically since the Cretaceous (Burokowsky, 1970). *P. merguiensis* in the Gulf of Carpentaria produces the bulk of its eggs during the season least appropriate for larval survival, while its major cohort is recruited from a minor spawning event each year.

The example of *Penaeus merguiensis* in the Gulf of Carpentaria confirms that the bimodal recruitment pattern is very basic to penaeids, since it is maintained under such apparently anomalous conditions, and the situation illustrated in Fig. 10.7 is probably valid for most species of penaeids: Their populations, at any one time, consist of two partly overlapping cohorts arising from the spring and autumn spawning peaks. Rothlisberg and his colleagues believe that for each cohort of *P. merguiensis* in the Gulf of Carpentaria, it is the first spawning event that produces the more eggs, the second representing the effort of only a small number of survivors. Garcia (1985), however, believes that the normal generation interval, the period between massive reproduction of parent to massive reproduction of progeny generation, is 1 year in *Penaeus,* based on his observations of *P. notialis = duorarum* in the Gulf of Guinea, and on theoretical grounds. Garcia allows two cohorts each year but does not believe that a major spawning event occurs at the end of the first 6 months of life.

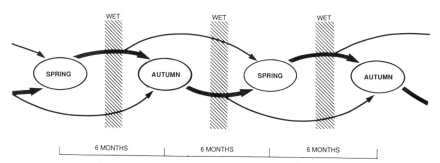

Fig. 10.7. "Interlocking" of cohorts in penaeid shrimp populations as illustrated by a generalization of life cycles in *P. merguiensis* throughout the Indo-Pacific (Rothlisberg *et al.,* 1985b).

Link between Coastal Wetlands and Recruitment

MacNae (1974) was among the first to suggest the existence of a causal link between the area of mangrove forests and the yield of adjacent shrimp fisheries, which are generally proportional to recruitment, given the strong exploitation they are usually subjected to and the short life span of penaeids. This idea was expanded in the form of correlative studies by Martsubroto and Naamin (1977) and Turner (1977), and the empirical relationships they obtained have been widely used for practical purposes, such as for environmental impact assessment studies, and by ecologists as illustrative of the general relationships between coastal organisms and the tidal wetlands.

Although such relationship can be expanded to account for more predictors of shrimp yields than intertidal vegetation area alone, e.g., latitude (Turner, 1977), they can also be viewed as a conceptual cul-de-sac because they do not tell us which attributes of vegetated intertidal areas are causal. More importantly, these simple relationships have one very basic constraint: They cannot model those areas that have very large shrimp catches, but relatively small areas of intertidal vegetation, such as the coast of northwest India and Pakistan.

As we have indicated in Chapter 4, there are penaeids that require brackish habitats for their juvenile development and those that do not. Off northwest India, at least two of the principal species in the shrimp stocks (*Metapenaeus affinis* and *Parapenaeopsis stylifera*) do not use estuarine regions for their juvenile development, so that their presence must modify any simple relationship between shrimp production and area of adjacent coastal wetland.

In fact there is no certainty that such a relationship is causal even when it can be demonstrated to exist; it seems just as likely that causation is through the general level of productivity of the continental shelf sustained by terrestrial organic material borne by rivers. In fact, it is noteworthy that the region of the tropical seas that produces the greatest biomass of shrimp is also that which receives the greatest discharges of terrestrial organic material, as a simple comparison of Fig. 2.1 with Table 10.2 will verify.

Various relationships have in fact been reported between shrimp catches, and rainfall or rate of river discharge, notably on the West African (Garcia and LeReste, 1981), East African (da Silva, 1986) and Australian coasts (Fig. 10.8). In some cases these relationships indicate that moderate rainfall and/or river discharge increases recruitment, while very strong rainfall decreases recruitment. Such dome-shaped relationships may imply that fresh water input of the right magnitude may, in a given

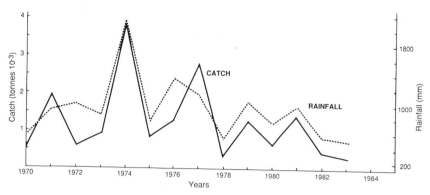

Fig. 10.8. Catch of *Penaeus merguiensis* and rainfall for the Karumba region of the southeastern Gulf of Carpentaria, Australia, 1970–1983 (Staples, 1985).

inshore system, maintain water of appropriate salinity in lagoons, estuaries, and other coastal systems where juvenile shrimp occur, while reduced or excessive freshwater inputs lead to the coastal system becoming limnic or fully marine, respectively. As Browder and Moore (1981) point out, juvenile shrimp require the matching of two sets of variables for good survival: a static set related to the benthic environment and a dynamic set related to water conditions. Match or mismatch of the two sets can determine survival or failure of estuarine juvenile stocks. Staples (1985) describes the complex factors determining emigration of juvenile *P. merguiensis* in the Gulf of Carpentaria as a function of river discharge. The proportion of the juvenile population that emigrates increases with increasing river discharge, though the size of the emigrants falls. Years of lowest rainfall are those in which only a few, large juveniles emigrate. Staples finds that the relationship between rainfall and recruitment shown in Fig. 10.7 depends not on the absolute size of the juvenile stock, but the proportion of it which emigrates.

Synthesis: Modeling Shrimp Recruitment

The various elements mentioned above, which have all been shown to affect shrimp recruitment, are but a few of those that have been hypothesized to have such effects and models have been formulated that attempt simultaneously to incorporate a number of these effects (Staples, 1985) and that may even include the added complexity of a fishery (Grant *et al.*, 1981). These models are variously successful, but some have a rather low predictive capability, especially where the economic vagaries of a fishery intervene. Predictive success probably requires very profound understanding of the biology of the species concerned, as well as of the economic background driving the fishery. The simple relationship between rainfall and shrimp catches shown in Fig. 10.8 has some predictive capability, and has been formulated as a mathematical multistage model described also by Staples (1985). This incorporates adult spawning stock, juvenile stock, and rainfall, and it is predictive of emigrants and the resulting adult catches sufficiently closely to be of use at least in long-term fishery management planning.

SQUIDS AS COMPONENTS OF TROPICAL MULTISPECIES SYSTEMS

Application of Fish Models to Squid Populations

Squids (Cephalopoda, Teuthoidea) are important components of tropical marine ecosystems, both neritic and oceanic, although the true magnitude of their biomass and level of production has become apparent on a world-wide basis only with the extension of the Japanese squid fisheries beyond the northwestern Pacific Ocean (Sato and Hatanaka, 1983; Worms, 1983; Lange and Sissenwine, 1983; Voss, 1983). Another phenomenon that has recently attracted scientific attention is the explosive growth of cephalopod populations after the reduction of the associated fish component of their ecosystem; this has occurred in the Gulf of Thailand and off Mauretania (Ritrasaga, 1976; Pauly, 1979; Boonyubol and Pramokchutima, 1982; Caddy, 1983). Squid inhabit a variety of habitats (Packard, 1972; Brandt, 1983; Boyle, 1983; also Fig. 10.9). We deal here only with neritic and shelf species occurring in the tropics and subtropics.

Because of their unusual interest to neurophysiologists and ethologists, the majority of the scientific literature on cephalopods is of little relevance to their population biology. Indeed, the preoccupation of most squid biologists with giant nerve fibers and ethology has led to the suggestion that "squids are not fish and that few principles of fish biology apply to these molluscs" (Juanico, 1983). On the contrary, we wish to suggest that most principles of fish population biology do apply to squids. In the following section, we shall discuss the role of squids in some tropical ecosystems through the application of fish population models.

Since major research programs devoted to the study of squid populations are relatively recent in origin, knowledge of the life cycle of individual species is often lacking, as can be seen clearly from the level of information available in the 22 species reviews presented by Boyle (1983).

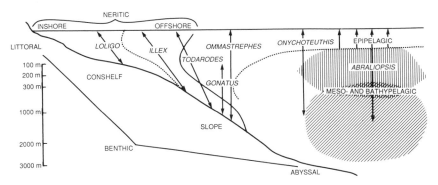

Fig. 10.9. The depth distribution of decapods in the major regions of the sea (Packard, 1972; Clarke and Lu, 1975; Roper and Young, 1975; Arnold, 1979).

However, the available evidence on growth, mortality, and other parameters of the population dynamics of squids is sufficient to suggest that they can be accomodated within fish population models (Pauly, 1985b). For example, Table 10.3 presents data for the maximum size of various species of squids in different parts of their range, and strongly suggests that squids conform to metabolic growth models for fish (Chapter 9), so that L_{max} for squid is a negative function of temperature. Further confirmation of the metabolic growth model can be provided by within-species comparisons of the same parameters as functions of temperature (Table 10.4).

Squids grow rapidly, and their growth is a function of temperature. Consequently, experimental and observational studies have produced inconsistent (and contradictory) growth models, including apparently linear growth in small species in the absence of food limitation, cyclic growth related to environmental temperature, sigmoid and asymptotic growth, and even expontential growth throughout life. Growth has been measured by analysis of daily growth rings in statoliths, experiments in tanks, and length–frequency analysis.

However, these seemingly contradictory statements about the growth of cephalopods can be resolved within a basic asymptotic growth pattern with superimposed seasonal growth oscillations. This conforms to a von Bertalanffy growth model modified to include a constant expressing the amplitude C of the growth oscillations (Pauly, 1985b, and Chapter 9). Figure 10.10 shows how length–frequency data for the tropical squid *Sepioteuthis lessioniana* can be analyzed quite simply with techniques developed for fish growth models, thus rendering superfluous the often complex schemes proposed by some authors to describe squid growth.

Figure 10.11 compares the growth parameters of a number of squid

TABLE 10.3. Maximum Size of Species of Squid in Relation to Water Temperature as Indicated by Latitude[a]

	Sepioteuthis lessoniana		Illex coindetii	Illex illecebrosus				Todarodes sagittarius		Nototodarus sloani		Dosidicus gigas	
			male	male		female		male	female				
	cm	kg	cm	cm	kg	cm	kg	cm	cm	cm	kg	cm	kg
High north latitude	36	1.4	37	31	1.6	27	0.4	49	64	—		—	
Low latitudes	33	1.2	26	20	0.2	18	0.1	32	37	32	0.6	100	18.8
High south latitudes	—		—	—		—		—	—	42	1.8	150	62.0

[a] Mantle lengths and weights from Roper et al., (1984). D. gigas weights from Ehrhardt et al., 1983.

TABLE 10.4. **Evidence for Applicability to *Loligo pealei* of a Fish Growth Model Which Gives a Limiting Role to O_2 Supply to Tissues**

Observation	Interpretation
Largest specimens in coldest parts of range	Table 10.3 for other squids, and see Chapter 9 for fish.
Maturation occurs at smaller sizes in warmer parts of range	As occurs in fish, see Chapter 9
Gill filaments grow relatively faster in warm parts of range	As required to elevate standard metabolism in warm, low O_2 water
Males have more gill filaments than females	Males reach 50 cm, females only 40 cm, off New England (Roper *et al.*, 1984)
L. pealei has relatively larger gills than *L. roperi*	*L. pealei* is larger species even when they co-occur

[a] All information extracted from Cohen (1976).

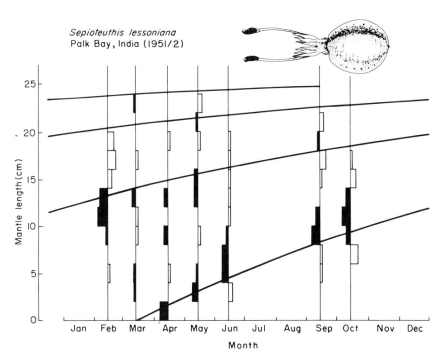

Fig. 10.10. Growth of the Palk Bay squid *Sepioteuthis lessoniana* based on length frequency data (in Rao, 1954) and the ELEFAN I program.

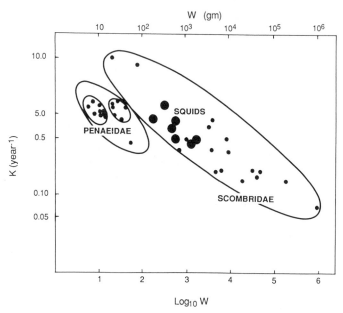

Fig. 10.11. Comparison of the growth performance of some squids with fast-growing fish (Scombridae) and other invertebrates (penaeid shrimps) by means of an auximetric grid, which leads to the conclusion that squids have as rapid growth as their pelagic competitors. (Data from Pauly, 1985a,b.)

species with those of penaeid shrimps and scombroid fish. As can be seen from this auximetric grid, the growth parameters of squids place them in the same league as the fast-growing, high-metabolism mackerel-like fishes, with which they share their pelagic habitats. Moreover, as Fig. 10.12 shows, the migration patterns for an oceanic squid, *Dosidicus gigas* of the eastern Pacific, show that it uses its environment very much in the same manner as large pelagic predatory teleosts, of which it is an evolutionary convergent ecological equivalent.

Analysis of mortality in cephalopods is not simple. There is excellent evidence that many, perhaps most, species of cephalopods undergo mortality in the same pattern as Pacific salmon, including a terminal, post-spawning mortality of both sexes. This is not a usual pattern of mortality for marine fish, and the application of fish population models must be sensitive to this difference. Spawning stress has been noted in many cephalopods as producing thin and watery flesh, damaged epithelium, opaque pupils, a cessation of feeding, and lack of germ cells in gonads. Of the 22 species synopses presented in Boyle (1983) of midge squid to octopuses, the 17 that include information on mortality describe some aspect of postspawning mortality.

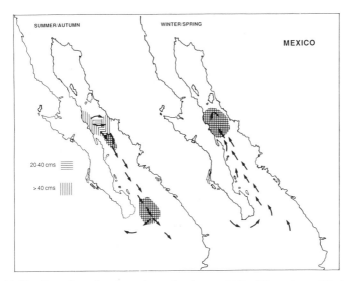

Fig. 10.12.   Tentative migration scheme for giant squid *Dosidicus gigas* off Baja California (Ehrhardt *et al.*, 1983).

However, cephalopod mortality is not as simple as this might seem to imply. First, because natural mortality must occur at some rate throughout life, it is only a small percentage of larvae that recruit, and of recruits that survive to die in terminal postspawning mortality. Second, because there is evidence in some species that some individuals survive their first spawning and undergo serial spawning with mortality only at the end of the process. Thus, Bakhayoko (1980) describes complex migrations by *Sepia officinalis* off Senegal which include a return of some females to a second spawning event; Boyle (1983) includes data suggesting that *Loligo vulgaris* of the North Atlantic behaves in the same manner, and Juanico (1983) believes that *Doryteuthis plei* of the Caribbean is another multiple spawner.

Pauly (1985b) presented length-converted catch curves for *Loligo pealei* of the Gulf of Mexico, and *Ommastrephes* off Japan, which apparently include no evidence of terminal postspawning mortality, though the same kind of plot for *Dosidicus gigas* seems to do so.

Figure 10.13 shows a similar plot for the tropical neritic squid *Sepioteuthis lessoniana* from southern India, which seems to show no evidence of a terminal postspawning mortality though Rao (1954) noted many of the symptoms of spawning stress in this species in Palk Bay where, he says, females have mostly empty stomachs during the breeding season in addition to being stressed in other ways. It seems likely that the

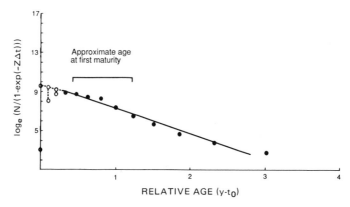

Fig. 10.13. Length-converted catch-curve for the squid *Sepioteuthis lessoniana* off southern India, with an estimated value of $Z = 2.56$ (data from Fig. 10.11).

Palk Bay squid will prove to have the same kind of life cycle as we have noted above for *Sepia officinalis* off Senegal.

Food Consumption of Squids

Because there are now several known instances of tropical squid outbursts following reduction of fish biomass by commercial fishing, the trophic relations between these two ecologically similar, but morphologically very different, groups are important to understand. We have already noted (Chapter 4) that cephalopods are capable of capturing unusually large prey, frequently including smaller individuals of the same species. Squids in tropical ecosystems act, as elsewhere, both as predators and prey (Fig. 10.14) and we shall discuss these two roles, using *Loligo duvauceli* and *Sepioteuthis lessoniana* as representative species, to illustrate the consequences of cephalopods as components of pelagic ecosystems.

The food intake of *L. duvauceli*, which like other squids is reported to feed on crustaceans, fishes, and squids, has not been estimated directly. However, we can apply the recently developed method of Pauly (1986b) and Chapter 9, based on food conversion data for *L. opalescens* (Fig. 10.15), which is a species very similar in size and morphology to *L. duvauceli*. The results (Table 10.5), which have been corrected for the different temperatures of the habitats of these two species, suggests that a population of *L. duvauceli* will eat approximately 9.4% of its own weight per day or 34 times its own weight per year. This is well within the range of food consumption estimated for other squid populations (Table 10.6),

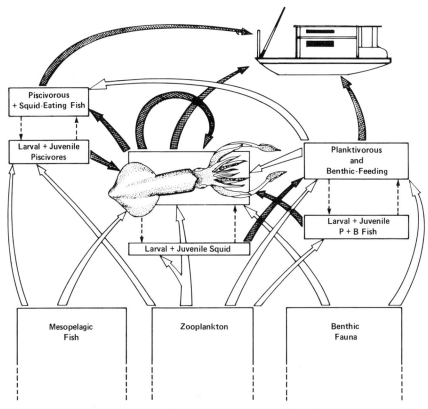

Fig. 10.14. Hypothesized trophic relations of squid with their biotic environment after intensive fishing for finfish and prior to a major squid fishery. Open arrows indicate flow of biomass, solid arrows are flows thought to be of key importance in determining abundance of squid and larger fish. Box size approximately corresponds to relative standing stocks of components. (Modified from Caddy, 1983.)

which covers an unrealistically wide range of 0.7–21.3% of body weight per day, or 4.4–78 times body weight per year. More interesting is the fact that our estimates for *L. duvaucelli* approach those for *Katsuwonus pelamis*, a very active pelagic fish.

In absolute terms, our estimates imply that the cephalopod biomass of the Gulf of Thailand (if one accepts that it behaves similarly to *D. duvauceli*, its major component species) consumed about 700,000 tons of prey in the early 1960s, and 500,000 tons in the early 1980s when the biomass of squid was about 20,000 and 10,000 tons, respectively (Pauly, 1985b). The lack of direct proportionality is due to the fact that a strongly

TABLE 10.5. Food Consumption as Percentage of Body Weight of Squids and Skipjack, *Katsuwonus pelamis*

Species	Area	Temp (°C)	Daily food intake (%)	Comment	Reference
Illex illecebrosus.	N.W. Atlantic	10°C	0.7 & 21.3	Spring (2 years)	Maurer and Bowman, 1985
		9°C	3.7 & 7.2	Summer (2 years)	Maurer and Bowman, 1985
		10°C	1.2 & 3.3	Fall (2 years)	Maurer and Bowman, 1985
Loligo pealei	N.W. Atlantic	10°C	8.3 & 20.9	Spring (2 years)	Maurer and Bowman, 1985
		13°C	3.6 & 9.9	Summer (2 years)	Maurer and Bowman, 1985
		14°C	1.4 & 18.6	Fall (2 years)	Maurer and Bowman, 1985
Loligo duvauceli	Gulf of Thailand	27°C	9.4	Annual mean	Table 10.5
Sepioteuthis lessoniana	Southern India	27°C	5.3	Annual mean	Table 10.5
Katsuwonus pelamis	East Tropical Pacific	26°C	5.9	Maintenance	Kitchell *et al.*, 1978
		26°C	7.9–19.0	Fast growth fish	Kitchell *et al.*, 1978
		26°C	30	Maximum intake	Kitchell *et al.*, 1978

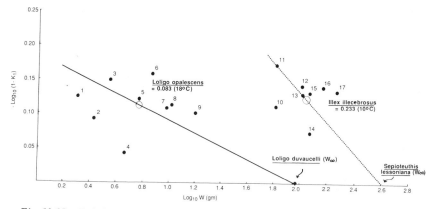

Fig. 10.15. Relationship between food conversion efficiency (K_1) and weight (W) in two species of squids, as used to estimate food conversion of squids of similar size (*L. duvauceli* and *Sepioteuthis lessoniana*, respectively). (Data from Hanlon *et al.*, 1983; O'Dor *et al.*, 1980.)

TABLE 10.6. Dimensions and Population Parameters
of Tropical Squids

Dimension or parameter	*Loligo duvauceli*	*Sepioteuthis lessoniana* (= *arctipinnis*)
ML max (cm)[a]	29.00	33.00
W max (g)[a]	150.00	1200.00[b]
ML inf (cm)	29.00[a]	27.00[d]
W inf (g)	150.00[c]	650.00[d]
$K(y^{-1})$	1.00[c]	0.73[d]
t (y)	0 (set value)	0 (set value)
$M(y^{-1})$	1.50[e]	1.10[e]
β	0.083[f]	0.23[g]
Q/B (exptl. temps.)[h]	4.25% (18°C)	1.22 (10°C)
Q/B (adjusted for higher metabolism)[i]	9.4% (27°C)	5.3% (27°C)

[a] Mantle length based on data of Roper *et al.* (1984).
[b] Based on length/weight relationship given by Rao (1954).
[c] From Pauly (1985b).
[d] See Fig. 10.12.
[e] Assumes M/K = 1.5 (see text, Chapter 10).
[f] Assumes data from Fig. 10.16 for *L. opalescens* are relevant.
[g] Assumes data from Fig. 10.16 for *I. illecebrosus* are relevant.
[h] Daily food consumption, % body weight, specific for temperature (see text).
[i] Temperature adjustment based on Winberg (1971, Table 3.3).

exploited stock, consisting of small individuals, has higher consumption per unit biomass than an exploited stock.

Sepioteuthis lessoniana in Southern India is often found associated with schools of juvenile fish (*Sardinella* spp., *Leiognathus* spp., etc) and the available evidence suggests that these fish constitute its main food. As in the case of *L. duvauceli,* food consumption has not been estimated directly. However, information is available on the gross food conversion efficiency for *Illex illecebrosus,* which reaches a similar weight to *S. lessoniana* off southern India (Fig. 10.15), and together with directly estimated growth parameters we can obtain a first approximation of the temperature-adjusted food consumption of *S. lessoniana.*

Our estimate (Table 10.5) is 5.3% body weight per day, so that a *S. lessoniana* population would consume 19 times its own weight per year. This is slightly less than estimated for *L. duvauceli,* which seems appropriate given that *S. lessoniana* is the larger organism. Our two estimates of food consumption by tropical squid suggest that these animals have the potential for a major impact on their ecosystems. We have presently no measure of the relevance of cannibalism in their role in the ecosystem, but it seems likely, as Caddy (1983) has suggested, that this may be a strategy to enable cephalopods to bridge gaps in the pelagic particle size spectrum and for the species as a whole to exploit indirectly, in an energy-efficient manner, food that is too small to be useful to the larger, maturing individuals. Nellen (1986) makes the same suggestion for fish, explaining the very high fecundity of fish principally as a mechanism for enabling large individuals to exploit small planktonic food at second hand. Alternatively, observations on squid in large tanks suggest that cannibalism is a mechanism by which a school survives a short period of food dearth; large individuals of *Illex* kill and share in the consumption of smaller individuals after only about 3 days food deprivation.

Predator–Prey Interactions between Squids and Fishes

Squids are exploited currently with demersal fishes and penaeid shrimps in a number of tropical fisheries throughout the world. These mixed fisheries are relatively well documented in Southeast Asia (Ritrasaga, 1976; Hernando and Flores, 1981), and some data are available (Pauly, 1979, 1985b; Larkin and Gazey, 1982; Chikuni, 1983) from the Gulf of Thailand which allow analysis of the cause of the squid outburst we have discussed already in Chapter 6.

Larkin and Gazey (1982) showed that a simple assumption of reduced trophic competition with pelagic fishes could not explain the increase of squids that took place in the 1970s when, moreover, demersal fish de-

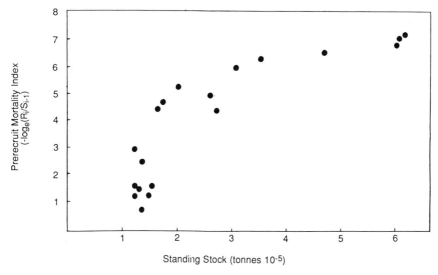

Fig. 10.16. Relationship between estimated prerecruit mortality of cephalopods in the Gulf of Thailand and total fish biomass in the inshore waters of the Gulf, 1961–1980 (Pauly, 1985b).

clined much more than the pelagic fish (Pauly, 1979). A more probable explanation, originally due to Jones (1982), is that the increase in squid was a direct consequence of a reduction of predation on their young.

Evidence along this line has been presented by Pauly (1985a) from which Fig. 10.16 is taken. This shows squid prerecruit mortality (e.g., of eggs and larvae) as a negative function of fish biomass and that there was a marked decline (of about one order of magnitude) in the daily prerecruit mortality as total fish biomass in the Gulf of Thailand declined. This confirms that the Gulf of Thailand squid outburst was due to a release of predator control on squid eggs and/or prerecruits.

This finding is in line with anecdotal evidence on the behavior of squid stocks (Gulland, 1982) and cephalopods in general (Caddy, 1983). It is also similar to what seems to happen with penaeid shrimp stocks, as we have discussed above. It is possible that a similar explanation lies behind the cephalopod outburst off Mauretania (Caddy, 1983) in which the decline in continental shelf sparid stocks after a decade of commercial fishing released first an outburst of *Sepia officinalis* and subsequently of *Octopus vulgaris*.

In this chapter we have shown that the population dynamics of tropical animals as diverse as sea urchins, shrimps, and squid can, at least in part, be investigated by means of models developed for the management of fish by an understanding of their population biology. This confirms an earlier suggestion (Pauly, 1985b) that nothing can be gained by emphasizing the differences between squids and other pelagic animals and by inferring that methods that have been developed to study fishes will not work when they are applied to other aquatic animals. Rather, fish models of known properties should be applied systematically to these animals and the deviations from such models studied in detail, because it is these deviations that will indicate how these animals are different from fishes.

References

Abele, L. G., and Walters, K. (1979). Marine benthic diversity, a critique and alternative explanation. *J. Biogeogr.* **6,** 115–126.

Abramson, N. J. (1970). Computer programmes for fish stock assessment. *FAO Fish. Tech. Pap.* **101,** 1–154.

Adey, W. H., and Steneck, R. S. (1985). Highly productive eastern Caribbean reefs: Synergistic effects of biological, chemical physical and geological factors. *Symp. Ser. Underwater Res. NOAA, U.S. Dept. Commer.* **3,** 163–187.

Alheit, J. (1982). Feeding interactions between coral reef fishes and the zoobenthos. *Proc. Int. Coral Reef Symp., 4th, 1981,* **2,** pp. 545–552.

Ali, R. M., and Thomas, P. J. (1979). Fish landing survey in Khorfakkan, November 1976– October 1977. *U.A.R. Dep. Fish., Tech. Rep.* **3,** 1–15.

Allen, G. R. (1975). "The Anemone Fishes. Their Classification and Biology." TFH Publications, Neptune City, New Jersey.

Allen, J. R. L. (1965). Late Quaternary Niger delta and adjacent areas: Sedimentary environments and lithofacies. *Am. Assoc. Pet. Geol. Bull.* **49,** 547–600.

Allen, J. R. L., and Wells, J. W. (1962). Holocene coral banks and subsidence in the Niger delta. *J. Geol.* **10,** 381–397.

Allen, K. R. (1971). Relation between production and biomass. *J. Fish. Res. Board Can.* **28,** 1573–1587.

Aller, R. C., and Dodge, R. E. (1974). Animal-sediment relationship in a tropical lagoon; Discovery Bay, Jamaica. *J. Mar. Res.* **32,** 209–232.

Alverson, F. (1963). The food of yellowfin and skipjack tunas in the eastern tropical Pacific Ocean. *Bull. Int.-Am. Trop. Tuna Comm.* **7,** 295–396.

Andrews, J. C., and Gentien, P. (1982). Upwelling as a source of nutrients for the Great Barrier Reef ecosystem: A solution to Darwin's question? *Mar. Ecol.: Prog. Ser.* **8,** 257–269.

Angel, M. V. (1969). Planktonic ostracods from the Canary Islands region. Their depth distributions, diurnal migration and community organisation. *J. Mar. Biol. Assoc. U.K.* **49,** 515–553.

Anonymous (1974a). "The Philippines Fisheries Industry." Media Systems Inc., Manila.

Anonymous (1974b). A review of the trawling industry of West Malaysia. Fisheries Division, Kuala Lumpur, Malaysia. *Proc., Indo-Pac. Fish. Counc.* **13** (3), 541–545.

Anonymous (1978a). Pelagic Resources Evaluation; Report on the Biology and Resources of Mackerels (*Rastrelliger* spp.) and Round Scads (*Decapterus* spp.) in the South China Sea, Part I, SCS/GEN/78/17, pp. 1–69. FAO-UNDP, Manila.

Anonymous (1978b). "Survey Results of the R/V Nansen in July 1978," Rep. J. NORAD/
Seychelles Proj., pp. 1–11. Inst. Mar. Res., Bergen, Norway.

Arnold, G. P. (1979). Squid: A review of their biology and fisheries. Minist. Agric. Fish.
Food (*G.B.*), *Fish. Res. Lab. Leafl.* **48**, 1–37.

Arntz, W., Landa, E., and Tarazona, J., eds. (1985). "El Niño: Su Impacto en la Fauna
Marina." Bol. Inst. Mar. Peru, Vol. Extr., Callao.

Arntz, W. E. (1986). The two faces of El Niño 1982–83. *Meeresforschung* **31** (1), 1–46.

Atkinson, M. J., and Grigg, R. W. (1984). Model of a coral reef ecosystem. II. Gross and net
benthic primary production at French Frigate Shoals, Hawaii. *Coral Reefs* **3**, 13–22.

Azam, F., Fenchel, T., Field, J. G., Gray, J. S., Meyer-Reil, L. A., and Thingstad, F.
(1983). The ecological role of water-column microbes in the sea. *Mar. Ecol.: Prog.
Ser.* **10**, 257–263.

Babcock, R. C., Bull, G. D., Harrison, P. L., Heyward, A. J., Oliver, J. K., Wallace, C. C.,
and Willis, B. L. (1986). Synchronous spawnings of 105 scleractinian coral species on
the Great Barrier Reef. *Mar. Biol. (Berlin)* **90**, 379–394.

Backus, R. H., and Craddock, J. E. (1977). Pelagic faunal provinces and sound scattering
levels in the Atlantic Ocean. *In* "Oceanic Sound Scattering Prediction" (N. R. Anderson and B. J. Zahuranec, eds.), pp. 529–547. Plenum, New York.

Bacolod, P. T., and Dy, D. T. (1986). Growth, recruitment pattern and mortality rate of the
sea uchin, *Tripneustes gracilla* Linnaeus in a seaweed farm at Danahon Reef, Central
Philippines. *Philipp. Sci.* (in press).

Bacon, P. R. (1971). Plankton studies in a Caribbean estuarine environment. *Caribb. J. Sci.*
11, 81–89.

Bain, K. H. (1965). Trawling operations in the Bay of Bengal. *Food Agric. Org. Extended
Progr. Tech. Assist. (FAO/EPTA)* **2121**, 1–21.

Bainbridge, V. (1960a). "The Plankton of the Inshore Waters off Freetown, Sierra Leone,"
Vol. 13, pp. 1–48. Colonial Office, Fishery Publ., London.

Bainbridge, V. (1960b). Occurrence of *Calanoides carinatus* in the plankton of the Gulf of
Guinea. *Nature (London)* **188**, 932–933.

Bainbridge, V. (1961). Early life history of the bonga *Ethmalosa dorsalis. J. Cons., Cons.
Int. Explor. Mer* **26**, 347–353.

Bainbridge, V. (1963). The food, feeding habits and distribution of the bonga, *Ethmalosa
dorsalis. J. Con., Cons. Int. Explor. Mer* **28** (2), 270–284.

Bainbridge, V. (1972). The zooplankton of the Gulf of Guinea. *Bull. Mar. Ecol.* **7**, 6–97.

Bakháyoko, M. (1980). Pêche et biologie des céphalopodes exploités sur les côtes du Séné-
gal. Thèse Doct. 3ème Cyle, No. 122, pp. 1–119. Université de Bretagne Occidentale,
Brest.

Bakun, A. (1986). Local retention of pelagic larvae in tropical demersal reef/bank systems:
The role of vertically structured hydrodynamic processes. *In* "Proceedings of the
IREP/OSLR Workshop on the Recruitment of Coastal Demersal Communities, Cam-
peche, Mexico" (A. Yáñez-Arancibía and D. Pauly, eds.), IOC/UNESCO Rep. Ser.
UNESCO, Paris **44** (in press).

Bakun, A., Beyer, J., Pauly, D., Pope, J. G., and Sharp, G. D. (1982). Ocean science in
relation to living resources. *Can. J. Fish. Aquat. Sci.* **39**, 1059–1070.

Bakus, G. J. (1967). The feeding habits of fishes and primary production at Eniwetok Atoll.
Micronesica **3**, 135–149.

Banse, K. (1968). Hydrography of the Arabian Sea shelf of Pakistan and India and effects on
demersal fish. *Deep-Sea Res.* **15**, 47–79.

Banse, K. (1982). Mass-scaled rates of respiration and intrinsic growth in very small invertebrates. *Mar. Ecol. Prog. Ser.* **9**, 281–297.

Banse, K., and Mosher, S. (1980). Adult body mass and annual production/biomass relationships of field populations. *Ecol. Monogr.* **50**, 355–379.

Bapat, S. V., Radhakrishnan, N., and Kartha, K. N. R. (1974). A survey of the trawl fish resources off Karwar, India. *Proc, Ind-Pac. Fish Counc.* **13** (3), 354–383.

Barber, R. T., and Smith, R. L. (1981). Coastal upwelling ecosystems. *In* "Analysis of Marine Ecosystems" (A. R. Longhurst, ed.), pp. 31–68. Academic Press, New York.

Bard, F. X., and Antoine, L. (1983). "Croissance du listao dans l'Atlantique Est," pp. 1–24. Document provisoire présentée à la Conférence ICCAT de l'année internationale du listao, Tenerife, juin 1983.

Bardach, J. E. (1961). Transport of calcareous fragments by reef fishes. *Science* **133**, 98.

Barham, E. G. (1971). Deep sea fishes: Lethargy and vertical orientation. *In* "Biological Sound Scattering in the Ocean" (G. B. Farquhar, ed.), pp. 100–118. Office of Naval Research, Washington, D.C.

Barkley, R. A. (1972). Johnston Atoll's wake. *J. Mar. Res.* **30**, 201–216.

Baslow, M. H. (1977). "Marine Pharmacology." R. E. Krieger Publ. Co., New York.

Bauer, J. C. (1976). Growth, aggregation and maturation of the echinoid, *Diadema antillarum*. *Bull. Mar. Sci.* **26**, 273–277.

Bauer, J. C. (1980). Observations on geographical variations in population density of the echinoid *Diadema antillarum* within the western Mid-Atlantic. *Bull. Mar. Sci.* **30**, 509–515.

Bauer, J. C. (1982). On the growth of laboratory-reared *Diadema antillarum*. *Bull. Mar. Sci.* **32**, 643–645.

Baumgartner, A., and Reidel, E. (1975). "The World Water Balance," 179 pp. Richard Lee, Amsterdam.

Bayagbona, E. O. (1966). Age determination and the Bertalanffy parameters of *P. typus* and *P. senegalensis* using the 'burnt otolith' technique. *In* "Proceedings of a Symposium on Oceanography and Fisheries Resources of the Tropical Atlantic," pp. 335–360. IOC/UNESCO, Paris.

Bayagbona, E. O. *et al.* (1963). "Quarterly Research Report," mimeo., pp. 1–8. Fed. Fish. Serv., Lagos, Nigeria.

Bayliff, W. H. (1963). The food and feeding of the anchoveta in the Gulf of Panama. *Bull. Int.-Am. Trop. Tuna Comm.* **7**, 397–498.

Beckman, W. (1984). Mesozooplankton on a transect from the Gulf of Aden to the central Red Sea during the winter monsoon. *Oceanol. Acta* **7**, 87–102.

Beers, J. R., and Stewart, G. L. (1971). Microzooplankters in the plankton communities of the upper waters of the eastern tropical Pacific. *Deep-Sea Res.* **18**, 861–883.

Ben-Avraham, Z., and Emery, K. O. (1973). Structural framework of the Sunda Shelf. *Bull. Am. Petrol. Geolog.* **57**, 2323–2366.

Berkeley, S. A., and Houde, E. D. (1978). Biology of two exploited species of half-beak, *Hemiramphus brasiliensis* and *H. balao*, from south-east Florida. *Bull. Mar. Sci.* **28**, 624–644.

Bertalanffy, L. von (1938). A quantitative theory of organic growth (Enquiries on growth laws, 2). *Hum. Biol.* **10**, 181–213.

Bertalanffy, L. von (1951). "Theoretische Biologie," vol. 2. A. Francke, A. G. Verlag, Bern.

Beverton, R. J. H. (1963). Maturation, growth and mortality of clupeid and engraulid stocks in relation to fishing. *Rapp. P.-V. Réun. Int. Cons. Explor. Mer.* **154**, 44–67.

Beverton, R. J. H., and Holt, S. J. (1956). A review of methods for estimating mortality rates in fish populations, with special references to sources of bias in catch sampling. *Rapp. P.-V. Réun. Int. Cons. Explor. Mer.* **140**, 67–83.

Beverton, R. J. H., and Holt, S. J. (1957). On the dynamics of exploited fish populations. *Fish. Invest. Ser. 2* **19**, 1–533.

Beverton, R. J. H., and Holt, S. J. (1959). A review of the lifespan and mortality rates of fish in nature and their relation to growth and other physiological characteristics. *Ciba Found. Colloq. Ageing* **5**, 142–180.

Bhattacharya, C. G. (1967). A simple method of resolution of a distribution into Gaussian components. *Biometrics* **23**, 115–135.

Birch, J. R., and Marko, J. R. (1986). The severity of the iceberg season on the Grand Banks of Newfoundland: An El Niño connection? *Trop. Atmos.-Ocean Newsl.* **36**, 18–24.

Bird, E. C. F. (1979). Geomorphology of the sea floor around Australia. *In* "Australia's Continental Shelf" (J. R. V. Prescott, ed.), pp. 1–22. Thomas Nelson, Melbourne.

Blaber, S. J. M. (1980). Fish of the Trinity Inlet System of North Queensland with notes on the ecology of fish faunas of tropical Indo-Pacific estuaries. *Aust. J. Mar. Freshwater Res.* **31**, 137–146.

Blaber, S. J. M., and Blaber, T. G. (1981). The zoogeographical affinities of estuarine fishes in South-East Africa. *S. Afr. J. Sci.* **77**, 305–307.

Blackburn, M. (1965). Oceanography and the ecology of tunas. *Oceanogr. Mar. Biol.* **3**, 299–322.

Blackburn, M. (1968). Micronekton of the eastern tropical Pacific Ocean; family composition, distribution, abundance and relations to tuna. *Fish. Bull.* **67**, 71–115.

Blackburn, M. (1977). Studies on pelagic animal biomasses. *In* "Oceanic Sound Scattering Prediction" (N. R. Anderson and B. J. Zahuranec, eds.), pp. 116–123. Plenum, New York.

Blackburn, M. (1981). Low latitude gyral regions. *In* "Analysis of Marine Ecosystems" (A. R. Longhurst, ed.), pp. 3–29. Academic Press, New York.

Blackburn, M., and Williams, F. (1975). Distribution and ecology of skipjack tuna, in an offshore area of the eastern tropical Pacific Ocean. *Fish. Bull.* **73**, 382–411.

Blackburn, M., Laurs, R. M., Owen, R. W., and Zeitzschel, B. (1970). Seasonal and areal changes in the standing stocks of phytoplankton, zooplankton and micronekton in the eastern tropical Pacific. *Mar. Biol. (Berlin)* **7**, 14–31.

Blay, J., and Eyeson, K. N. (1982). Feeding activity and habits of the shad, *Ethmalosa fimbriata,* in the coastal waters of Cape Coast, Ghana. *J. Fish Biol.* **21**, 403–410.

Boely, T., and Fréon, P. (1980). Coastal pelagic fisheries. *FAO Fish. Tech. Pap.* **186**, 13–76.

Boonyubol, M., and Pramokchutima, S. (1982). Demersal trawling in the Gulf of Thailand. *Thai Fish. Rep. Bangkok* **9**, 1–23 (ICLARM Transl. No. 4, 1–12, 1984).

Boradatov, V. A., and Karpetchenko, Y. L. (1958). Fisheries investigations off West Africa. *Rybn. Khoz. (Moscow)* **34**, 10–20.

Boto, K., and Isdale, P. (1985). Fluorescent bands in massive corals result from terrestrial fulvic acid inputs to nearshore zone. *Nature (London)* **315**, 396–397.

Boyle, P. R., ed. (1983). "Cephalopod Life Cycles," Vol. 1, Species Accounts," pp. 1–475. Academic Press, New York.

Boysen-Jensen, P. (1919). Valuation of the Limfjord. I. Studies on the fish food of the Limfjord 1909–1917. *Rep. Dan. Biol. Stn.* **26**, 1–44.

Bozeman, E. L., and Dean, J. M. (1980). The abundance of estuarine larval and juvenile fish in a South Carolina intertidal creek. *Estuaries* **3**, 89–97.

Bradbury, M. G. (1971). Studies on the fauna associated with the DSL in the equatorial Indian Ocean during October and November, 1964. *In* "Biological Sound Scattering in the Ocean" (G. B. Farquhar, ed.), pp. 409–447 Office of Naval Research, Washington, D.Ċ.

Brandt, S. B. (1983). Pelagic squid association with a warm core eddy of the East Australia Current. *Aust. J. Mar. Freshwater Res.* **34**, 573–585.

Breder, C. M., and Rosen, D. E. (1966). "Modes of Reproduction in Fishes: How Fish Breed." TFH Publications, Jersey City, New Jersey.

Brey, T., and Pauly, D. (1986). Electronic length-frequency Analysis: A revised and expanded users guide to ELEFAN 0,1 and 2. *Ber. Inst. Meereskd. Christian-Albrechts Univ., Kiel* (149), 1–77.

Brinca, L., Rey, F., Silva, C., and Saetre, R. (1981). "A Survey of the Marine Fish Resources of Mozambique," Rep. Surv. with R/V Nansen, pp. 1–58. Inst. Mar. Res., Bergen, Norway.

Brock, V. E. (1954). Some aspects of the biology of the aku, *Katsuwonus pelamis,* in the Hawaiian Islands. *Pacif. Sci* **8,** 94–104.

Brock, R. E. (1979). An experimental study on the effects of grazing by parrotfishes and the role of refuges in benthic community structures. *Mar. Biol. (Berlin)* **51,** 381–388.

Broom, M. J. (1982). Structure and seasonality in a Malaysian mudflat community. *Estuarine Coastal Shelf Sci.* **15,** 135–150.

Brothers, E. B. (1981). What can otolith microstructure tell us about daily and subdaily events in the early life history of fish? *Rapp. P.-V. Réun., Cons. Int. Explor. Mer* **178,** 393–394.

Brothers, E. B., and McFarland, W. N. (1981). Correlations between otolith microstructure, growth and life history transition in newly recruited French grunts (*Haemulon flavolineatum*) *Rapp. P.-V. Réun, Cons. Int. Explor. Mer* **178,** 369–374.

Brouard, F., Grandperrin, R., and Cillaurin, E. (1984). Croissance des jeunes thons jaunes (*Thunnus albacares*) et des bonites (*Katsuwonus pelamis*) dans le Pacifique tropical occidental," Notes Doc. Océanogr. No. 10. Mission ORSTOM, Nouméa.

Browder, J., and Moore, D. (1981). A new approach to determining the quantitative relationship between fishing production and flow of freshwater to estuaries. *In* "Freshwater Inflow to Estuaries" (R. Cross and D. Williams, eds.), FSW/OBS-81/04; pp. 403–430. USFW Office Biol. Serv., Washington, D.C.

Bruce, A. J. (1975). Shrimps and prawns of coral reefs with special reference to commensalism. *In* "Biology and Geology of Coral Reefs" (O. A. Jones and R. Endean, eds.), Vol. 3, pp. 37–94. Academic Press, New York.

Bruce, J. G., Keeling, J. L., and Beatty, W. H. (1985). On the north Brazilian eddy field. *Prog. Oceanogr.* **14,** 57–63.

Buchanan, J. B. (1957). Benthic fauna of the continental edge off Accra, Ghana. *Nature (London)* **179,** 634–635.

Buchanan, J. B. (1958). The bottom fauna across the continental shelf off Accra, Gold Coast. *Proc. Zool. Soc. London* **130,** 1–56.

Buchanan, J. B., Sheader, M., and Kingston, P. R. (1978). Sources of variability in the benthic macrofauna off the Northumberland coast. *J. Mar. Biol. Assoc. U.K.* **58,** 191–210.

Burggren, W. W., and Cameron, J. N. (1980). Anaerobic metabolism, gas exchange and acid-based balance during hypoxic exposure in the channel fish *Ictalurus punctatus. J. Exp. Zool.* **213,** 405–416.

Burokowsky, R. N. (1970). Origin of the amphiatlantic range of the shrimps *Penaeus duorarum. Okeanologiya (Moscow)* **6,** 1086–1089.

Cabanban, A. S. (1984). Some aspects of the biology of *Pterocaesio pisang* in the Central Visayas. M.Sc. Thesis, pp. 1–69. College of Science, University of the Philippines, Quezon City.

Caddy, J. F. (1983). The cephalopods: Factors relevant to their population dynamics and to the assessment and management of stocks. *FAO Fish. Tech. Pap.* **231,** 1–452.

Cadenat, J. A. (1954). Notes d'Ichthyologie ouest-Africaine. 7. Biologie. Régime alimentaire. *Bull. Inst. Fr. Afr. Noire, Ser. A* **16,** 564–583.

Carey, F. G., and Robison, B. H. (1981). Daily patterns in the activities of swordfish, observed by acoustic telemetry. *Fish. Bull.* **79**, 281–292.

Carpenter, E. J., and McCarthy, J. J. (1975). Nitrogen fixation and uptake of combined nitrogen nutrients by *Oscillatoria spp.* in the western Sargasso Sea. *Limnol. Oceanogr.* **20**, 389–401.

Carpenter, E. J., and McCarthy, J. J. (1978). Benthic nutrient regeneration and high rate of primary production in continental shelf waters. *Nature (London)* **274**, 188–189.

Cassie, R. M. (1954). Some uses of probability paper in the analysis of size-frequency distributions. *Aust. J. Mar. Freshwater Res.* **5**, 513–522.

Cavarivière, A. (1982). Les balistes des côtes africaines (*Balistes carolinensis*). Biologie, prolifération et possibilités d'exploitation. *Oceanol. Acta* **5**, 453–460.

Cervignon, F. (1965). Exploratory fishing off the Orinoco delta. *Proc. Annu. Gulf Caribb. Fish. Inst.* **17**, 20–23.

Chacko, P. I., Thomas, S. D., and Pillay, T. V. R. (1967). Scombroid fishes of Madras State, India. *Mar. Biol. Assoc. India, Symp. Ser.* **1**, 1006–1008.

Chalker, B. E., Dunlap, W. C., and Oliver, J. K. (1983). Bathymetric adaptations of reef-building corals at Jarvis Reef, Great Barrier Reef. II. Light saturation curves for photosynthesis and respiration. *J. Exp. Mar. Biol. Ecol.* **73**, 37–56.

Chang, J. H. (1974). The result of experimental trawl fishing in South China Sea. *Proc., Indo-Pac. Fish. Comm.* **13** (3), 585–596.

Chao, N. L., Viera, J. P., and Barbieri, L. R. R. (1987). Lagoa dos Patos as a nursery ground for shore fishes off Southern Brazil. *In* "Proceedings of the IREP Workshop on Recruitment in Coastal Demersal Communities, Campeche, Mexico" (A. Yañez-Arancibia and D. Pauly, eds.) IOC/UNESCO Workshop Series **44** (in press).

Chevey, N. (1934). Method of reading the scales from fish of the intertropical zone. *Proc. Pac. Sci. Congr., 5th, 1933,* pp. 3817–3829.

Chi, K. S., and Yang, R. T. (1973). Age and growth of skipjack tuna in the waters around the southern part of Taiwan. *Acta Oceanogr. Taiwan.* **3**, 199–222.

Chikuni, S. (1983). Cephalopod resources of the Indo-Pacific region. *FAO Fish. Tech. Pap.* **231**, 264–305.

Christensen, J. M. (1964). Burning of otoliths, a technique for age determination of soles and other fish. *J. Cons., Cons. Int. Explor. Mer* **29**, 73–81.

Clarke, M. R., and Lu, C. C. (1975). Vertical distribution of cephalopods at 18N 25W in the North Atlantic. *J. Mar. Biol. Assoc. U.K.* **55**, 165–182.

Clarke, T. (1973). Some aspects of the ecology of lanternfishes (Myctophidae) in the Pacific Ocean near Hawaii. *Fish. Bull.* **71**, 401–434.

Cloud, P. E. (1959). Geology of Saipan, Mariana Islands. Part 4. Submarine topography and shoal water ecology. *Gol. Surv. Prof. (U.S.)* **28-K**, 361–445.

Cohen, A. C. (1976). The systematics and distribution of *Loligo* in the western North Atlantic, with description of two new species. *Malacologia* **15**, 299–367.

Collette, B. B., and Rützler, K. (1978). Reef fishes over sponge bottoms off the mouth of the Amazon River. *Proc. Int. Coral Reef Symp., 3rd, 1977,* Vol. 1, pp. 306–310.

Committo, J. A., and Ambrose, W. G. (1985). Multiple trophic levels in soft-bottom communities. *Mar. Ecol.: Prog. Ser.* **26**, 289–293.

Connell, J. H. (1978). Diversity in tropical rain forests and coral reefs. *Science* **199**, 1302–1310.

Connor, E. F., and McCoy, E. D. (1979). The statistics and biology of the species-area relationship. *Am. Nat.* **113**, 791–833.

Conover, R. J. (1975). Production in marine plankton communities. *In* "Unifying Concepts in Ecology'' (W. H. Van Dobben and R. H. Lowe-McConnell, eds.), pp. 159–164. Junk, the Hague.

Cowen, R. K. (1985). Large scale pattern of the recruitment by the labrid, *Semicossyphus pulcher*: Causes and implications. *J. Mar. Res.* **14**, 719–742.

Cromwell, T., Montgomery, R. B., and Stroup, E. D. (1954). Equatorial undercurrent in Pacific Ocean revealed by new methods. *Science* **119**, 648–649.

Crosnier, A. (1963). Fonds de pêche le long de la République fédérale du Cameroun. Mimeo, *ORSTOM*, pp. 1–66.

Crosnier, A., and Berrit, G. R. (1966). Fonds de pêche le long de la côte des Répliques du Dahomey et du Togo. Mimeo, *ORSTOM*, pp. 1–94.

Cruiz, R., Coyula, R., and Ramirez, A. T. (1981). Crecimiento y mortalidad de la langosta espinosa en la plataforma de Cuba. *Rev. Cubana Invest. Pesqui.* **6**, 89–119.

Csirke, J. (1980). Recruitment in the Peruvian anchovy and its dependence on the adult population. *Rapp. P.-V. Réun., Cons. Int. Explor. Mer* **33**, 340–362.

Cullen, J. J. (1982). The deep chlorophyll maximum: Comparing vertical profiles of chlorophyll *a*. *Can. J. Fish. Aquat. Sci.* **39**, 791–803.

Cushing, D. H. (1971a). A comparison of production in temperate seas and the upwelling areas. *Trans. R. Soc. S. Afr.* **40**, 17–33.

Cushing, D. H. (1971b). ''Survey of Resources in the Indian Ocean and Indonesian Area,'' IOFC/DEV/71/2, pp. 1–123. Indian Ocean Fishery Commission, Rome.

Cushing, D. H. (1981). ''Fisheries Biology: A Study in Population Dynamics.'' Univ. of Wisconsin Press, Madison.

Daan, N. (1973). A qualitative analysis of the food intake of North Sea cod. *Neth. J. Sea Res.* **6**, 279–517.

Daget, J., and Ecoutin, J. M. (1976). Models mathématiques de production applicable aux poissons tropicals subissant un arrêt annuel prolongé de croissance. *Cah. ORSTOM, Ser. Hydrobiol.* **10**, 59–69.

Dalzell, P. (1983). ''The Distribution and Production of Anchovies in Papua-New Guinea Waters,'' Res. Rep. 85-03, p. 24. Fish. Res. Surv. Branch, Dep. Primary Ind., Port Moresby.

Dalzell, P. (1984a). ''The Influence of Rainfall on Catches of Stolephorid Anchovies in Papua-New Guinea Waters,'' Res. Rep. 84-04, pp. 1–18. Fish. Res. Surv. Branch, Dep. Primary Ind., Port Moresby.

Dalzell, P. (1984b). ''The Population Biology and Management of Bait-fish in Papua-New Guinea Waters,'' Res. Rep. 84-05, p. 59. Fish. Res. Surv. Branch, Dep. Primary Ind., Port Moresby.

Dandonneau, Y., and Charpy, L. (1985). An empirical approach to the island mass effect in the south tropical Pacific. *Deep-Sea Res.* **32**, 707–721.

Danil'chenko, P. G. (1960). ''Bony Fishes on the Maikop Beds of the Caucasus,'' No. 78. Akad. Nauk SSSR Tr. Paleontol. Inst., Moscow.

Darwin, C. (1842). ''The Structure and Distribution of Coral Reefs'' (Univ. of California Press, Berkeley, 1962).

da Silva, A. J. (1986). River run-off and shrimp abundance in a tropical coastal ecosystem—the example of Sofala Bank, Mozambique. *In* ''The Role of Freshwater Outflow in Coastal Marine Ecosystems'' (S. Skreslet, ed.), pp. 124–141. Springer-Verlag, Berlin and New York.

da Sylva, J. (1984). Hydrology and fish distribution at the Sofala Bank (Mozambique). *Rev. Invest. Pesqu. (Maputo, Mozambique)* **12**, 5–36.

D'Aust, B. G. (1971). Physiological constraints on vertical migration by mesopelagic fishes.

In "Biological Sound Scattering in the Ocean" (G. B. Farquhar, ed.), pp. 86–96. Office of Naval Research, Washington, D.C.

Davis, W. M. (1982). The coral reef problem. *Am. Geogr. Soc., Spec. Publ.* **9**, 1–596.

Davis, W. S. (1957). Ova production of American shad in Atlantic coast rivers. *Res. Rep.— U.S. Fish. Wildl. Serv.* **49**, 1–5.

Dawes, C. J. (1981). "Marine Botany." Wiley, New York.

Day, F. (1865). "Fishes of Malabar," pp. 1–231. B. Quaritch, London.

Day, J. H., Blaber, S. J. M., and Wallace, J. H. (1981). Estuarine fishes. *In* "Estuarine Ecology, with Special Reference to South Africa" (J. H. Day, ed.), pp. 197–221. A. A. Balkema, Rotterdam.

Deegan, L. A., and Day, J. W. (1984). Estuarine fishery habitat requirements. *In* "Proceedings of a Conference on Research for Managing the Nations Estuaries," pp. 315–335. UNC Sea Grant Programme, North Carolina State University, Raleigh.

Deegan, L. A., Day, J. W., Gosselink, J. G., Yáñez-Arancibia, A., Chavez, G. S., and Sanchez-Gil, P. (1986). Relationship among physical characteristics, vegetation distribution and fisheries yield in Gulf of Mexico estuaries. *In* "Estuarine Variability," pp. 83–100. Academic Press, New York.

de Groot, S. J. (1971). On the interrelationships between morphology of the alimentary tract, food and feeding behaviour in flatfishes. *Neth. J. Sea Res.* **5**, 121–196.

de Groot, S. J. (1973). Gaps in the studies on behaviour of Indian Ocean flatfishes belonging to the Psettodidae and Cynoglossidae. *J. Mar. Biol. Assoc. India* **15**, 251–261.

D'Elia, C., Webb, K., and Porter, J. (1981). Nitrate rich ground water inputs to Discovery Bay, Jamaica: A significant source of nitrogen to local reefs? *Bull. Mar. Sci.* **31**, 903–910.

Dessier, A. (1985). Dynamique et production d'*Eucalanus pileatus* à Pointe-Noire, Congo. *Océanogr. Trop.* **20**, 3–18.

de Sylva, D. P. (1963). Systematics and life history of the Great Barracuda, *Sphyraena barracuda* (Walbaum). *Stud. Trop. Océanogr.* **1**, 1–179.

Dhebtaron, Y., and Chotiyaputta, K. (1974). Review of the mackerel fishery in the Gulf of Thailand. *Proc., Indo-Pac. Fish Counc.* **15**(3), 265–286.

Diamond, J. (1985). Filter-feeding on a grand scale. *Nature (London)* **316**, 679–680.

Dickson, R. R. (1983). Global summaries and intercomparisons: Flow statistics from long-term current meter moorings. *In* "Eddies in Marine Science" (A. R. Robinson, ed.), pp. 278–353. Springer-Verlag, Berlin and New York.

Dietrich, G., Kalle, K., Krauss, W., and Siedler, G. (1980). "General Oceanography, an Introduction." Wiley, New York.

Doty, M. S. (1973). Farming the red seaweed, *Euchema,* for carrageen. *Micronesica* **9**, 59–73.

Doty, M. S., and Oguri, M. (1956). The island mass effect. *J. Cons., Cons. Int. Explor. Mer* **22**, 33–37.

Druzhinin, A. D., and Hlaing, U. P. (1974). Observations on the trawl fishery of southern Burma. *Proc., Indo-Pac. Fish. Counc.* **13**, 151–209.

Dugdale, R. C., and Goering, J. J. (1967). Uptake of new and regenerated forms of nitrogen in primary productivity. *Limnol. Oceanogr.* **12**, 196–206.

Durbin, E. G., and Durbin, A. G. (1975). Grazing rates of the Atlantic menhaden *Brevoortia tyrannus* as a function of particle size and concentration. *Mar. Biol. (Berlin)* **33**, 265–277.

Durbin, E. G., and Durbin, A. G. (1980). Standing stock and estimated production rates of phytoplankton and zooplankton in Narragansett Bay, R.I. *Estuaries* **4**, 24–41.

Dwiponggo, A. (1972). The fishery for the preliminary study on the growth rate of "lemuru" (oil sardine) at Muntjar, Bali Strait. *Proc., Indo.-Pac. Fish. Counc.* **15**, 221–240.

Ebert, T. A. (1975). Growth and mortality of post-larval echinoids. *Am. Zool.* **15**, 755–775.

Ebert, T. A. (1983). Recruitment in Echinoderms. *In* "Echinoderm Studies" (M. Jangoux and J. M. Lawrence, eds.), pp. 169–203. A. A. Balkema, Rotterdam.

Eckert, G. J. (1984). Annual and spatial variation in recruitment of labroid fishes among reefs, Great Barrier Reef. *Mar. Biol. (Berlin)* **78**, 123–127.

Eggleston, D. (1972). Patterns of biology in the Nemipteridae *J. Mar. Biol. Assoc. India* **14**, 357–364.

Ehrhardt, N. M., Jacquemin, P. S., Garcia, F., Conzalez, G., Lopez, J. M., Ortiz, B., and Solis, A. (1983). On the fishery and biology of the giant squid *Dosidicus gigas* in the Gulf of California, Mexico. *FAO Fish. Tech. Pap.* **231**, 306–340.

Ekman, S. (1953). "Zoogeography of the Sea." Sidjwick & Jackson, London.

Elliott, J. M., and Persson, L. (1978). The estimation of daily rates of food consumption for fish. *J. Anim. Ecol.* **47**, 977–991.

Elton, C. (1927). "Animal Ecology." Macmillan, New York.

Emery, K. O. (1967). Estuaries and lagoons in relation to continental shelves. In "Estuaries" (G. H. Lauff, ed.), Publ. No. 83, pp. 9–11. Am. Assoc. Adv. Sci., Washington, D.C.

Emery, K. O., Uchupi, E., Sunderland, J., Uktolseja, H. L., and Young, E. M. (1972). Geological structure and some water characteristics of the Java Sea and adjacent continental shelf. UN ECAFE/CCOP *Tech. Bull.* **6**, 197–223.

Eppley, R. W. (1981). Autotrophic production of particulate matter. *In* "Analysis of Marine Ecosystems" (A. R. Longhurst, ed.), pp. 343–362. Academic Press, New York.

Evans, G., Murray, J. W., Biggs, H. E., Bate, R., and Bush, P. R. (1973). The oceanography, ecology, sedimentology, and geomorphology of parts of the Trucial Coast barrier island complex, Persian Gulf. *In* "The Persian Gulf" (B. H. Purser, ed.), pp. 233–278. Springer-Verlag, Berlin and New York.

Everitt, B. S., and Hand, D. J. (1981). "Finite Mixture Distributions." Chapman & Hall, London.

Fagade, S. O., and Olaniyan, C. I. O. (1972). The biology of the West African shad *Ethmalosa fimbriata* in the Lagos lagoon, Nigeria. *J. Fish Biol.* **4**, 519–533.

Fager, E. W., and Longhurst, A. R. (1968). Recurrent group analysis of species assemblages of demersal fish in the Gulf of Guinea. *J. Fish. Res. Board Can.* **25**, 1405–1421.

Fasham, M. J. R., ed. (1984). "Flows of Energy and Materials in Marine Ecosystems." Plenum, New York.

Fernando, E. F. W. (1974). Species composition of fish captured by trawlers in the Wadge Bank. *Proc., Ind-Pac. Fish. Counc.* **13** (3), 521–531.

Fischer, W., and Whitehead, P. J. P., eds. (1974). "FAO Species Identification Sheets for Fishery Purposes. Eastern Indian Ocean, and Western Central Pacific" (Note: similar sets exist also under similar titles for the Western Indian Ocean and the tropical Atlantic, East and West), 4 vols. FAO, Rome.

Flint, R. W., and Rabelais, N. N. (1981). Gulf of Mexico shrimp production: A food web hypothesis. *Fish. Bull.* **79**, 737–748.

Fontana, A., and LeGuen, J. C. (1969). Étude de la maturité sexuelle et de la fecondité de *Pseudotolithus (fonticulus) elongatus*. *Cah. ORSTOM, Ser. Oceanogr.* **7**, 9–19.

Food and Agricultural Organization (1982). Stock assessment of the Kenyan demersal offshore resources. *Offshore Trawling Surv., Work Rep* **8**, 1–58 (KEN/74/023).

Forbes, A. M. G. (1984). The contributions of local processes to seasonal hydrography of the Gulf of Carpentaria. *Océanogr. Trop.* **19**, 193–201.

Fricke, H. W. (1975). Lösen einfacher Probleme bei einem Fisch (Freiwasserversuch an *Balistes fuscus*). *Z. Tierpsychol.* **38**, 18–33.

Fricke, H. W., and Fricke, S. (1977). Monogamy and sex change by aggressive dominance in coral reef fishes. *Nature (London)* **66**, 830–832.

Gallardo, V. A. (1975). On a benthic sulphide system on the continental shelf of north and central Chile. *Proc. Int. Semin. Upwelling, Coquimbo, Chile, Nov. 1975.*

Gallardo, V. A., Castillo, J. G., Retamal, M. A., Yañez, A., Moyano, H. I., and Hermosa, J. G. (1977). Quantitative studies on the soft-bottom macrobenthic animal communities of shallow antarctic bays. *In* "Adaptation within Antarctic Ecosystems. Proceedings of the Third SCAR Symposium on Antarctic Biology" (G. A. Llano, ed.), pp. 361–387. Smithsonian Institution, Washington, D.C.

Garcia, D., and LeReste, L. (1981). Life cycles, dynamics, exploitation and management of coastal penaeid shrimp stocks. *FAO Fish. Tech. Pap.* **203**, 1–215.

Garcia, S. (1983). The stock-recruitment relationship in shrimps: reality or artefacts and misinterpretations? *Océanogr. Trop.* **18**, 25–48.

Garcia, S. (1985). Reproduction, stock assessment models and population parameters in exploited penaeid shrimp populations. *In* "Proceedings of the Second Australian National Prawn Seminar" (P. C. Rothlisberg, D. J. Staples, and P. J. Crocos, eds.), pp 139–158. NPS2, Cleveland, Australia.

Garrett, C., and Munk, W. H. (1979). Internal waves in the ocean. *Annu. Rev. Fluid Mech.* **11**, 339–369.

Garzoli, S. L., and Katz, E. J. (1983). The forced annual reversal of the Atlantic North Equatorial Counter Current. *J. Phys. Oceanogr.* **13**, 2082.

Gerber, R. P., and Gerber, M. B. (1979). Ingestion of natural particulate organic matter and subsequent assimilation by tropical lagoon zooplankton. *Mar. Biol.* **52**, 33–43.

George, K., and Banerji, S. K. (1964). Age and growth studies on the Indian mackerel *Rastrelliger kanagurta* (Cuvier) with special reference to length-frequency data collected at Cochin. *Indian J. Fish.* **11**, 621–638.

George, P. C. (1964). Food and feeding habits of *Rastrelliger kanagurta*. *Mar. Biol. Assoc. India, Symp. Ser.* **1**, 569–573.

Gill, A. E., and Clarke, A. J. (1974). Wind-induced upwelling, coastal currents, and sea level changes. *Deep-Sea Res.* **21**, 325–345.

Gjøsaeter, J., and Kawaguchi, K. (1980). A review of the world resources of mesopelagic fish. *FAO Fish. Tech. Pap.* **193**, 1–151.

Gjøsaeter, J. G., Dayaratne, P., Bergstad, O. A., Gjøsaeter, H., Sousa, M. I., and Beck, I. M. (1984). Aging tropical fish by growth rings in their otoliths. *FAO Fish. Circ.* **776**, 1–54.

Glynn, P. W. (1985). El Niño-associated disturbance to coral reefs and post-disturbance mortality by *Acanthaster planci*. *Mar. Ecol.: Prog. Ser.* **26**, 295–300.

Golikov, A. N., and Scarlato, O. A. (1973). Comparative characteristics of some ecosystems of the upper regions of the shelf in tropical, temperate and arctic waters. *Helgol. Wiss. Meeresunters* **24**, 219–234.

Gomez, E. D., Gueib, R. A., and Aro, E. (1983). Studies on the predators of commercially important seaweeds. *Fish. Res. J. Philipp.* **8**, 1–17.

Goodwin, C. D., and Nacht, M. (1986). Decline and renewal: Causes and cures of the decay among foreign-trained intellectuals and professionals in the Third World. *Inst. Int. Educ. Rep.* **9**, 1–75.

Gordon, A. L. (1969). Circulation of the Caribbean Sea. *J. Geophys. Res.* **42**, 6207–6223.

Gordon, I. D. M. (1977). The fish populations of the inshore waters of the west of Scotland. The biology of the Norway pout (*Trisopterus esmarkii*). *J. Fish Biol.* **10**, 417–430.

Gorshkov, S. G., ed. (1978). "World Ocean Atlas: Atlantic and Indian Oceans." Pergamon, Oxford.

Grandperrin, R., and Rivaton, J. (1966). Individualisation de plusieurs ichthyofaunes le long de l'équateur. *Cah. ORSTOM, Sér. Océanogr.* **5**, 69–77.

Grant, W. E., Isaksen, K. G., and Griffin, W. L. (1981). A general bioeconomic model for annual-crop marine fisheries. *Ecol. Model.* **13**, 195–219.

Grassle, J. F. (1973). Variety in coral reef communities. *In* "Biology and Geology of Coral Reefs" (O. A. Jones and R. Endean, eds.), Vol. 2, pp. 247–270. Academic Press, New York.

Gray, I. E. (1954). Comparative study of the gill area of marine fishes. *Biol. Bull. (Woods Hole, Mass.)* **107**, 219–225.

Gray, J. S. (1985). Nitrogenous excretion by meiofauna from coral reef sediments. *Mar. Biol. (Berlin)* **80**, 31–35.

Green, C. F., and Birtwhistle, W. (1927). "Report on the Steam Trawler Tongkol for the Period 28 May–31 Dec, 1926." Govt. Printing Office, Singapore.

Greze, V. N. (1978). Production in animal populations. *In* "Marine Ecology" (O. Kinne, ed.), Vol. 4, pp. 89–114. Wiley (Interscience) London.

Grice, G. D., and Hulseman, K. (1965). Abundance, vertical distribution and taxonomy of calanoid copepods at selected stations in the north central Atlantic. *J. Zool.* **146**, 213–262.

Griffiths, R. C., and Simpson, J. G. (1967). Temperature structure of the Gulf of Carioca, Venezuela, from August 1959–August, 1961. *Ser. Recurs. Explot. Pesq. Venez.* **1**, 117–169.

Grigg, R. W., Polovina, J. J., and Atkinson, M. J. (1984). Model of a coral reef ecosystem. III. Resource limitation, fisheries yield, and resource management. *Coral Reefs* **3**, 23–27.

Gulland, J. A., ed. (1971). "The Fish Resources of the Ocean." Fishing New Books, West Byfleet, England.

Gulland, J. A. (1982). The management of tropical multi-species fisheries. *In* "Theory and Management of Tropical Fisheries" (D. Pauly and G. I. Murphy, eds.), pp. 287–298. ICLARM, Manila.

Gulland, J. A. (1983). "Fish Stock Assessment: A Manual of Basic Methods." Wiley, New York.

Gulland, J. A., and Holt, S. J. (1959). Estimation of growth parameters for data at unequal time intervals. *J. Cons., Cons. Int. Explor. Mer* **25**, 47–49.

Gunter, G. (1967). Some relationships of estuaries to the fisheries of the Gulf of Mexico. *In* "Estuaries" (G. H. Lauff, ed.), Publ. No. 83, pp. 621–638. Am. Assoc. Adv. Sci., Washington, D.C.

Hamner, W. M., and Carleton, J. H. (1979). Copepod swarms: Attributes and role in coral reef ecosystems. *Limnol. Oceanogr.* **24**, 1–14.

Hammond, L. S., and Wilkinson, C. R. (1985). Exploitation of sponge exudates by coral reef holothuroids. *J. Exp. Mar. Biol. Ecol.* **94**, 1–9.

Hampton, J., and Majkowski, J. (1987). An examination of the ELEFAN computer programmes for length-based stock assessment. *In* "Length-based Methods in Fisheries Research" (D. Pauly and G. R. Morgan, eds.), ICLARM Conf. 14 (in press).

Hanlon, R. T., Nixon, R. F., and Hulet, W. H. (1983). Survival, growth and behaviour of the loliginid squids, *Loligo plei, L. peali,* and *Lolliguncula brevis* in closed sea-water systems. *Biol. Bull. (Woods Hole, Mass.)* **165**, 637–685.

Harden-Jones, F. R. (1968). "Fish Migration." Arnold, London.

Harding, J. P. (1949). The use of probability paper for the graphical analysis of polymodal frequency distributions. *J. Mar. Biol. Assoc. U.K.* **28**, 141–153.

Hargraves, P. E., and Viquez, R. (1985). Spatial and temporal distribution of phytoplankton in the Gulf of Nicoya. *Bull. Mar. Sci.* **37**, 577–585.

Harkantra, S. N., Rodrigues, C. L., and Parulekar, A. H. (1982). Macrobenthos of the shelf of north-eastern Bay of Bengal. *Indian J. Mar. Sci.* **11**, 115–121.

Harmelin-Vivien, M. L. (1981). Trophic relationships of reef fishes in Tuléar (Madagascar). *Oceanol. Acta* **4**, 365–374.

Hartman, W. D., and Goreau, T. F. (1970). Jamaican coralline sponges: Their morphology, ecology and fossil relatives. *Symp. Zool. Soc. London* (25), 205–243.

Hatcher, B. G. (1983). Grazing in reef ecosystems. *In* "Perspectives on Coral Reefs" (D. J. Barnes, ed.), pp. 164–179. Aust. Inst. Mar. Sci., Townsville, Qd.

Hay, M. E. (1984a). Pattern of fish and urchin grazing on Caribbean coral reefs: Are previous results typical? *Ecology* **65**, 446–454.

Hay, M. E. (1984b). Predictable special escape from herbivory: How do these affect the evolution of herbivore resistance in tropical marine communities? *Oecologia* **64**, 396–407.

Hedgepeth, J. W. (1969). An intertidal reconnaissance of rocky shores of the Galapagos. *Wasmann J. Biol.* **27**, 1–24.

Heincke, F. (1913). Investigation on the plaice. General Report I. Plaice fishery and protective regulations. Part I. *Rapp. P.-V. Réun., Cons. Int. Explor. Mer* **17A**, 1–153.

Heinrich, A. K. (1970). "Production Cycles in the Marine Pelagic Zone. Principles of Biological Productivity in the Ocean and its Utilisation." Nauka Press, Moscow.

Heinrich, A. K. (1977). Communities of the tropical areas of the ocean. *In* "Oceanology: Biology of the Ocean" (M. E. Vinogradov, ed.), Vol. 2, pp. 91–103. Nauka Press, Moscow.

Herbland, A. (1978). Heterotrophic activity in the Mauretanian upwelling. *In* "Upwelling Ecosystems" (R. Boje and M. Tomczak, eds.), pp. 156–166. Springer-Verlag, Berlin and New York.

Herbland, A., and Voituriez, B. (1979). Hydrological structure analysis for estimating the primary production in the Atlantic Ocean. *J. Mar. Res.* **37**, 87–101.

Herman, A., and Platt, T. (1983). Numerical modelling of diel carbon production and zooplankton and zooplankton grazing on Scotian Shelf, based on observational data. *Ecol. Modell.* **18**, 55–72.

Hernando, A. M., and Flores, E. E. C. (1981). The Philippines squid fishery: A review. *Mar. Fish. Rev.* **43**, 13–20.

Heron, A. C. (1972). Population ecology of colonising species: The pelagic tunicate *Thalia democratica*. 2. Population growth rate. *Oecologia* **10**, 294–312.

Hiatt, R. W., and Strasburg, D. W. (1960). Ecological relationships of the fish fauna on coral reefs of the Marshall Islands. *Ecol. Monogr.* **30**, 65–127.

Hickling, C. F. (1930). The natural history of the hake. Part 3. Seasonal changes in the condition of the hake. *Fish. Invest. Ser.* **12**, 1–78.

Hickling, C. F. (1931). The structure of the otolith of the hake. *Q. J. Microsc. Sci.* (N.S.) **74**, 547–561.

Hida, T., and Pereyra, W. (1966). Results of bottom trawling in Indian seas by RV "Anton Bruun" in 1963. *Proc. Indo.-Pac. Fish Counc.* **11**, 156–171.

Hirota, J., Ferguson, R., Finn, J. A., Shuman, R. F., and Taguchi, S. (1984). Primary productivity, the cycling of nitrogen and spatiotemporal variability in components of the epipelagic ecosystem in Hawaiian waters. *In* "Second Symposium on Resource

Evaluation of the Northwest Hawaiian Islands'' (W. Grigg and K. Y. Tan, eds.), Sea Grant MR-8401; pp. 344-374. University of Hawaii, Honolulu.

Hobson, E. S. (1974). Feeding relationships of teleostean fishes on coral reefs in Kona, Hawaii. *Fish. Bull.* **72**, 915-1031.

Hobson, E. S., and Chess, J. R. (1978). Trophic relations among fishes and plankton in the lagoon of Eniwetok atoll. *Fish. Bull.* **76**, 133-153.

Hochachka, P. W., Hulbert, W. C., and Guppy, M. (1978). The tuna power plant and furnace. *In* ''The Physiological Ecology of Tunas'' (G. D. Sharp and A. E. Dizon, eds.), pp. 153-174. Academic Press, New York.

Hogg, N. G., Katz, E. J., and Sanford, T. D. (1978). Eddies, islands and mixing. *JGR, J. Geophys. Res.* **83**, 2921-2938.

Holligan, P. M. (1981). Biological implications of fronts on the northwestern European continental shelf. *Philos. Trans. R. Soc. London, Ser.* **A302**, 547-562.

Holm, R. F. (1978). The community structure of a tropical marine lagoon. *Estuarine, Coastal Shelf Sci.* **7**, 329-345.

Holthuis, L. B. (1980). Shrimps and prawns of the World—an annotated catalogue of species of interest to fisheries. *FAO Fish. Synopsis* **124**, 1-271.

Hopkins, T. L., and Baird, R. C. (1977). Aspects of the feeding ecology of mid-water fishes. *In* ''Oceanic Sound Scattering Prediction'' (N. R. Anderson and B. J. Zahuranec, eds.), pp. 325-360. Plenum, New York.

Hopkins, T. L., and Baird, R. C. (1985). Aspects of the trophic ecology of the mesopelagic fish *Lampanyctus alatus* in the Gulf of Mexico. *Biol. Oceanogr.* **3**, 285-312.

Houghton, R. W., and Beer, T. (1984). Wave propagation during the Ghana upwelling. *J. Geophys. Res., C: Oceans* **89**, 4423-4429.

Howard, G. V., and Landa, A. (1958). A study of the age, growth, sexual maturity and spawning of the anchoveta in the Gulf of Panama. *Bull. Int.-Am. Trop. Tuna Comm.* **2**, 389-437.

Hunter, J. R. (1981). The feeding ecology and predation of marine fish larvae. *In* ''Marine Fish Larvae'' (R. Lasker, ed.), pp. 33-71. U. Wash. Press.

Hunter, J. R. (1984). Tidal and stratification/mixing model of Kuwait waters. *Kuwait Bull. Mar. Sci.* (5), 11-35.

Hunter, J. R., and Sharp, G. D. (1983). Physics and fish populations: Shelf sea fronts and fisheries. *FAO Fish. Rep.* **291**, 659-682.

Hunter, J. R., Argue, A. W., Bayliff, W. H., Dizon, A. E., Fontenau, A., Goodman, D., and Seckel, G. R. (1987). The dynamics of tuna movements: An evaluation of past and present research. *FAO Fish. Tech. Pap.* **277**, 1-78.

Hussain, A. G., Burney, M. A., and Mohiuddin, S. Q. (1974). Analysis of demersal catches taken from the exploratory fishing off the coast of West Pakistan. *Proc., Indo-Pac. Fish. Counc.* **13** (3), 61-84.

Huston, M. (1979). A general hypothesis of species diversity. *Am. Nat.* **113**, 81-101.

Ikeda, T. (1985). Metabolic rates of epipelagic marine zooplankton as a function of body mass and temperature. *Mar. Biol. (Berlin)* **85**, 1-11.

Iles, T. D. (1974). The tactics and strategy of growth in fishes. *In* ''Sea Fisheries Research'' (F. R. Harden-Jones, ed.), pp. 331-345. Elek Science, London.

Ingles, J., and Pauly, D. (1981). Aspects of the growth and mortality of exploited coral reef fishes. *Proc. Int. Coral Reef Symp., 4th, 1981,* Vol. 1, pp. 89-98.

Ingles, J., and Pauly, D. (1984). An atlas of the growth, mortality and recruitment of Philippine fishes. *ICLARM Tech. Rep.* **13**, 1-127.

International Development & Research Council (IDRC/FAO) (1982). ''Fish by-Catch— Bonus from the Sea,'' Can. Int. Dev. & Res. Counc., Ottawa.

Isarankura, A. (1978). Assessment of stocks of demersal fishes off the west coast of Thailand and Malaysia. FAO & Indo Pacific Fishery Commission/DEV/71, **20**, 1–20.

Isdale, P. (1984). Fluorescent bands in massive corals record centuries of coastal rainfall. *Nature (London)* **310**, 578–579.

Iturriaga, R., and Mitchell, B. G. (1986). Chroococcoid cynobacteria: A significant component in the food web dynamics of the open ocean. *Mar. Ecol.: Prog. Ser.* **28**, 291–297.

Jangoux, M., and Lawrence, J. M., eds. (1983). "Echinoderm Studies." A. A. Balkema, Rotterdam.

Jenkins, W. J. (1982). Oxygen utilisation rates in the North Atlantic sub-tropical gyre and primary production in oligotrophic systems. *Nature (London)* **300**, 246–248.

Jenkins, W. J., and Goldman, J. C. (1985). Seasonal oxygen cycles in the Sargasso Sea. *J. Mar. Res.* **43**, 465–491.

Johannes, R. E. (1978). Reproductive strategies of coastal marine fishes in the tropics. *Environ. Biol. Fishes* **3**, 65–84.

Johannes, R. E. (1981). "Words of the Lagoon: Fishing and Marine Lore in the Palau District of Micronesia." Univ. of California Press, Berkeley.

Johannes, R. E., and Tepley, L. (1974). Examination of the reef coral *Porites lobata* in situ, using time-lapse photography. *Proc. Coral Reef Symp., 2nd, 1973,* Vol. 1, pp. 127–131.

Johns, B., Dube, S. K., Mohanty, V. C., and Sinha, P. C. (1981). Numerical simulation of the surge generated by the 1977 Andrha cyclone. *Q. J. Meteorol. Soc.* **107**, 919–934.

Johnson, P. W., Xu, H.-S., and Sieburth, J. M. (1983). Utilisation of chroococcoid Cyanobacteria by marine protozooplankton but not by calanoid copepods. *Ann. Inst. Oceanogr. (Paris)* **58**, 297–308.

Joiris, C., Lancelot, C., Daro, M., Mommaerts, J. P., Bertels, A., Bossicart, M., Nijs, J., and Hecq, J. H. (1982). A budget of carbon cycling in the Belgian coastal zone: relative roles of zooplankton, bacteroplankton, and benthos in the utilisation of primary production. *Neth. J. Sea Res.* **16**, 260–275.

Jones, C. (1986). Determining age of larval fish with the otolith increment technique. *Fish. Bull.* **84**, 91–103.

Jones, N. S. (1950). Marine bottom communities. *Biol. Rev. Cambridge Philos. Soc.* **25**, 283–313.

Jones, R. (1976). Growth of fishes. *In* "The Ecology of the Seas" (D. H. Cushing and J. J. Walsh, eds.), pp. 251–279. Blackwell, Oxford.

Jones, R. (1982). Ecosystems, food chains and fish yields. *In* "Theory and Management of Tropical Fisheries" (D. Pauly and G. I. Murphy, eds.), pp. 195–239. ICLARM, Manila.

Jordán, R. S. (1980). Biology of the anchoveta. I. Summary of the present knowledge. *In* "Proceedings of the Workshop on the Phenomenon known as El Niño," pp. 1–284. UNESCO, Paris.

Joseph, J., and Calkins, T. (1969). Population dynamics of the skipjack tuna (*Katsuwonus pelamis*) in the eastern Pacific Ocean. *Bull. Int. Am. Trop. Tuna Comm.* **13**, 1–273.

Joseph, P. S. (1985). Linkage of mudbank and upwelling off the south-west coast of India. *Oceanol. Acta* (in press).

Josse, E., LeGuen, J. C., Kearney, R. E., Lewis, A. D., Smith, A., Marec, L., and Tomlinson, P. K. (1979). Growth of skipjack. *Occas. Pap.—South Pac. Comm.* **11**, 1–83.

Juanico, M. (1983). Squid maturity scales for population analysis. *FAO Fish. Tech. Pap.* **231**, 341–378.

Kamimura, T., and Honma, M. (1963). Distribution of the yellowfin tuna in the tuna longline fishing grounds of the Pacific Ocean. *Rep. Nankai Reg. Fish. Res. lab.* **17**, 31–53.

Kaufman, L. (1978). The three spot dameslfish: Effects on benthic biota of Caribbean coral reefs. *Proc. Int. Coral Reef Symp., 3rd, 1977,* Vol. 1, pp. 559–564.

Kearney, R. E. (1974). The research methods employed in the study of the Papua-New Guinea skipjack fishery. *Papua New Guinea Agric. J.* **25,** 31–37.

Keenleyside, M. H. A. (1979). "Diversity and Adaptation in Fish Behaviour." Springer-Verlag, Berlin and New York.

Kerr, R. A. (1985). Small eddies are mixing the ocean. *Science* **230,** 793.

Kesteven, G. L. (1971). "Report on the Fisheries of Pakistan," FAO files, MS rep., pp. 1–9. FAO, Rome.

Kesteven, G. L., Nakken, O., and Strømme, T. (1981). "The Small Pelagic and Demersal Fish Resources of the NW Arabian Sea," pp. 1–56. Inst. Mar. Res., Bergen, Norway.

Ketchum, B. H. (1962). Regeration of nutrients by zooplankton. *Rapp. P.-V. Réun., Cons. Int. Explor. Mer* **153,** 142.

Kimmerer, W. J. (1983). Direct measurement of the production : biomass ratio of the sub-tropical calanoid copepod *Acrocalanus inermis. J. Plankton Res.* **5,** 1–14.

Kimor, B. (1973). Plankton relations of the Red Sea, Persian Gulf and Arabian Sea. *In* "Biology of the Indian Ocean" (B. Zeitschel, ed.), pp. 221-232. Springer-Verlag, Berlin and New York.

Kimor, B. (1981). The role of phagotrophic dynoflagellates in marine ecosystems. *Kiel. Meeresforsch., Sonderh.* **5,** 164–173.

King, M. G. (1986). The fishery resources of Pacific island countries. Part 1. Deep-water shrimp resources. *FAO Fish. Tech. Pap.* **272** (1), 1–45.

Kinsey, D. W. (1979). Carbon turnover and accumulation by coral reefs. Ph.D. Thesis, University of Hawaii, Honolulu.

Kinsey, D. W., and Domm, A. (1974). Effect of fertilisation on a coral reef environment—primary production studies. *Proc. Coral Reef Symp., 2nd, 1973,* Vol. 1, pp. 49–66.

Kinzer, K. (1977). Observations on the feeding habits of mesopelagic fish *Benthosema glaciale* off North-west Africa. *In* "Oceanic Sound Scattering Prediction" (N. R. Anderson and B. J. Zahuranec, eds.), pp. 381–392. Plenum, New York.

Kitchell, J. F., Neill, W. H., Dizon, A. E., and Magnuson, J. J. (1978). Bioenergetics spectra of skipjack and yellowfin. *In* "The Physiological Ecology of Tunas" (G. D. Sharp and A. E. Dizon, eds.), pp. 357–368. Academic Press, New York.

Kraljevic, M. (1984). On the experimental feeding of the sea-bream (*Sparus aurata*) under aquarium conditions. *Acta Adriat.* **25,** 183–204.

Krishnamurti, T. N. (1981). Cooling of the Arabian Sea and the onset-vortex during 1979. *In* "Recent Progress in Equatorial Oceanography. Final Report of SCOR W/G47" (J. P. McCreay, D. W. Moore, and J. M. Witte, eds.), pp. 1–12. UNESCO, Paris.

Kröger, A., and Remmert, H. (1984). Temperature and teleosts. *Oecologia* **61,** 426–427.

Krumbein, W. E. (1981). Biogeochemsitry and geomicrobiology of lagoons and lagoonary environments. *UNESCO Tech. Pap. Mar. Sci.* **33,** 97–110.

Krumbein, W. E., and Cohen, Y. (1977). Solar Lake, Sinai. 4. Stomatolithic cyanobacterial mats. *Limnol. Oceanogr.* **22,** 635–656.

Kühlmorgen-Hille, G. (1976). Preliminary study of the life history of the flatfish *Psettodes erumei. In* "Fisheries Resources Management in Southeast Asia" (K. Tiews, ed.), pp. 261–268. German Foundation for International Development, Berlin.

Kurian, C. V. (1971). Distribution of benthos on the south-west coast of India. *In* "Fertility of the Sea (J. D. Costlow, ed.), pp. 225–239. Gordon & Breach, New York.

Kutkuhn, J. H. (1966). The role of estuaries in the development and perpetuation of commercial shrimp resources. *Am. Fish. Soc., Spec. Publ.* **3,** 16–36.

Lambert, T., and Ware, D. (1984). Reproductive strategies of demersal and pelagic spawning fish. *Can. J. Fish. Aquat. Sci.* **41,** 1565–1569.

Lambshead, P. J. D., Plat, H. M., and Shaw, K. M. (1983). The detection of differences among assemblages of marine benthic species based on an assessment of dominance and diversity. *J. Nat. Hist.* **17**, 859–874.

Lange, A. M. T., and Sissenwine, M. P. (1983). Squid resources of the Northwest Atlantic. *FAO Fish. Tech. Pap.* **231**, 21–54.

Larkin, P. A., and Gazey, W. (1982). Applications of ecological simulations models to management of tropical multispecies fisheries. *In* "Theory and Management of Tropical Fisheries" (D. Pauly and G. I. Murphy, eds.), pp. 123–140. ICLARM, Manila.

Lasker, R. (1975). Field criteria for survival of anchovy larvae: The relation between the onshore chlorophyll layers and successful feeding. *Fish. Bull.* **71**, 453–462.

Lasker, R. (1978). The relation between oceanographic conditions and larval anchovy food in the California Current: Identification of factors contributing to recruitment failure. *Rapp. P.-V. Réun., Cons. Int. Explor. Mer* **173**, 212–230.

Lasker, R., ed. (1985). An egg production method for estimating spawning biomass of pelagic fish: Application to the northern anchovy, *Engraulis mordax. NOAA Tech. Rep., NMFS* **36**, 1–99.

Laurs, R. M., and Lynn, R. J. (1977). Seasonal migration of north Pacific albacore *Thunnus alalunga* in North American coastal waters. *Fish. Bull.* **75**, 795–822.

Lawrence, A. L. (1985). Marine shrimp culture in the western hemisphere. *In* "Second Australian Shrimp Seminar" (P. C. Rothlisberg, B. J. Hill, and D. J. Staples, eds.), pp. 327–336. NPS2, Cleveland, Australia.

Lawson, G. W. (1966). The littoral ecology of West Africa. *Oceanogr. Mar. Biol.* **4**, 405–448.

LeBorgne, R. (1978). Evaluation de la production secondaire planctonique en milieu océanique par la methode des rapports CNP. *Oceanol. Acta* **1**, 107–118.

LeBorgne, R. (1981). Relationships between the hydrological structure, chlorophyll and zooplankton biomasses in the Gulf of Guinea. *J. Plankton Res.* **3**, 577–592.

LeBorgne, R. (1982). Zooplankton production in the eastern tropical Atlantic Ocean: net growth efficiency and P:B in terms of carbon, nitrogen and phosphorus. *Limnol. Océanogr.* **27**, 681–698.

LeBorgne, R., and Dufour, P. (1979). Premiers résultats sur l'excrétion et la production du mesozooplankton de la lagune Ebrié (Côte d'Ivoire). *Doc. Sci. Cent. Pech. Océanogr., Abidjan* **10**, 1–39.

LeBorgne, R., Herbland, A., LeBouteiller, A., and Roger, C. (1983). Biomasse, excrétion et production du zooplancton-micronekton—relations avec le phytoplancton et les particules. *Océanogr. Trop.* **18**, 419–460.

LeBorgne, R., Dandonneau, Y., and Lemasson, L. (1985). The problem of the island mass effect on chlorophyll and zooplankton standing crops around Mare Island and New Caledonia. *Bull. Mar. Sci.* **37**, 450–459.

LeDanois, E. (1948). "Les profondeurs de la mer, trente ans de récherches sur la faune sous-marine au large des côtes de France." Payot, Paris.

Leetma, A., Quadfasel, D. R., and Wilson, D. (1981). Somali Current: Observations and theory. *In* "Recent Progress in Equatorial Oceanography. Final Report of SCOR W/ G47" (J. P. McCreay, D. W. Moore, and J. M. Witte, eds.), pp. 393–404. UNESCO, Paris.

Legand, M., and Rivaton, F. (1969). Cycles biologiques des poissons mesopelagiques: Action prédatrice des poissons micronekton. *Cah. ORSTOM, Ser. Océanogr.* **7**, 29–45.

Legand, M., Bourret, P., Fourmanoir, P., Grandperrin, R., Gueredrat, J. A., Michel, A., Rancurel, P., Repelin, R., and Roger, C. (1972). Relations trophiques et distributions verticales en milieu pélagiques dans l'océan Pacifique inter-tropicale. *Cah. ORSTOM, Ser. Océanogr.* **10,** 303–393.

Le Guen, J.-C. (1970). Dynamique des populations de *Pseudotolithus elongatus.* Thèse Doct. Sci. 3ème Cycle, Nat. Fac. Sci., Université de Paris (ORSTOM, Paris, 1970, pp. 1–94).

Le Guen, J.-C. (1976). Utilisation des otolithes pour la lecture de l'âge de Sciaenid's intertropicaux. Marques saisonières et journalières. *Cah. ORSTOM Ser. Océanogr.* **14,** 331–340.

LeLoeuff, P., and Intes, A. (1973). Note sur le regime alimentaire de quelques poissons demersaux de Côte d'Ivoire. *Doc. Sci. Cent. Rech. Océanogr., Abidjan* **4,** 17–44.

Lessios, H. A., Robertson, D. R., and Cubit, J. D. (1984a). Spread of Diadema mass mortality through the Caribbean. *Science* **226,** 335–337.

Lessios, H. A., Cubit, J. D., Robertson, D. R., Shulmen, M. K., Partker, M. R., Garrity, S. D., and Levins, S. C. (1984b). Mass mortality of *Diadema antillarum* on the Caribbean coast of Panama. *Coral Reefs* **3,** 173–182.

Lévêque, C., Durand, J.-R., and Ecoutin, J.-M. (1977). Relations entre le rapport P/B et la longevité des organismes. *Cah. ORSTOM Ser. Hydrobiol.* **11,** 17–31.

Lewis, J. B. (1977). Processes of organic production on coral reefs. *Biol. Rev. Cambridge Philos. Soc.* **52,** 305–347.

Lewis, J. B. (1981). Coral reef ecosystems. *In* "Analysis of Marine Ecosystems" (A. R. Longhurst, ed.), pp. 127–159. Academic Press, New York.

Li, K. M. (1960). Synopsis of the biology of *Sardinella* in the tropical eastern Indo-Pacific area. *FAO Species Synopsis* **5,** 175–212.

Li, W. K., Subba Rao, D. V., Harrison, W. G., Smith, J. C., Cullen, J. J., Irwin, B., and Platt, T. (1983). Autotrophic picoplankton in the tropical ocean. *Science* **219,** 292–295.

Lim, L. C. (1975). Record of an offshore upwelling in the southern South China Sea, Singapore. *Singapore J. Primary Ind.* **3,** 53–61.

Lindroth, A. (1950). Die Assoziationen der marinen Weichböden. *Zool. Bidr. Uppsala* **15,** 331.

Longhurst, A. R. (1957a). "An Ecological Survey of the West African Marine Benthos," No. 11, pp. 1–102. Colonial Office, Fishery Publ., London.

Longhurst, A. R. (1957b). The food of the demersal fish of a West African estuary. *J. Anim. Ecol.* **26,** 369–387.

Longhurst, A. R. (1959). Benthos densities off tropical West Africa. *J. Cons., Cons. Int. Explor. Mer* **25,** 21–28.

Longhurst, A. R. (1960a). A summary survey of the food of West African demersal fish. *Bull. Inst. Fr. Afr. Noire, Ser. A* **22,** 267–282.

Longhurst, A. R. (1960b). Local movements of *Ethmalosa fimbriata* off Sierra Leone from tagging data. *Bull. Inst. Fr. Afr. Noire, Ser. A* **22,** 1337–1340.

Longhurst, A. R. (1963). "The Bionomics of the Fishery Resources of the Eastern Tropical Atlantic," No. 20, pp. 1–66. Colonial Office, Fishery Publ., London.

Longhurst, A. R. (1964a). The coastal oceanography of Western Nigeria. *Bull. Inst. Fr. Afr. Noire, Ser. A* **26,** 337–402.

Longhurst, A. R. (1964b). Bionomics of the Sciaenidae of tropical West Africa. *J. Cons., Cons. Int. Explor. Mer* **29,** 93–114.

Longhurst, A. R. (1964c). A review of the present situation in benthic synecology. *Bull. Inst. Oceanogr.* **63,** 1–54.

Longhurst, A. R. (1965). A survey of the fish resources of the eastern Gulf of Guinea. *J. Cons. Cons. Int. Explor. Mer* **29,** 302–334.

Longhurst, A. R. (1966). Species assemblages in tropical demersal fisheries. *In* "Proceedings of the Symposium on Oceanography and Fisheries Resources in the Tropical Atlantic," pp. 147–170. UNESCO, Paris.

Longhurst, A. R. (1967a). Vertical distribution of zooplankton in relation to the eastern Pacific oxygen minimum. *Deep-Sea Res.* **14,** 51–64.

Longhurst, A. R. (1967b). Diversity and trophic structure of zooplankton communities in the California Current. *Deep-Sea Res.* **14,** 393–408.

Longhurst, A. R. (1971a). Curstacean resources. *In* "The Fish Resources of the Ocean" (J. A. Gulland, ed.), pp. 206–245. Fishing News Books, West Byfleet, England.

Longhurst, A. R. (1971b). The clupeid resources of tropical seas. *Oceanogr. Mar. Biol.* **9,** 349–385.

Longhurst, A. R. (1976). Interactions between zooplankton and phytoplankton profiles in the eastern tropical Pacific Ocean. *Deep-Sea Res.* **23,** 720–754.

Longhurst, A. R. (1983). Benthic-pelagic coupling and export of organic carbon from a tropical Atlantic continental shelf—Sierra Leone. *Estuarine, Coastal Shelf Sci.* **17,** *261–285.*

Longhurst, A. R. (1984). Importance of measuring rates and fluxes in marine ecosystems. *In* "Flows of Energy and Materials Marine in Ecosystems" (M. J. R. Fasham, ed.), pp. 1–32. Plenum, New York.

Longhurst, A. R. (1985a). The structure and evolution of plankton communities. *Prog. Oceanogr.* **15,** 1–35.

Longhurst, A. R. (1985b). Relationship between diversity and the vertical structure of the upper ocean. *Deep-Sea Res.* **32,** 1535–1569.

Longhurst, A. R., and Herman, A. W. (1981). Do oceanic zooplankton aggregate at, or near, the deep chlorophyll maximum? *J. Mar. Res.* **39,** 353–356.

Lopez, R. B. (1964). "Problemas de la distribución geografica de los peces marinos suramericanos," pp. 57–62. Bol. Inst. Biol. Mar., Argentina.

Love, C. M., ed. (1971). EASTROPAC Atlas 6 (fig. 40.PP.ei). *U. S. Dep. Commer. Fish Circ.* **330.**

Lowe-McConnell, R. H. (1962). The fishes of the British guiana continental shelf, with notes on their natural history. *J. Linn. Soc. London, Zool.* **44,** 669–700.

Luckhurst, B. E., and Luckhurst, K. (1977). Recruitment patterns of coral reef fishes on the fringing reef of Curacao, Netherlands Antilles, *Can. J. Zool.* **55,** 681–689.

Lutjeharms, J. R. E., Bang, N. D., and Duncan, C. P. (1981). Characteristics of the currents south and east of Madagascar. *Deep-Sea. Res.* **28,** 879–899.

Lutz, R. A., and Rhoads, D. C. (1977). Anaerobiosis and a theory of growth line formation. *Science* **198,** 1222–1227.

MacDonald, J. S., and Green, R. H. (1983). Redundancy of variables used to describe importance of prey species in fish diets. *Can. J. Fish. Aquat. Sci.* **40,** 635–637.

McGowan, J. A. (1972). The nature of ocean ecosystems. *In* "The Biology of the Oceanic Pacific" (C. B. Miller, ed.), pp. 9–28. Oregon Univ. Press, Eugene.

McGowan, J. A., and Hayward, T. L. (1978). Mixing and ocean productivity. *Deep-Sea Res.* **25,** 771–793.

MacIsaac, J. J., Dugdale, R. C., Barber, R. J., Blasco, D., and Packard, T. T. (1985). Primary production cycle in an upwelling centre. *Deep-Sea Res.* **32,** 503–529.

MacLeish, W. H. (1976). Ocean eddies. *Oceanus* **19,** 1–86.

McManus, J. W. (1985a). Marine speciation, tectonics, and sea-level changes in south-east Asia. *Proc. Int. Coral Reef. Symp. 5th, 1984,* Vol. 4, pp. 133–138.

McManus, J. W. (1985b). Descriptive community dynamics: Background and application to tropical fisheries management. Ph.D. Thesis, pp. 1–217. Univ. of Rhode Island, Kingston.

Macnae, W. (1968). A general account of the fauna and flora of mangrove swamps and forests in the Indo-west Pacific Region. *Adv. Mar. Biol.* **6,** 73–270.

Macnae, W. (1974). "Mangrove Forests and Fisheries," Indian Ocean Fishery Commission IOFC/DEV/74.34, Rome, 35 pp.

Macnae, W., and Kalk, M. (1962). The fauna and flora of sand flats at Inhaca Island, Mozambique. *J. Anim. Ecol.* **31,** 93–122.

McPherson, G. R. (1981). Investigations of Spanish mackerel in Queensland waters. *In* "Northern Pelagic Fish Seminar, Darwin" (C. J. Grant and D. G. Walter, eds.), pp. 51–58. Aust. Govt. Publ. Serv., Townsville, Qd.

McRoy, P. C., and Lloyd, D. S. (1981). Comparative function and stability of macrophyte-based ecosystems. *In* "Analysis of Marine Ecosystems" (A. R. Longhurst, ed.), pp. 473–490. Academic Press, New York.

Madhupratap, M., and Haridas, P. (1986). Epipelagic calanoid copepods of the northern Indian Ocean. *Oceanol. Acta* **9,** 105–117.

Madhupratap, M., Rao, T. S. S., and Haridas, P. (1977). Secondary production in the Cochin backwaters, a tropical monsoonal estuary. *Proc. Symp. Warm Water Zooplankton, 1976*, pp. 515–519.

Mahon, R., Oxenford, H. A., and Hunte, W. (1986). "Development Strategies for Flyingfish Fisheries of the Eastern Caribbean. Proceedings of a Workshop at University of the West Indies," IDRC-MR128e, pp. 1–148. Can. Int. Dev. Agency, Ottawa.

Malone, T. C. (1980). Size-fractionated primary production of marine phytoplankton. *In* "Primary Productivity of the Sea" (P. G. Falkowsky, ed.), pp. 301–319. Plenum, New York.

Malovitskaya, L. M. (1971). Production of species of mass-occurrence copepods in the Gulf of Guinea. *Tr., Atl. Nauchno-Issled. Inst. Rybn. Khoz. Okeanogr.* **37,** 401–405.

Manacop, P. R. (1955). Commercial trawling in the Philippines. *Philipp. J. Fish.* **3,** 117–188.

Mandelli, E. (1981). On the hydrography and chemistry of some coastal lagoons of the Pacific coast of Mexico. *UNESCO Tech. Pap. Mar. Sci.* **33,** 81–96.

Mann, K. H. (1977). Destruction of kelp beds by sea-urchins: Cyclic phenomenon or irreversible degradation? *Helgol. Wiss. Meeresunters.* **30,** 455–467.

Mann, K. H. (1982). "Ecology of Coastal Waters." Blackwell, Oxford.

Mann, K. H. (1984). Fish production in open ocean ecosystem. *In* "Flows of Energy and Materials in Marine Ecosystems" (M. J. R. Fasham, ed.), pp. 435–458. Plenum, New York.

Marcille, J., and Stequert, B. (1976). Croissance des jeunes albacores *Thunnus albacares* et patudos, *T. obesus*, dans la côte nord-ouest de Madagascar. *Cah. ORSTOM, Ser. Océanogr.* **14,** 153–162.

Marr, J. C. (1982). The realities of fishery management in the Southeast Asia region. *In* "Theory and Management of Tropical Fisheries" (D. Pauly and G. I. Murphy, eds.), p. 360. ICLARM, Manila.

Martsubroto, P., and Naamin, N. (1977). Relationship between tidal forests and commercial shrimp production in Indonesia. *Mar. Res. Indonesia* **8,** 81–86.

Masuda, K., Nakane, S., Saito, S., and Fujii, T. (1964). Survey of trawl grounds off the north-west coast of Australia with species reference to hydrographic conditions of the ground. *Bull. Fac. Fish. Hokkaido Univ.* **15** (2), 77–88.

Mathews, C. P. (1985). The present state of Kuwait's shrimp fishery. *In* "Proceedings of the 1984 Shrimp and Fin Fisheries Management Workshop" (C. P. Mathews, ed.), pp. 3–31. Kuwait Inst. Sci. Res., Safat.

Mathews, C. P. (1987). Fisheries management in a tropical country: The most appropriate balance of size and age/length related methods for practical assessments. *In* "Length-based Methods in Fisheries Research'' (D. Pauly and G. R. Morgan, eds.), ICLARM Conf. Proc. 14 (in press).

Mathews, C. P., Al-Hossaini, M., Abdul Ghaffar, A. R., and Al-Shoshanni, M. (1987). Assessment of short-lived stocks with special reference to Kuwaits shrimp fisheries. *In* "Length-based Methods in Fisheries Research" (D. Pauly and G. R. Morgan, eds.), ICLARM Conf. Proc. 14 (in press).

Maurer, D., and Vargas, J. A. (1979). Diversity of soft-bottom benthos in a tropical estuary. *Mar. Biol. (Berlin)* **81,** 97–106.

Maurer, R. O., and Bowman, R. E. (1985). Food consumption of squids (*Illex illecibrosus* and *Loligo pealei*) off the NE United States. *NAFO Sci. Counc. Stud.* **9,** 117–124.

Menaché, M. (1961). Decouverte d'un phénomène de remontée d'eaux profondes au sud du canal de Mozambique. *Mém. Inst. Sci. Madagascar, Ser. F* **4,** 167–173.

Menasveta, D. (1970). Potential demersal fish resources of the Sunda Shelf. *In* "The Kuroshio" (J. W. Marr, ed.), pp. 525–555. East-West Center, Honolulu.

Menasveta, D., and Isrankura, A. P. (1968). Country Report, Thailand. *In* "Proceedings of the International Seminar on Fishery Development in Southeast Asia," pp. 281–309. German Foundation for International Development, Berlin.

Mensah, M. A. (1974). The reproduction and feeding of the marine copepod *Calanoides carinatus* in Ghanaian waters. *Ghana J. Sci.* **14,** 167–191.

Menzel, D. W. (1960). Utilisation of food by a Bermudan reef fish *Epinephelus guttatus*. *J. Cons., Cons. Int. Explor. Mer* **25,** 216–222.

Merle, J. (1980). Seasonal heat budget in the equatorial Atlantic Ocean. *J. Phys. Oceanogr.* **10,** 464–469.

Merle, J., and Arnault, S. (1985). Seasonal variability of the surface dynamic topography in the tropical Atlantic Ocean. *J. Mar. Res.* **43,** 267–288.

Merrett, N. R., and Roe, H. S. J. (1974). Patterns and selectivity in the feeding of certain mesopelagic fishes. *Mar. Biol. (Berlin)* **28,** 115–126.

Methot, R. D. (1984). Seasonal variation in survival of larval *Engraulis mordax* estimated from the age distribution of juveniles. *Fish. Bull.* **81,** 741–750.

Miller, C. B. Johnson, J. K., and Heinle, D. R. (1977). Growth rules in the marine copepod genus *Acartia*. *Limnol. Oceanogr.* **22,** 326–335.

Milliman, J. D., and Meade, R. H. (1983). World-wide delivery of river sediment to the oceans. *J. Geol.* **91** (1), 1–21.

Mistakidis, M. N. (1973). The Crustacean Resources and Related Fisheries in the Countries Bordering the South China Sea," SCS/DEV/73/7, pp. 1–39. FAO South China Sea Development and Coordinating Programme, Rome.

Mitchell, W. G., and McConnell, R. H. (1959). The trawl survey carried out by the r/v Cape St. Mary off British Guiana. *Br. Guiana Fish. Div., Dep. Agric., Bull.* **2,** 1–53.

Mittelbach, G. G. (1981). Foraging efficiency and body size: A study of optimal diet and habitat use by bluegill. *Ecology* **62,** 1370–1386.

Mohr, E. (1921). Altersbestimmung bei tropischen Fischen. *Zool. Anz.* **53,** 87–95.

Mosieev, P. A. (1969). "The Living Resources of the World Ocean." Izd. Pishch. Prom-., Moskva.

Monod, T. (1927). Contribution à l'étude de la faune du Cameroun. Pisces I. *Faune Col. Fr.* **1,** 643-742.

Monod, T. (1961). *Brevoortia* Gill, 1861 et *Ethmalosa* Regan 1871. *Bull. Inst. Fr. Afr. Noire, Ser. A* **23,** 506–547.

Morgan, G. R. (1987). Incorporating age data into length-based stock assessment methods. *In* "Length-based Methods in Fisheries Research" (D. Pauly and G. R. Morgan, eds.), ICLARM Conf. Proc. 14 (in press).

Morgans, J. F. C. (1964). "A Preliminary Fishery Survey of Bottom Fishing on the North Kenya Banks," No. 21, pp. 1–64. Colon. Office Fish. Publ., London.

Muir, B. S. (1969). Gill dimensions as a function of fish size. *J. Fish Res. Board Can.* **26** (1), 165–170.

Munk, W., and Sargent, M. C. (1948). Adjustment of Bikini Atoll to ocean waves. *Trans. Am. Geophys. Union* **29**, 855-860.

Munro, J. L. (1980). Stock assessment models: Applicability and utility in tropical small-scale fisheries. *In* "Stock Assessments for Tropical Small-scale Fisheries" (P. M. Roedel and S. Saila, eds.), pp. 35–47. Int. Cent. Mar. Res. Dev., University of Rhode Island, Kingston.

Munro, J. L., ed. (1983). Caribbean coral reef fishery resources. *ICLARM Stud. Rev.* **7**, 1–276.

Munro, J. L., and Gwyther, J. (1981). Growth rates and mariculture potential of tridacnid clams. *Proc. Int. Coral Reef Symp. 5th, 1980*, Vol. 2, pp. 633–636.

Munro, J. L., and Pauly, D. (1985). A simple method for comparing the growth of fish and invertebrates. *Fishbyte* **1** (1), 5–6.

Munro, J. L., and Thompson, R. (1983). Areas investigated, objectives and methodology. *In* "Caribbean coral reef fishery resources" (J. L. Munro, ed.). *ICLARM Stud. Rev.* **7**, 15–25.

Munro, J. L., and Williams, D. McB. (1985). Assessment and management of coral reef fisheries: Biological, environmental and socioeconomic aspects. *Proc. Int. Coral Reef Symp. 5th, 1984*, Vol. 4, pp. 545–581.

Murphy, G. I. (1982). Recruitment of tropical fishes. *In* "Theory and Management of Tropical Fisheries" (D. Pauly and G. I. Murphy, eds.), pp. 141–148. ICLARM, Manila.

Muscatine, L., and Porter, J. W. (1977). Reef corals: Mutualistic symbioses adapted to nutrient-poor environments. *BioScience* **27**, 454–460.

Musick, J. A. (1987). Seasonal recruitment of sub-tropical sharks in Chesapeake Bight. *In* "Proceedings of the IREP Workshop on Recruitment in Coastal Demersal Communities, Campeche, Mexico," (A. Yañez-Aranciba and D. Pauly, eds.). IOC/UNESCO Workshop Series **44** (in press).

Mysak, L. A. (1986). El Niño, interannual variability and fisheries in the northeast Pacific Ocean. *Can. J. Fish. Aquat. Sci.* **43**, 464–497.

Nair, R. V., and Subrahmanyan, R. (1955). The diatom, *Fragillaria oceanica*, an indicator of abundance of the Indian Oil sardine. *Curr. Sci.* **24** (2), 41–42.

Narasimham, K. A., Rao, G. S., Sastry, Y. A., and Venugopalam, W. (1979). Demersal fishery resources off Kakinda with a note on economics of commercial trawling. *Indian J. Fish.* **26**, 90–100.

Nees, J., and Dugdale, R. C. (1959). Computation of production for populations of aquatic midge larvae. *Ecology* **40**, 425–430.

Nellen, W. (1966). Horizontal and vertical distribution of plankton production in the Gulf of Guinea and adjacent areas, February-May, 1961. *In* "Proceedings of the Symposium on Oceanography and Fishery Resources of the Tropical Atlantic," pp. 265–268. UNESCO, Paris.

Nellen, W. (1986). A hypothesis on the fecundity of bony fish. *Meeresforschung* **31**, 75–89.

New, M. B., and Rabanal, H. R. (1985). A review of the status of penaeid aquaculture in Southeast Asia. *In* "Second Australian National Prawn Seminar" (P. C. Rothlisberg, B. J. Hill, and D. J. Staples, eds.), pp. 307–326. NPS2, Cleveland, Australia.

Neyman, A. A., Sokolova, M. N., Viogradova, N. G., and Pasternak, F. A. (1973). Some patterns of the distribution of bottom fauna in the Indian Ocean. *In* "Biology of the Indian Ocean" (B. Zeitzschel, ed.), pp. 467–473. Springer-Verlag, Berlin and New York.

Nichols, M., and Allen, G. (1981). Sedimentary processes in coastal lagoons. *UNESCO Tech. Pap. Mar. Sci.* **33**, 27–80.

Nixon, S. W. (1982). Nutrient dynamics, primary production and nutrient yields of lagoons. *Oceanol. Acta, Vol. Spec.* **4**, Suppl. 357–373.

O'Day, W. T., and Nafpaktitis, B. G. (1967). A study of the effects of expatriation on the gonads of two myctophid fishes in the North Atlantic Ocean. *Bull. Mus. Comp. Zool.* **136**, 77–89.

O'Dor, R. K., Durwood, R. D., Vessey, E., and Amaratunga, T. (1980). Feeding and growth in captive squid, *Illex illecibrosus*, and the influence of food availability on growth in the natural population. *ICNAF Sel. Pap.* **6**, 15–21.

Odum, H. T., and Odum, E. P. (1955). Trophic structure and productivity of a windward coral reef community on Eniwetok Atoll. *Ecol. Monogr.* **25**, 291–320.

Odum, W. E. (1970). Utilisation of the direct grazing and plant detritus food chains by the striped mullet *Mugil cephalus*. *In* "Marine Food Chains" (J. H. Steele, ed.), pp. 222–240. Oliver & Boyd, Edinburgh.

O'Gorman, R., Barwick, D. H., and Bowen C. A. (1987). Discrepancies between otolith and scale age determinations for alewives from the Great Lakes. *In* "Age and Growth of Fishes" (R. C. Summerfeldt and G. E. Hall, eds.), pp. 203–210. Iowa Univ. Press, Ames.

Okera, W. (1976). Observations on some population parameters of exploited stocks of *Senilia* (=*Arca*) *senilis* in Sierra Leone. *Mar. Biol. (Berlin)* **38**, 217–229.

Okera, W. (1982). "Organization of Fish Assemblages on the Northern Australian Continental Shelf." CSIRO, Cronulla (unpublished).

Olson, R., and Boggs, C. H. (1986). Apex predation by yellowfin tuna: Independent estimates from gastric evacuation and stomach contents, bioenergetics and caesium concentration. *Can. J. Fish. Aquat. Sci.* **43**, 1760–1775.

Ommaney, F. D. (1961). "Malayan Offshore Trawling Grounds—The Experimental and Exploratory Fishing of the FRV Manahine in Malayan and Borneo Waters, 1955–56," No. 18, pp. 1–95. Colon Off. Fish. Publ., London.

Omori, M. (1975). The systematics, biogeography and fishery of epipelagic shrimps of the genus *Acetes* (Sergestidae). *Bull. Ocean Res. Inst.,* Univ. Tokyo, No. 7, 1–91.

Orsi, J. J. (1986). Interaction between diel vertical migration of a mysidacean shrimp and two-layered estuarine flow. *Hydrobiologia* **137**, 79–87.

Osborn, T. R. (1978). Measurements of energy dissipation adjacent to an island. *JGR, J. Geophys. Res.* **83**, 2939–2957.

Owen, R. W., and Zeitzschel, B. (1970). Phytoplankton production: Seasonal changes in the oceanic eastern tropical Pacific. *Mar. Biol. (Berlin)* **7**, 32–36.

Packard, A. (1972). Cephalods and fish: The limits of convergence. *Biol. Rev. Cambridge Philos. Soc.* **47**, 241–307.

Paine, R. T. (1966). Food complexity and species diversity. *Am. Nat.* **100**, 65–75.

Palomares, M. L. (1985). Developing countries dominate the shrimp scene. *ICLARM Newsl.* **8**, 3–5.

Panella, G. (1971). Fish otoliths: Daily growth layers and periodical patterns. *Science* **173**, 1124.

Panella, G. (1974). Otolith growth patterns: An aid in age determination in temperate and tropical fishes. *In* "Ageing of Fish" (T. Bagenal, ed.), pp. 28–39. Unwin Bros., Old Woking.

Parekular, A. H., Harkantra, S. N., and Ansari, Z. A. (1982). Benthic production and assessment of demersal fishery resources of the Indian Seas. *Indian J. Mar. Sci.* **11**, 107–114.

Parin, N. V. (1970). "Ichthyofauna of the Epipelagic Zone." Nauka, Moscow (Isr. Program Sci. Transl., Jerusalem, 1970).

Parin, N. V. (1977). Midwater fishes in the western tropical Pacific Ocean and the seas of the Indonesian-Australian archipelago. *Tr. Inst. Okeanol. im. P.P. Shirshova, Akad. Nauk SSSR* **107**, 68–188.

Parker, R. R., and Larkin, P. A. (1959). A concept of growth in fishes. *J. Fish. Res. Board Can.* **16**, 721–745.

Parsons, T. R., Takahashi, M., and Hargrave, B. (1977). "Biological Oceanographic Processes." Pergamon, Oxford.

Pati, S. (1982). The influence of temperature and salinity on the pelagic fishery in the northern part of the Bay of Bengal. *J. Cons., Cons. Int. Explor. Mer* **40**, 220–225.

Pauly, D. (1975). On the ecology of a small west African Lagoon. *Ber. Dtsch. Wiss. Komm. Meeresforsch.* **24**, 46–62.

Pauly, D. (1976). The biology, fishery and potential for aquaculture of *Tilapia melanotheron* in a small West African lagoon. *Aquaculture* **7**, 33–49.

Pauly, D. (1978). A preliminary compilation of fish length growth parameters. *Ber. Inst. Meereskd. Christian-Albrechts Univ., Kiel* **55**, 1–200.

Pauly, D. (1979). Theory and management of tropical multispecies stocks: A review, with emphasis on the southeast Asian demersal fisheries. *ICLARM Stud. Rev.* **1**, 1–35.

Pauly, D. (1980a). On the interrelationships between natural mortality, growth parameters and mean environmental temperature in 175 fish stocks. *J. Cons., Cons. Int. Explor. Mer* **39**, 175–192.

Pauly, D. (1980b). A new methodology for rapidly acquiring basic information on tropical fish stocks: Growth, mortality and stock recruitment relationships. *In* "Stock Assessments for Tropical Small-scale Fisheries" (P. M. Roedel and S. Saila, eds.), pp. 154–172. Int. Cent. Mar. Res. Dev., University of Rhode Island, Kingston.

Pauly, D. (1981). The relationship between gill surface area and growth performance in fish: A generalization of von Bertalanffy's theory of growth. *Ber. Dtsch. Wiss. Komm. Meeresforch.* **28**, 252–282.

Pauly, D. (1982a). The fishes and their ecology. *In* "Small-scale fisheries of San Miguel Bay Philippines: Biology and stock assessment" (D. Pauly and A. N. Mines, eds.). *ICLARM Tech. Rep.* **7**, 15–33.

Pauly, D. (1982b). Studying single-species dynamics in a multi-species context. *In* "Theory and Management of Tropical Fisheries" (D. Pauly and G. I. Murphy, eds.), pp. 33–70. ICLARM, Manila.

Pauly, D. (1982c). A method to estimate the stock-recruitment relationship of shrimps. *Trans. Am. Fish. Soc.* **111**, 13–20.

Pauly, D. (1984a). Fish population dynamics in tropical waters: A manual for use with programmable calculators. *ICLARM Stud. Rev.* **8**, 1–325.

Pauly, D. (1984b). A mechanism for the juvenile-to-adult transition in fishes. *J. Cons., Cons. Int. Explor. Mer* **41**, 280–284.

Pauly, D. (1984c). Reply to comments on prerecruit mortality in Gulf of Thailand shrimps. *Trans. Am. Fish. Soc.* **113**, 404–406.

Pauly, D. (1985a). Zur Fischereibiologie tropischer Nutztiere: Eine Bestandsaufnahme von Konzepten und Methoden. *Ber. Inst. Meereskd. Christian-Albrechts Univ. Kiel* **147**, 1–55.

Pauly, D. (1985b). Population dynamics of short-lived species, with emphasis on squids. *NAFO Sci. Counc. Stud.* **9**, 143–154.

Pauly, D. (1986a). On improving operation and use of ELEFAN programmes (III): Correcting length–frequency data for effects of gear selection and/or incomplete recruitment. *Fishbyte* **4,** 11–13.

Pauly, D. (1986b). A simple method for estimating the food consumption of fish populations from growth data and food conversion experiments. *Fish Bull.* **84,** 829–842.

Pauly, D. (1987). Fisheries Research and the demersal fisheries of Southest Asia. *In* "Fish Population Dynamics" (J. A. Gulland, ed.), 2nd ed. (in press).

Pauly, D., and Calumpong, H. (1984). Growth and mortality of the sea-hare *Dolabella auricula* in the central Visaya, Philippines. *Mar. Biol. (Berlin)* **79,** 289–293.

Pauly, D., and David, N. (1981). ELEFAN I, a BASIC program for the objective extraction of growth parameters from length-frequency data. *Meeresforschung* **28,** 205–211.

Pauly, D., and Gaschütz, G. (1979). A simple method for fitting oscillating length growth data with a programme for pocket calculators. *Cons. Int. Explor. Mer,* C.M. 1979/G: 24, Demersal Fish Committee.

Pauly, D., and Ingles, J. (1986). The relationship between shrimp yields and intertidal vegetation areas: A reassessment. *In* "Proceedings of the IREP/OSLR Workshop on the Recruitment of Coastal Demersal Communities, Campeche, Mexico" (A. Yañez-Arancibia and D. Pauly, eds.) IOC-UNESCO Rep. Ser. (44), UNESCO, Paris (in press).

Pauly, D., and Morgan, G. R., eds. (1987). "Length-based Methods in Fishery Research," ICLARM Conf. Proc. 14 (in press).

Pauly, D., and Murphy, G. I. (1982). "Theory and Management of Tropical Fisheries." ICLARM, Manila.

Pauly, D., and Navaluna, N. A. (1983). Monsoon-induced seasonality in the recruitment of Philippine fishes. *FAO Fish Rep.* **291,** 823–833.

Pauly, D., and Palomares, M. L. (1985). Shrimp consumption by fish in Kuwait waters: A methodology, preliminary results and their indications for management and research. *In* "Proceedings of the 1985 Shrimp and Fin Fisheries Management Workshop" (C. P. Mathews, ed.), pp. 1–270. Kuwait Inst. Sci. Res., Safat. pp. 1–270.

Pauly, D., and Pullin, R. S. V. (1987). On the relationship between hatching time, egg diameter and temperature in marine eggs. *Environ. Biol. Fish.* (in press).

Pauly, D., Aung, S., Rijavec, L., and Htein, H. (1984a). The marine living resources of Burma: A short review. *Indo-Pac. Fish. Comm. FAO Fish Rep.* **318,** 96–107.

Pauly, D., Ingles, J., and Neal, R. (1984b). Application to shrimp stocks of objective methods for the estimation of growth, mortality and recruitment-related parameters from length–frequency data (ELEFAN I and II). *In* "Penaeid Shrimps—Their Biology and management" (J. A. Gulland and B. Rothschild, eds.), pp. 220–234. Fishing News Books, Farnham.

Pearcy, W. G., Krygier, E. E., Mesecar, R., and Ramsey, F. (1977). Vertical distribution and migration of oceanic micronekton off Oregon. *Deep-Sea Res.* **24,** 223–245.

Pearson, J. C. (1929). Natural history and conservation of the redfish and other common sciaenids on the Texas coast. *Fish. Bull.* **44,** 129–144.

Penchaszadeh, P. E., and Salaya, J. J. (1984). Estructura y ecologia trofica de las communidaes demersales en el Golfo Triste, Venzuela. *In* "Recursos Pesqueros Potenciales de Mexicó: La Pesca Acompañante del Camerón" (A. Yáñez-Arancibia, ed.), pp. 571–598. Univ. Nac. Auton., Mexico.

Pérès, J.-M., and Picard, J. (1958). Manuel de biolonomie benthique de la mer Méditerranée. *Recl. Trav. Stn. Mar. Endoume* **23,** 5–122.

Pérès, J.-M., and Pichon, M. (1962). Note préliminaire générale sur le benthos littoral de la région de Tuléar (Madagascar). *Ann. Fac. Sci. Tech. Madagascar,* pp. 145–152.

Petersen, C. G. J. (1892). Fiskensbiologiske Forhold; Holboek Fjord, 1890–91. *Beret. Dan. Biol. Stn.*, pp. 121–183.

Petersen, C. G. J. (1918). The sea-bottom and its production of fish food. *Rep. Dan. Biol. Stn.* **25**, 1–62.

Petit, D. (1982). *Calanoides carinatus* sur le plateau continental congolais. III. Abondance, tailles et temps de génération. *Océanogr. Trop.* **17**, 155–175.

Petrie, B. D. (1983). Current response at the shelf break to transient wind forcing. *JGR, J. Geophys. Res.* **88**, 9567–9578.

Philander, S. G. (1985). Tropical oceanography. *Adv. Geophys.* **28A**, 461–477.

Pianka, E. R. (1966). Latitudinal gradients in species diversity: A review of concepts. *Am. Nat.* **100**, 33–46.

Pichon, M. (1962). Note preliminaire sur la répartition et le peuplement des sables fins et des sables vaseux non-fixés de la zone inter-tidale de Tuléar. *Recl. Trav. Stn. Mar. Endoume, Fasc. Hors Ser., Suppl.* **1**, 220–235.

Pillay, T. V. R. (1958). Biology of the hilsa, *Hilsa ilisha,* of the Rivery Hooghly. *Indian J. Fish.* **5**, 201–257.

Pillay, T. V. R. (1967a). Estuarine fisheries of West Africa. *In* "Estuaries" (G. H. Lauff, ed.), Publ. No. 83, pp. 639–646. Am. Assoc. Adv. Sci., Washington, D.C.

Pillay, T. V. R. (1967b). Estaurine fisheries of the Indian coasts. *In* "Estuaries" (G. H. Lauff, ed.), Publ. No. 83, pp. 644–657. Am. Assoc. Adv. Sci., Washington, D.C.

Pingree, R. D., and Mardell, G. T. (1981). Slope turbulence, internal waves and phytoplankton growth at the Celtic Sea shelf-break. *Philos. Trans. R. Soc. London, Ser. A* **302**, 663–682.

Pingree, R. D., Mardell, G. T. Holligan, P. M., Griffiths, D. K., and Smithers, J. (1982). Celtic Sea and Armorican Current structure and the vertical distribution of temperature and chlorophyll. *Cont. Shelf Res.* **1**, 99–116.

Piton, B., and Fusey, F.-X. (1982). Trajectories of satellite-tracked buoys in the Gulf of Guinea, July 1978–July 1979. *Trop. Ocean-Atmos. Newsl.* **10**, 5–7.

Plante, R. (1967). Etude quantitative du benthos dans la région de Nosy-Bé. *Cah. ORSTOM, Ser. Océanogr.* **5**, 7–108.

Platt, T. (1985). Structure of the marine ecosystem: Its allometric basis. *Can. Bull. Fish. Aquat. Sci.* **21**, 1–260.

Platt, T., Subba Rao, D. V., and Irwin, B. (1983). Photosynthesis of picoplankton in the oligotrophc ocean. *Nature (London)* **300**, 702–704.

Poinsard, F., and Troadec, J-P. (1966). Détermination de l'age par la lecture des otolithes chez deux espèces de sciaenidés Ouest-Africains. *J. Cons., Cont. Int. Explor. Mer* **30**, 291–307.

Poll, M. (1951). Poissons, généralités. *Res. Sci. Exp. Belge Eaux Côtieres Afr. Atl. Sud.* **4**, 1–154.

Polovina, J. J. (1984). Model of a coral reef ecosystem. I. The ECOPATH model and its application to French Frigate Shoals. *Coral Reefs* **3**, 1–11.

Por, F. D. (1978). Lessepsian migration: The influx of Red Sea biota onto the Mediterranean by way of the Suez Canal. *Ecol. Stud.* **23**, 1–228.

Porter, J. W. (1974). Zooplankton feeding by the Caribbean coral *Montastrea cavernosa. Proc. Coral Reef Symp., 2nd, 1973,* Vol. 1, pp. 111–125.

Postel, E. (1955). Les faciés bionomiques des côtes de Guinée française. *Rapp. P.-V. Réun., Cons. Int. Explor. Mer* **136**, 11–13.

Postel, E. (1960). Rapport sur la sardinelle (*Sardinella aurita* V.). *In* "Proceedings of the

World Scientific Meeting on Biology of Sardines and Related Species, Rome, 1959,'' pp. 55–95. FAO, Rome.

Postel, E. (1965). Apercu général sur les langoustes de la zone intertropicale africaine et leur exploitation. *Pêche Marit.* (1045), 313–323.

Postel, E. (1966). Langoustes de la zone intertropicale africaine. *Mem. Inst. Fondam. Afr. Noire, Ser. A* **77**, 399–474.

Potter, M. A., and Dredge, M. C. L. (1985). Deepwater prawn resources off southern and central Queensland. *In* "Second Australian National Prawn Seminar" (P. C. Rothlisberg, B. J. Hill, and D. J. Staples, eds.), pp. 221–229. NPS2, Cleveland, Australia.

Potts, D. C. (1985). Sea-level fluctuations and speciation in Scleractinia. *Proc. Int. Coral Reef Symp. 5th, 1984,* Vol. 4, pp. 127–132.

Potts, G. W., Clark, P. F., and Shand, J. (1986). Behavioral ecology of underwater organisms. *Prog. Underwater. Sci.* **11**, 1–182.

Purser, B. H., and Evans, G. (1973). Regional sedimentation along the Trucial Coast, S. E. Persian Gulf. *In* "The Persian Gulf" (N. H. Purser, ed.), pp. 211–231. Springer-Verlag, Berlin and New York.

Pütter, G. W. (1920). Studien über Physiologische Ähnlichkeit. VI. Wachstumsähnlichkeit. *Pflüegers Arch. Gesamte Physiol. Menschen Tiere* **180**, 298–340.

Qasim, S. Z. (1970). Some problems related to the food chain in a tropical estuary. *In* "Marine Food Chains" (J. H. Steele, ed.), pp. 45–51. Oliver & Boyd, Edinburgh.

Qasim, S. Z. (1972). The dynamics of food and feeding habits of some marine fishes. *Indian J. Fish.* **19**, 11–28.

Qasim, S. Z. (1973). Productivity of backwaters and estuaries. *IBP Ecol. Stud.* **3**, 143–154.

Quader, O., Pramanik, M. A. H., Khan, F. A., and Polcyn, F. C. (1986). Mangrove ecosystem study of Chakaria Sunderbans at Chittagong with special emphasis on shrimp ponds by remote sensing techniques. *In* "Marine Interfaces Hydrodynamics" (J. C. J. Nihoul, ed.), pp. 645–653. Elsevier, Amsterdam.

Quinn, N. J., and Kojis, B. L. (1986). Annual variation in the nocturnal nekton assemblage of a tropical estuary. *Estuarine, Coastal Shelf Sci.* **22**, 63–90.

Qureshi, M. R. (1955). "Marine Fishes of Karachi and the Coasts of Sind and Makram." Govt. Pakistan.

Rainer, S. F. (1984). Temporal changes in a demersal fish and cephalopod community of an unexploited coastal area in Northern Australia. *Aust. J. Mar. Freshw. Res.* **35**, 747–768.

Raja, B. T. A. (1969). The Indian oil sardine. *Bull. Cent. Mar. Fish. Res. Inst.* (*Mandapam Camp, India*) **16**, 1–128.

Raja, B. T. A. (1972a). Estimation of the age and growth of the Indian oil-sardine. *Indian J. Fish.* **17**, 84–88.

Raja, B. T. A. (1972b). A forecast for the ensuing oil-sardine fishery. *Seafood Export J.* **4** (10), 27–33.

Raja, B. T. A. (1973). Possible explanation for the fluctuations in abundance of the Indian oil-sardine. *Proc., Indo-Pac. Fish Comm.* **15**, 241–252.

Ralston, S. (1981). A study of the Hawaiian deepsea handline fishery with special reference to the population dynamics of Opakapaka *Pristipomoides filamentosus.* Ph.D. Thesis, University of Washington, Seattle.

Ralston, S. (1985). A novel approach to aging tropical fish. *ICLARM Newsl. Manila* **8**, 14–15.

Ralston, S. (1986). Mortality rates of snapper and groupers. *In* "Proceedings of a Workshop on the Biology of Tropical Groupers and Snappers, Honolulu, Hawaii, 1985." (J. J. Polovina and S. Ralston, eds.), pp. 16–24.

Ralston, S., and Polovina, J. R. (1982). Multispecies analysis of the commercial deep-sea handline fishery in Hawaii. *Fish. Bull.* **80,** 435–448.

Ramage, C. S. (1986). El Niño. *Sci. Am.* **254,** 77–83.

Randall, J. E. (1962). Tagging reef fishes in the Virgin Islands. *Proc. Annu. Gulf Caribb. Fish. Inst.* **14,** 201–241.

Randall, J. E. (1967). Food habits of reef fishes of the West Indies. *Stud. Trop. Oceanogr* (5), 665–847.

Randall, J. E. (1985). This week's citation classic "Food Habits of Reef Fishes of the West Indies." *Curr. Contents* **19,** 22.

Rao, K. V. (1954). Biology and fishery of the Palk Bay squid, *Sepioteuthis arctipinnis. Indian J. Fish.* **1,** 37–67.

Rao, K. V. (1969). Distribution pattern of the major exploited marine fishery resources of India. *Bull. Cent. Mar. Fish. Res. Inst. (Cochin, India)* **6,** 1–69.

Rao, K. V., Dorairaj, K., and Kagwade, P. V. (1974). Results of the exploratory fishing operations of the Government of India vessels at Bombay Base for the period 1961–1971. *Proc., Indo-Pac. Fish Counc.* **13** (3), 402–430.

Rao, K. V. S. (1981). Food and feeding habits of lizard fishes (*Saurida spp.*) from the northwestern part of Bay of Bengal. *Indian J. Fish.* **28,** 47–64.

Reeson, P. H. (1983). The biology, ecology and bionomics of the parrot fishes. Scaridae. *In* "Caribbean coral reef fishery resources" (J. L. Munro, ed.). *ICLARM Stud. Rev.* **7,** 166–190.

Reeve, M. R., and Baker, L. D. (1975). Production of two planktonic carnivores in S. Florida inshore waters. *Fish. Bull.* **73,** 238–248.

Reid, J. L., Brinton, E. Fleminger, A., Venrick, E., and McGowan, J. A. (1978). Ocean circulation and marine life. *In* "Advances in Oceanography" (H. Charnock and G. Deacon, eds.), pp. 65–130. Plenum, New York.

Reinthal, P. N., Kensley, B., and Lewis, S. M. (1984). Dietary shifts in the queen triggerfish *Balistes vetula* in the absence of its primary food item, *Diadema antillarum. Mar. Ecol.* **5,** 191–195.

Richards, A. R. (1955). "Trawl-fishing in the south-east Caribbean. A report prepared for the Government of Trinidad and Tobago and the Caribbean Commission," pp. 1–147. Caribbean Commission, Port-of-Spain, Trinidad.

Richards, W. J. (1969). An hypothesis on yellowfin migrations in the eastern Gulf of Guinea. *Cah. ORSTOM, Ser. Océanogr.* **7,** 3–7.

Richardson, P. L. (1984). Moored current meter measurements in the Atlantic North Equatorial Counter Current during 1983. *Geophys. Res. Lett.* **11,** 749–752.

Ricker, W. E. (1954). Stock and recruitment. *J. Fish. Res. Board Can.* **1,** 559–623.

Ricker, W. E. (1975). Computation and interpretation of biological statistics of fish populations. *Bull. Fish. Res. Board Can.* (191), 1–382.

Ritrasaga, S. (1976). Results of the studies on the status of demersal fish resources in the Gulf of Thailand from trawling surveys, 1963–1972 *In* "Fisheries Resources Management in Southeast Asia" (K. Tiews, ed.), pp. 198–223. German Foundation for International Development, Berlin.

Rivas, L. R. (1978). Preliminary models of annual life history cycles of North Atlantic bluefin tuna. *In* "The Physiological Ecology of Tunas" (G. D. Sharp and A. E. Dizon, eds.), pp. 369–394. Academic Press, New York.

Robertson, A. I. (1979). The relationship between annual production/biomass ratios and lifespans for marine macrobenthos. *Oecologia* **38,** 193–202.

Robinson, A. R., ed. (1983). "Eddies in Marine Science." Springer-Verlag, Berlin and New York.

Robinson, M. H. (1978). Is tropical biology real? *Trop. Ecol.* **9**, 30–52.

Roden, G. I. (1961). On the wind-driven circulation in the Gulf of Tehuantepec and its effects upon surface temperatures. *Geofis. Int.* **1**, 55–72.

Roe, H. S. J. (1983). Vertical distributions of euphausiids and fish in relation to light intensity in the Northeastern Atlantic. *Mar. Biol. (Berlin)* **77**, 287–298.

Roe, H. S. J., and Babcock, J. (1984). The diel migrations and distributions within a mesopelagic community in the north-east Atlantic. 5. Vertical migration and feeding of fish. *Prog. Oceanogr.* **13**, 389–424.

Roff, D. A. (1980). A motion for the retirement of the von Bertalanffy function. *Can. J. Fish. Aquat. Sci.* **37**, 127–129.

Roger, C. (1977). The use of a group of macroplanktonic organisms (euphausiid crustaceans) in the study of the warm water pelagic food webs. *Proc. Symp. Warm Water Zooplankton, 1976*, pp. 309–318.

Rollet, B. (1981). "Bibliography on Mangrove Research, 1600–1875." UNESCO, Paris.

Ronquillo, I. A. (1974). A review of the roundscad fishery in the Philippines. *Proc., Indo-Pac. Fish. Counc.* **15** (3), 351–375.

Roper, C. F. E., and Young, R. E. (1975). Vertical distribution of pelagic cephalopods. *Smithson. Contrib. Zool.* **209**, 61–87.

Roper, C. F. E., Sweeney, M. J., and Nauen, C. E. (1984). "An Annotated and Illustrated Catalogue of Species of Interest to Fisheries," FAO Species Catalogue No. 3. FAO, Rome.

Rose, A. E., ed. (1967). "Thermobiology." Academic Press, New York.

Rosen, B. R. (1981). The tropical high diversity enigma—the coral's eye view. *In* "Chance, Change, and Challenge, the Evolving Biosphere" (P. H. Greenwood and P. L. Forey, eds.), pp. 103–129. Br. Mus. (Nat. Hist.), London.

Rosen, B. R. (1984). Reef coral biogeography and climate through the late Caenozoic: Just islands in the sun or a critical pattern of islands. *In* "Fossils and Climate" (P. Benchley, ed.), pp. 201–259. Wiley, New York.

Rothlisberg, P. C. (1982). Vertical emigration and its effect on dispersal of penaeid shrimp larvae in the Gulf of Carpentaria, Australia. *Fish. Bull.* **80**, 541–554.

Rothlisberg, P. C., Church, J. A., and Forber, A. M. G. (1983). Modelling the advection of vertically migrating shrimp larvae. *J. Mar. Res.* **41**, 511–554.

Rothlisberg, P. C., Hill, B. J., and Staples, D. J., eds. (1985a). "Second Australian National Prawn Seminar." NPS2, Cleveland, Australia.

Rothlisberg, P. C., Staples, D. J., and Crocos, P. J. (1985b). A review of the life history of the banana prawn, *Penaeus merguiensis*, in the Gulf of Carpentaria. *In* "Second Australian National Prawn Seminar" (P. C. Rothlisberg, B. J. Hill, and D. J. Staples, eds.), p. 125–136. NPS2, Cleveland, Australia.

Rothschild, B. (1967). Estimates of the growth of the skipjack tuna (*Katsuwonus pelamis*) in the Hawaiian Islands. *Proc., Indo-Pac. Fish Counc.* **12**, 100–111.

Rothschild, B. J. (1965). Hypotheses on the origin of exploited skipjack tuna in the eastern and central Pacific Ocean. *U. S., Fish Wildl. Serv., Spec. Rep—Fish.* **512**, 1–20.

Rougerie, F., and Wauthy, B. (1986). Le concept d'endo-upwelling dans le fonctionnement des atolls-oasis. *Oceanol. Acta* **9**, 133–148.

Rowe, G. T., and Smith, K. L. (1977). Benthic-pelagic coupling in the mid-Atlantic Bight. *In* "Ecology of Marine Benthos" (B. C. Couall, ed.), pp. 55–65. Univ. of South Carolina Press, Columbia.

Rowe, G. T., Smith, S., Falkowski, P., Whitledge, T., Theroux, R., Phoel, W., and Ducklow, H. (1986). Do continental shelves export organic matter? *Nature (London)* **324**, 559–561.

Ruamrasaga, S., and Isarankura, A. P. (1965). "An Analysis of Demersal Fish Catches Taken from the Otterboard Trawling Survey in the Gulf of Thailand," Vol. 3, pp. 1–51. Cont. Dep. Fisheries, Thailand.

Rudnick, D. J., Elmgren, R., and Frithson, J. B. (1985). Meiofunal prominence and benthic seasonality in a coastal marine ecosystem. *Oecologia* 67, 157–168.

Russ, G. (1984). A review of coral reef fisheries. *UNESCO Rep. Mar. Sci.* 27, 74–92.

Russell, E. S. (1931). Some theoretical considerations of the overfishing problem. *J. Cons., Cons. Int. Explor. Mer* 6, 3–20.

Ryther, J. H. (1959). Potential productivity of the sea. *Science* 130, 602–608.

Ryther, J. H., and Yentsch, C. S. (1957). The estimation of phytoplankton production in the ocean from chlorophyll and light data. *Limnol. Oceanogr.* 2, 281–286.

Ryther, J. H., Menzel, D., and Corwin, N. (1967). Influence of the Amazon River outflow on the ecology of the western tropical Atlantic. *J. Mar. Res.* 25, 69–83.

Saeger, J., and Gayanilo, F. C. (1986). A revised and graphics-orientated version of ELEFAN 0, I and II basic program for use on HP 86/87 microcomputers. Univ. Philippines *Dep. Mar. Fish., Tech. Rep.* 8, 1–233.

Saetre, R. (1985). Surface currents in the Mozambique Channel. *Deep-Sea Res.* 32, 1457–1467.

Saetre, R., and da Silva, A. J. (1982). Water masses and circulation of the Mozambique Channel. *Rev. Invest. Pesq., Maputo, Mozambique* 3, 1–83.

Saetre, R., and de Paula e Silva, R. (1979). "The Marine Fish Resources of Mozambique." Inst. Mar. Res., Norway.

Sale, P. F. (1978). Maintenance of high diversity in coral reef fish communities. *Am. Nat.* 111, 337–359.

Sale, P. E. (1980). The ecology of fishes on coral reefs. *Oceanogr. Mar. Biol.* 18, 367–421.

Sale, P. F., and Dybahl, R. (1975). Determinants of community structure for coral reef fishes in an experimental habitat. *Ecology* 56, 1343–1355.

Salzen, E. A. (1957). A trawling survey off the Gold Coast. *J. Cons., Cons. Int. Explor. Mer* 23, 72–82.

Salzen, E. A. (1958). Observations on the biology of the West African shad, *Ethmalosa fimbiata. Bull. Inst. Fr. Afr. Noire, Ser A* 20, 1388–1426.

Sanders, H. L. (1968). Marine benthic diversity: A comparative study. *Am. Nat.* 102, 243–282.

Sanders, H. L. (1969). Benthic marine diversity and the stability-time hypothesis. *Brookhaven Symp. Biol.* 22, 71–81.

Sandstrom, H., and Elliott, J. A. (1984). Internal tide and solitons on the Scotian shelf: A nutrient pump at work. *JGR, J. Geophys. Res.* 89, 6415–6426.

Santander, H., and Zuzunaga, J. (1984). Impact of the 1982–83 El Niño on the pelagic resources off Peru. *Trop. Ocean-Atmos. Newsl.* 28, 10–12.

Sargent, M. C., and Austin, T. S., (1949). Biological economy of coral reefs. *Geol. Surv. Prof. Pap. (U.S.)* 260-E, 293–300.

Sato, T., and Hatanaka, H. (1983). A review of assessment of Japanese distant-water fisheries for cephalopods. *FAO Fish. Tech. Pap.* 231, 145–180.

Sauskan, V. I., and Ryzhov, V. M. (1977). Investigation of communities of demersal fish of Campeche Bank. *Oceanology* 17, 223–227.

Saville, A. (1977). Survey methods of appraising fishery resources. *FAO Fish. Tech. Pap.* 171, 1–76.

Schaefer, M. B. (1961). Report on the investigations of the IATTC for the year 1959. *Annu. Rep. Int.-Am. Trop. Tuna Comm., 1959*, pp. 39–156.

Scheibling, R. E., and Stevenson, R. L. (1984). Mass mortality of *Strongylocentrotus droebachiensis* off Nova Scotia, Canada. *Mar. Biol. (Berlin)* 78, 153–164.

Scholander, P. F., Flags, W., Walters, V., and Irving, L. (1953). Climatic adaptations in arctic and tropical poikilotherms. *Phys. Zool.* **26**, 67–92.

Schwinghamer, P. (1981). Characteristic size distributions of integral benthic communities. *Can. J. Fish. Aquat. Sci.* **38**, 1255-1263.

Scoffin, T. P., Stearns, C. W., Boucher, D., Frydl, P., Hawkins, C. M., Hunter, I. G., and MacGeachy, J. K. (1980). Calcium carbonate budget of a fringing reef on the west coast of Barbados. Part II. Erosion, sediments and internal structure. *Bull. Mar. Sci.* **30**, 475–508.

Scoffin, T. P., Stoddart, D. R., Tudhope, A. W., and Woodroffe, C. (1985). Rhodoliths and Coraliths of Muri Lagoon, Rarotonga, Cook Islands. *Coral Reefs* **4**, 125–134.

Selye, H. (1980). The stress concept today. *In* "Handbook of Stress and Anxiety" (E. L. Kutash *et al.*, eds.), pp. 1–580. Jossey-Bass, San Francisco, California.

Seshappa, G. (1953). Observations on the biological and physical features of the inshore seabottom along the Malabar coast. *Proc. Natl. Inst. Sci. India* **19**, 257–279.

Sharp, G., and Francis, R. C. (1976). Energetics model for the exploited yellowfin tuna population in the eastern Pacific Ocean. *Fish. Bull.* **77**, 36–51.

Sharp, G. D., and Csirke, J. (1983). "Proceedings of the Expert Consultation to Examine Changes in Abundance and Species Composition of Neritic Fish Resources," FAO Fish. Rep. No. 291, 3 vol. FAO, Rome.

Sharp, G. D., and Dizon, A. E., eds. (1978). "The Physiological Ecology of Tunas." Academic Press, New York.

Shephard, F. A. (1973). "Submarine Geology," 3rd ed. Harper & Row, New York.

Shepherd, J. G. (1982). A versatile new stock-recruitment relationship for fisheries and the construction of sustainable yield curves. *J. Cons., Cons. Int. Explor. Mer* **40**, 67–75.

Shih, C.-T. (1979). East-west diversity. *In* "Zoogeography and Diversity in Plankton (S. van der Spoel and A. C. Pierrot-Bults, eds.), pp. 87–102. Bunge Sci. Publ., Utrecht.

Shin, P. K. S., and Thompson, G. B. (1982). Spatial distribution of the infaunal benthos of Hong Kong. *Mar. Ecol.: Prog.-Ser.* **10**, 37–47.

Shomura, R. S., and Nakamura, E. L. (1969). Variations in marine zooplankton from a single location in Hawaiian waters. *Fish. Bull.* **68**, 87–100.

Shomura, R. S., Menasvata, D., Suda, A., and Talbot, F. (1967). The present status of fisheries and assessment of potential resources of the Indian Ocean and adjacent seas. *FAO Fish. Rep.* **54**, 1–32.

Shubnikov, D. A. (1976). Some problems in the study of commercial ocean ichthyofauna in low latitudes. *J. Ichthyol. (Engl. Transl.)* **16**, 190–194.

Shushkina, E. A., and Kisliakov, I. I. (1975). An estimation of the zooplankton production in an equatorial part of the Pacific Ocean in the Peruvian upwelling. *Tr. Inst. Okeanol. im. P. P. Shrishova, Akad. Nauk SSSR.* **102**, 384–395.

Shushkina, E. A., Vinogradov, M. E., Sorokin, Y. I., Lebedova, L. P., and Mikhleyev, V. N. (1978). Functional characteristics of plankton communities in the Peruvian upwelling region. *Oceanol.* **18**, 579–589.

Sibert, J. R., Kearney, R. E., and Lawson, T. A. (1983). Variations in growth increments of tagged skipjack (*Katsuwonus pelamis*). *South Pac. Comm., Tuna Billfish Assess. Programme Tech. Rep.* **10**, 1–43.

Siebert, L. M., and Popova, O. A. (1982). Particularidades de la alimentación el civil (*Caranx ruber*) en la región suroccidental de la plataforma cubana. *Rep. Invest. Inst. Oceanol. Ac. Cienc. Cuba* **3**, 1–19.

Sieburth, J. M. (1984). Protozoan bacterivory in pelagic marine waters. *In* "Heterotrophic Activity in the Sea" (J. E. Hobbie and P. J. leB. Williams, eds.), pp. 405–444. Plenum, New York.

Sieburth, J. M., and Davis, P. G. (1983). Role of heterotrophic nanoplankton in grazing and nurturing planktonic bacteria. *Ann. Inst. Océanogr. (Paris)* **58**, 285–296.

Silvert, W., and Pauly, D. (1987). On the compatibility of a new expression for gross conversion efficiency with the von Bertalanffy growth equation. *Fish. Bull.* **85**, (in press).

Simpson, A. C. (1982). A review of the database on tropical multispecies stocks in the southeast Asian region. *In* "Theory and Management of Tropical Fisheries" (D. Pauly and G. I. Murphy, eds.), pp. 5–32. ICLARM, Manila.

Simpson, J. G., and Griffiths, R. C. (1967). The fishery resources of Venezuela and their exploitation. *Ser. Rec. Explot. Pesq. Minist. Agric., Venez.* **1**, 175–206.

Simpson, J. H., and Hunter, J. R. (1974). Fronts in the Irish Sea. *Nature (London)* **250**, 404–406.

Sivalingham, S., and Medcof, J. C. (1955). Study of the Wadge Bank fishery. *Dep. Fish, Ceylon, Prog. Rep. Biol. Tech.* **1**, 10–12.

Skillman, R. (1981). "Estimates of von Bertalanffy Growth Parameters of Skipjack Tuna from Capture-recapture Experiments in Hawaiian Islands." NMFS/NOAA, SWFC, La Jolla, California.

Smayda, T. J. (1966). A quantitative analysis of the phytoplankton of the Gulf of Panama. *Bull. Int. Am. Trop Tuna Comm* **11**, 355–612.

Smetacek, V. (1985). Role of sinking in diatom life-history cycles: Ecological, evolutionary and geological significance. *Mar. Biol. (Berlin)* **84**, 239–251.

Smith, I. R., and Pestaño-Smith, R. (1980). A fishing community's response to seaweed farming. *ICLARM Newsl.* **3**, 6–8.

Smith, S. L. (1984). Biological indications of active upwelling in the northwestern Indian Ocean in 1964 and 1979 and a comparison with Peru and NW Africa. *Deep-Sea Res.* **31**, 951–968.

Smith, S. V. (1981). The Houtman Abrolhos Islands: Carbon metabolism of coral reefs at high latitude. *Limnol. Oceanogr.* **26**, 612–621.

Snedaker, S. C., and Snedaker, J. G. (1984). The mangrove ecosystem: Research methods. *UNESCO Monogr. Oceanogr. Methodol.* **8**, 1–251.

Soberón-Chávez, G., and Yáñez-Arancibia, A. (1984). Control ecológica de los peces demersales: Variabilidad ambiental de la zona costera y su influencia en la produccion natural de los recursos pesqueras. *In* "Recursos Pesqueras Potenciales de México: La Pesca Acompañente del Camarón" (A. Yáñez-Arancibia, ed.), pp. 399–485. Univ. Nac. Autonom., Mexico.

Soegiarto, V., and Birowo, S. (1975). "Atlas Oseanologi Perairan Indonesia dan Sekitarnya," Vols. 1 and 2. Lembaga Oseanologi Nasional, Jakarta.

Somjaiwong, D., and Chullasorn, S. (1974). Tagging experiments on the Indo-Pacific Mackerel in the Gulf of Thailand. *Proc. Indo.-Pac. Fish. Counc.* **15** (3), 287–296.

Sorokin, Yu. I. (1981). Microheterotrophic organisms in marine ecosystems. *In* "Analysis of Marine Ecosystems" (A. R. Longhurst, ed.), pp. 293–342. Academic Press, New York.

Sorokin, Yu. I., Kopylov, A. I., and Mamaeva, N. V. (1985). Abundance and dynamics of microplankton in the central tropical Indian Ocean. *Mar. Ecol.: Prog. Ser.* **24**, 27–41.

Soutar, A., and Isaacs, J. (1969). History of fish populations inferred from fish scales in anaerobic sediments off California. *Rep. Calif. Coop. Fish. Invest.* **13**, 63–70.

Sparre, P. (1987). A method for estimation of growth, mortality and gear selection/recruitment parameters from multiple c.p.u.e. length frequency data. *In* "Length-based Methods in Fishery Research" (D. Pauly and G. R. Morgan, eds.), ICLARM Conf. Proc. 14 (in press).

Springer, V. G. (1982). Pacific Plate biogeography with special reference to shore-fishes. *Smithson. Contrib. Zool.* **367**, 1–182.

Staples, D. J. (1985). Modelling the recruitment processes of the banana prawn, *Penaeus merguiensis,* in the Southeastern Gulf of Carpentaria *In* "Second Australian National Prawn Seminar" (P. C. Rothlisberg, B. J. Hill, and D. J. Staples, eds.), pp. 175–184. NPS2, Cleveland, Australia.

Stehli, F. G., and Wells, J. W. (1971). Diversity and age patterns in the hermatypic corals. *Syst. Zool.* **20**, 115–126.

Stephenson, T. A., and Stephenson, A. (1949). On the universal features of zonation between tidemarks on rocky coasts. *J. Ecol.* **40**, 1–49.

Stephenson, T. A., and Stephenson, A. (1972). "Life Between Tide Marks on Rocky Shores." Freeman, San Francisco, California.

Stephenson, W., and Searle, R. B. (1960). Experimental studies on the ecology of intertidal environments at Heron Island. *Aust. J. Mar. Freshwater Res.* **11**, 241–267.

Stephenson, W., and Williams, W. T. (1970). A study of the benthos of soft bottoms, Sek Harbour, New Guinea, using numerical analysis. *Aust. J. Mar. Freshwater Res.* **22**, 11–34.

Stephenson, W., Williams, W. T., and Lance, G. N. (1971a). The macrobenthos of Moreton Bay. *Ecol. Monogr.* **40**, 459–494.

Stephenson, W., Williams, W. T., and Cook, S. D. (1971b). Computer analysis of Petersen's original data on bottom communities. *Ecol. Monogr.* **42**, 387–408.

Stevenson, M. R. (1981). Seasonal variations in the Gulf of Guayaquil, a tropical estuary. *Biol. Cient. Tecnic. Inst. Nac. Pesca* **4**, 1–133.

Struhsaker, P. (1969). Demersal fish resources of the continental shelf of the southeastern United States. *Fish. Ind. Res.* **4**, 261–300.

Sutton, M. (1983). Relationships between reef fishes and coral reefs. *In* "Perspectives on Coral Reefs" (D. J. Barnes, ed.), pp. 248–255. Aust. Inst. Mar. Sci.

Suzuki, Z., Tomlinson, P. K., and Honma, M. (1978). Population structure of Pacific yellowfin tuna. *Bull. Int.-Am. Trop. Tuna Comm.* **17**, 277–441.

Sverdrup, H. U., Johnson, M. W., and Fleming R. H. (1942). "The Oceans: Their Physics, Chemistry and General Biology." Prentice-Hall, Englewood Cliffs, New Jersey.

Sweatman, H. P. A. (1985). The influence of adults of some coral reef fishes on larval recruitment. *Ecol. Monogr.* **55**, 469–485.

Talbot, F. H. (1965). A description of the coral structure of Tutia Reef and its fish fauna (Tanganyika). *Proc. Zool. Soc. London* **145**, 431–470.

Talbot, E. H., and Penrith, M. J. (1963). The white marlin from the seas around South Africa. *S. Afr. J. Sci.* **59**, 28–63.

Tandog, D. D. (1984). State of exploitation and population dynamics of skipjack tuna in waters off Misamis Oriental. M.Sc. Thesis, pp. 1–93. College of Fisheries, Univ. of Philippines in the Visayas, Quezon City.

Taylor, C. C. (1958). Cod growth and temperature. *J. Cons., Cons. Int. Explor. Mer* **23**, 366–370.

Taylor, C. C. (1962). Growth equations with metabolic parameters. *J. Cons., Cons. int. Explor. Mer* **27**, 270–286.

Taylor, D. L. (1969). The nutritional relationship of *Anemonia sulcata* and its dinoflagellate symbiont. *J. Cell Sci.* **4**, 751–762.

Taylor, R. E. (1959). Temperature and growth—the Pacific razor clam. *J. Cons., Cons. Int. Explor. Mer* **25**, 93–101.

Terchunian, A., Klemas, V., Segovia, A., Lavarez, A., Vasconez, B., and Guerrero, L. (1986). Mangrove mapping in Ecuador: The impact of shrimp pond construction. *Environ. Manage.* **10**, 345–350.

Tham, A. K. (1950). "Food and Feeding Relationships of the Fishes of Singapore Strait," pp. 1–35. Col. Office Fish. Publ., London.

Tham, A. K. (1968). Synopsis of biological data on the Malaya anchovy, *Stolephorus pseudoheterolobus*. *FAO Fish. Synopses* **37**.

Thiel, M. (1928). Madreporia. *Beitr. Meeresfauna Westafr.* **3**, 253–350.

Thompson, R., and Munro, J. L. (1983). The biology, ecology and bionomics of the hinds and groupers. *In* "Caribbean coral reef fishery resources" (J. L. Munro, ed.). *ICLARM Stud. Rev.* **7**, 39–81.

Thorpe, J. E. (1986). Age at first maturity in Atlantic salmon: freshwater period influences and conflicts with smolting. *Can. Fish. Aquat. Sci. Spec. Publ.* **89**, 7–14.

Thorson, G. (1957). Bottom communities. *Mem. Geol. Soc. Am.* **67**, 461–534.

Thorson, G. (1966). Some factors influencing the recruitment and establishment of marine benthic communities. *Neth. J. Sea Res.* **3**, 267–293.

Tiews, K. (1962). Experimental trawl fishing in the Gulf of Thailand and its results regarding the possibilities of trawl fisheries development in Thailand. *Veroeff. Inst. Kuesten-Binnenfisch., Hamburg* **25**, 1–53.

Tiews, K., Divino, P., Ronquillo, I. A., and Marques, J. (1972). On the food and feeding of eight species of *Leiognathus* found in Manila Bay and San Miguel Bay. *Proc., Indo-Pac. Fish Counc.* **13**, 93–99.

Tranter, D. J. (1973). Seasonal studies of a pelagic ecosystem (110E). *In* "Biology of the Indian Ocean" (B. Zeitzschel, ed.), pp. 476–520. Springer-Verlag, Berlin and New York.

Tricas, T. C. (1986). The economics of foraging in coral-feeding butterfly fishes of Hawaii. *Proc. Int. Coral Reef Symp., 5th, 1985*, Vol. 5, pp. 409–419.

Trono, G. C., Rabanal, H. R., and Santiko, I. (1980). "Seaweed farming." South China Sea Development & Coordinating Programme SCS/80/WP/91, Manila.

Turner, R. E. (1977). Intertidal vegetation and commercial yields of penaeid shrimps. *Trans. Am. Fish. Soc.* **106**, 411–416.

Uchiyama, J. H., and Struhsaker, P. (1981). Age and growth of skipjack and yellowfin tuna, as indicated by daily growth increments of sagittae. *Fish. Bull.* **79**, 151–162.

Ursin, E. (1967). A mathematical model of some aspects of fish growth, respiration and mortality. *J. Fish. Res. Board Can.* **24**, 2455–2453.

Ursin, E. (1973). On the prey size preference of cod and dab. *Medd. Dan. Fisk. Havunders* [N.S.] **7**, 85–98.

van Andel, T. H. (1967). The Orinoco delta. *J. Sediment. Petrol.* **37**, 297–310.

van Campen, W. G. (1950). Translation of "Poisonous Fishes of the South Seas" compiled by T. Kumada, Tokyo, 1943. *U.S., Fish Wildl. Serv., Spec. Sci. Rep.—Fish.* **25**, 1–221.

Vandermeulen, J. H., and Gillfillan, E. S. (1985). Petroleum pollution, corals and mangroves. *Mar. Tech. Soc. J.* **18**, 62–72.

van Thielen, R. (1977). The food of juvenile *Sardinella aurita* and of juvenile and adult *Anchoa guinensis* in the near-shore waters off Ghana, Africa. *Meeresforschung* **25**, 46–53.

Vareschi, E., and Fricke, H. (1986). Light response of a scleractinian coral. *Mar. Biol. (Berlin)* **90**, 395–402.

Victor, B. C. (1982). Daily otolith increments and recruitment in two coral reef wrasses, *Thalassoma bifasciatum* and *Halichoeres bivittatus*. *Mar. Viol. (Berlin)* **71**, 203–208.

Vinogradov, M. E. (1981). Ecosystems of equatorial upwellings. *In* "Analysis of Marine Ecosystems" (A. R. Longhurst, ed.), pp. 69–93. Academic Press, New York.

Vinogradov, M. E., Krapivi, P. F., Menshutkin, V. V., Fleyshman, V. S., and Shushkina,

E. A. (1973). Mathematical model of the functions of the pelagic ecosystem of tropical regions. *Oceanology (Engl. Transl.)* **13**, 704–717.

Vinogradov, M. E., Shushkina, E. A., and Kukina, I. N. (1976). Functional characteristics of planktonic communities of the equatorial upwelling. *Oceanology (Engl. Transl.)* **16**, 122–138.

Voituriez, B., and Herbland, A. (1982). Comparaison des systèmes productifs de l'Atlantique tropical est: Dômes thermiques, upwellings côtièrs et upwelling équatorial. *Rapp. P.-V. Réun., Cons. Int. Explor. Mer* **180**, 107–123.

Voss, G. L. (1969). The pelagic midwater fauna of the eastern tropical Atlantic with special reference to the Gulf of Guinea. *In* "Proceedings of a Symposium on Oceanography and Fisheries of the Eastern Tropical Atlantic," pp. 91–99. UNESCO, Paris.

Voss, G. L. (1983). A review of cephalopod fishery biology. *Mem. Natl. Mus. Victoria, Melbourne* **44**, 229–241.

Wade, B. A. (1972). A description of a highly diverse soft-bottom community in Kingston Harbour, Jamaica. *Mar. Biol. (Berlin)* **13**, 57–69.

Walsh, G. E. (1974). Mangroves: A review. *In* "Ecology of Halophytes" (R. J. Reimold and W. H. Queen, eds.), pp. 51–174. Academic Press, New York.

Walsh, J. J. (1977). A biological sketchbook for an eastern boundary current. *In* "The Sea: Ideas and Observations on Progress in the Study of the Sea" (E. D. Goldberg, ed.), pp. 923–968. Wiley, New York.

Walsh, J. J. (1983). Death in the sea: Engimatic phytoplankton losses. *Prog. Oceanogr.* **12**, 1–86.

Wanders, J. B. W. (1976). The role of benthic algae in the shallow reef of Curaçao (Netherlands Antilles). II. Primary productivity of the *Sargassum* beds on the north-east coast submarine plateau. *Aquat. Bot.* **2**, 327–335.

Wankowski, J. W. J. (1981). Estimated growth of surface-schooling skipjack tuna, from the Papua-New Guinea region. *Fish. Bull.* **79**, 517–532.

Wardle, C. S. (1978). Non-release of lactic acid from anaerobic swimming muscle of plaice *Pleuronectes platessa:* A stress reaction. *J. Exp. Biol.* **77**, 141–156.

Warfel, H. E., and Manacop, P. R. (1950). Otter trawl explorations in Philippine waters. *Res. Rep.—U.S. Fish Wildl. Serv.* **25**, 5–60.

Warwick, R. M. (1982). The partitioning of secondary production among species in benthic communities. *Neth. J. Sea Res.* **16**, 1–16.

Warwick, R. M., and George, C. L. (1980). Annual macrofauna production in an *Abra* community. *In* "Industrialized Embayments and their Environmental Problems" (M. B. Collins, ed.), pp. 517–538. Pergamon, Oxford.

Warwick, R. M., and Ruswahyuni (1987). A comparative study of the structure of some tropical and temperate marine soft-bottom macrobenthic communities. *Mar. Biol. (Berlin)* (submitted for publication).

Warwick, R. M., George, C. L., and Davies, J. R. (1978). Animal production in a *Venus* community. *Estuarine Coastal Mar. Sci.* **7**, 215–241.

Watts, J. C. D. (1959). Some observations on the marking of demersal fish in the Sierra Leone River estuary. *Bull. Inst. Fr. Afr. Noire, Ser. A* **21**, 1237–1252.

Wauthy, B. (1983). Introduction à la climatologie du Golfe de Guinée. *Océanogr. Trop.* **18**, 103–138.

Webb, J. E. (1958). The ecology of Lagos lagoon. (1) The lagoons of the Guinea Coast. *Philos. Trans. R. Soc. London, Ser. B* **241**, 307–318.

Weber, W. (1976). The influence of hydrographic features on the spawning time of tropical fish. *In* "Fisheries Resources Management in Southeast Asia" (K. Tiews, ed.), pp. 269–281. German Foundation for International Development, Berlin.

Weikert, H. (1980). Oxygen minimum layer in the Red Sea: Ecological implications of zooplankton occurrence. *Meeresforschung* **28**, 1–9.

Wells, J. T. (1983). Dynamics of coastal fluid muds in low-, moderate-, and high-tide-range environments. *Can. J. Fish. Aquat. Sci.* **40**, Suppl. 1, 130–142.

Wells, J. W. (1957). Coral reefs. *Mem. Geol. Soc. Am.* **67**, 609–632.

Wethey, D. S., and Porter, J. W. (1976). Sun and shade differences in productivity of reef corals. *Nature (London)* **262**, 281–282.

Wheeler, J. F. G., and Ommaney, F. D. (1953). "Report on the Mauritius-Seychelles Fishery Survey," No. 3, pp. 1–98. Colon. Off. Fish. Publ., London.

White, T. F. (1982). "The Philippine Tuna Fishery and Aspects of the Population Dynamics of Tuna in Philippine Waters," SCS/82/WP/114, pp. 1–64. South China Sea Fisheries Development and Coordinating Programme, Manila.

Whiteleather, R. T., and Brown, H. H. (1945). "An Experimental Fishery Survey in Trindad, Tobago and British Guiana." Anglo-Am. Caribb. Comm., Washington, D.C.

Whittaker, R. H. (1960). Vegetation of the Siskiyou Mountains, Oregon and California. *Ecol. Monogr.* **30**, 279–338.

Wiebe, W. J., Johannes, R. E., and Webb, K. L. (1975). Nitrogen fixation in a coral reef community. *Science* **188**, 257–259.

Wilkinson, C., and Sammarco, P. (1983). Effects of fish grazing and damselfish territoriality on coral reef algae. II. Nitrogen fixation. *Mar. Ecol.: Prog.-Ser.* **3**, 15–19.

Williams, D. B. (1983). Daily, monthly and yearly variability in recruitment of a guild of coral reef fishes. *Mar. Ecol.: Prog. Ser.* **10**, 231–237.

Williams, D. M., and Hatcher, A. (1983). Structure of fish communities on outer slopes of inshore, mid-shelf and outer-shelf reefs of the Great Barrier Reef. *Mar. Ecol.: Prog. Ser.* **10**, 239–250.

Williams, F. (1968). Report on the Guinean Trawling Survey. Vols. I, II, and III. *Publ., Sci., Sci. Tech. Res. Comm., Organ. Af. Unity* **99**, 1–529, 1–551).

Williams, F. (1972). Consideration of three proposed models of the migration of young skipjack tuna into the eastern Pacific Ocean. *Fish. Bull.* **70**, 741–762.

Williams, P. J. LeB. (1981). Incorporation of microheterotrophic processes into the classical paradigm of the planktonic food web. *Kiel. Meeresforsch., Sonderh.* **5**, 1–28.

Winberg, G. G. (1971). "Methods for the Estimation of Production in Aquatic Animals." Academic Press, New York.

Wise, J. P., and Davis, C. W. (1973). Seasonal distribution of tunas and billfish in the Atlantic. *NOAA Tech. Rep., NMFS SSRF* **NMFS SSRF-662**, 1–24.

Wolanski, E., Drew, E., Abel, K. M., and O'Brien, J. (1986). Tidal jets, nutrient upwelling, and their influence on the alga *Halimeda* in the ribbon reefs, Great Barrier Reef. *Symp. Mar. Sci. West. Pac. Indo-Pac Convergence, Townsville, Qd. Dec. 1986.* p. 47.

Wongratana, T. (1983). Diagnoses of 24 new species and a proposal for a new name of Indo-Pacific clupeid fishes. *Jpn. J. Ichthyol.* **24**, 375–385.

Worms, J. (1983). World fisheries for cephalopods: A synoptic overview. *FAO Fish. Tech. Pap.* **231**, 1–20.

Wyman, M., Gregory, R. P. F., and Carr, N. G. (1985). Novel role for phycoerythrin in a marine cyanobacterium *Synechococcus*. *Science* **230**, 818–820.

Wyrtki, K. (1961). The physical oceanography of the southeast Asian waters. *Naga Rep.* **2**, 1–195.

Wyrtki, K. (1962a). The oxygen minima in relation to ocean circulation. *Deep-Sea Res.* **9**, 11–23.

Wyrtki, K. (1962b). The upwelling in the region between Java and Australia during the south-east monsoon. *Aust. J. Mar. Freshwater Res.* **13**, 217–225.

Wyrtki, K. (1964). Upwelling in the Costa Rica Dome. *Fish. Bull.* **63,** 355–372.

Wyrtki, K. (1967). Circulation and water masses in the eastern Equatorial Pacific Ocean. *Int. J. Oceanol. Limnol.* **1,** 117–147.

Yamanaka, H. (1984). The relationship between El Niño episodes and fish emigrations and yields in the western Pacific. *Trop. Atmos.-Ocean Newsl.* **25,** 2–4.

Yáñez-Arancibia, A. (1978). Taxonomia, ecologia y estructura de las comunidades de peces en lagunas costeras del Pacifico de Mexico. *Inst. Cienc. del Mar y Limnol. Univ. Nat. Auton. Mexico. Publ. Esp.* **2.**

Yáñez-Arancibia, A. (1985). Recursos demersales de alta diversitad en las costas tropicales: Perspectiva ecologica. *In* "Recursos Pesqueros Potenciales de México: La Pesca Acompañente del Camarón" (A. Yáñez-Arancibia, ed.), pp. 17–38. Univ. Nacl. Autonom., Mexico.

Yáñez-Arancibia, A. (1986). Ecologia de la zona costera: Analisis de siete topicos. *AGT Editonal, Mexico D. F.*

Yáñez-Arancibia, A., and Day, J. W. (1982). Ecological characteristics of Terminos Lagoon, a tropical lagoon-estuarine system in the southern Gulf of Mexico. *Oceanol. Acta, Spec. Suppl.,* pp. 431–440.

Yáñez-Arancibia, A., Sanchez-Gil, P., Villalobos-Zapata, G. J., and Rodriguez-Capatillo, R. (1985). Distribucion y Abondancia de las especias doinantes en las poblaciones de peces de la plataforma continental Méxicana del Golfo de México. *In* "Recursos pesqueros de México" (A. Yáñez-Arancibia, ed.), pp. 315–397. Univ. Auton. Méx.

Yang, Jiming (1981). Trophic levels of North Sea fish. *J. Cons., Cons. Int. Explor. Mer.* **17,** 1–6.

Yong, M. Y., and Skillman, R. A. (1975). A computer program for the analysis of polymodal frequency distributions (ENORMSEP), FORTRAN IV. *Fish. Bull.* **73,** 681.

Zabi, S. G. (1982). Repartition et abondance des espèces de la macrofaune benthique de la lagune Ebrié. *Doc. Sci. Cient. Rech. Oceanogr. Abidjan ORSTOM* **13,** 73–96.

Zahn, W. (1984). Influence of bottom topography on currents in the Mozambique Channel. *Trop. Atmos. Océan Newsl.* (26), 22–23.

Zann, L. P., and Bolton, L. (1985). Distribution, abundance and ecology of the blue coral *Heliopora coerulea. Coral Reefs* **4,** 125–134.

Zei, M. (1966). Sardines and related species of the eastern tropical Atlantic. *In* "Proceedings of the Symposium on Oceanography and Fishery Resources of the Tropical Atlantic," pp. 101–108. UNESCO, Paris.

Zeitzschel, B. (1981). Field experiments on benthic ecosystems. *In* "Analysis of Marine Ecosystems" (A. R. Longhurst, ed.), pp. 607–626. Academic Press, New York.

Ziman, J. (1976). "The Force of Knowledge." Cambridge Univ. Press, London and New York.

Zurbrigg, R. E., and Scott, W. B. (1972). Evidence for expatriate populations of the lanternfish *Myctophum punctatum* in the Northwest Atlantic. *J. Fish. Res. Board Can.* **29,** 1679–1683.

Author Index*

A

Abele, L. G., 66
Abel, K. M., 92
Aboussouan, A., 291
Abramson, N. J., 272
Adey, W. H., 141
Alheit, J., 81
Ali, R. M., 161
Allen, G. R., 15, 279
Allen, J. R. L., 9, 10, 16
Allen, K. R., 237
Aller, R. C., 75, 79
Alverson, F., 252, 301
Amaratunga, T., 331
Ambrose, W. G., 236
Andrews, J. C., 48, 144
Anonymous, 161, 170, 204
Ansari, Z. A., 78, 128
Antoine, L., 265
Argue, A. W., 209
Arnault, S., 55
Arnold, G. P., 323
Arntz, W., 61
Aro, E., 22, 308
Atkinson, M. J., 108, 142, 247, 248
Aung, S., 161, 164, 273, 274
Austin, T. S., 108
Azam, F., 115, 117

B

Babcock, R. C., 142, 222, 223
Backus, R. H., 220, 221
Bacolod, P. T., 308
Bacon, P. R., 93
Bain, K. H., 161
Bainbridge, V., 93, 94, 131, 189, 190, 232, 292
Baird, R. C., 224, 254
Baker, L. D., 114
Bakhayoko, M., 327
Bakun, A., 31, 124, 259, 293, 299
Bakus, G. J., 246
Banerji, S. K., 259
Bang, N. D., 45
Banse, K., 33, 45, 112
Bapat, S. V., 161
Barber, R. C., 197
Barber, R. J., 129, 148
Barber, R. T., 49–51
Barbieri, L. R. R., 182
Bard, F. X., 265
Bardach, J. E., 241
Barham, E. G., 224
Barkley, R. A., 124
Bartels, A., 132
Barwick, D. H., 270
Baslow, M. H., 250

* Prepared by Victor Sambilay, Jr., College of Fisheries, University of the Philippines.

Geographic Index*

* Prepared by Victor Sambilay, Jr., College of Fisheries, University of the Philippines.

Taxonomic Index*

A

Abalistes, see Balistidae
Abra, 109, 115, *see also* Bivalve
Abraliopsis, 323,*see also* Squid
Acantharia, 93, 98, *see also* Protozoa
Acanthaster plancii, 60, 90, *see also* Aster-
    oidea
Acanthocybium, see Scombridae
Acanthopagrus, see Sparidae
Acanthostracion, see Ostraciontidae
Acanthuridae, 142, 177, 242
  Acanthurus, 2, 242
    bahianus, 264
    monroviae, 152, 232, 233
    Zanclus, 243
Acartia, 79, 114, *see also* Copepoda
Acetes, see Sergestidae
Achirus, see Myctophidae
Achroonema, see Bacteria
Acrocalanus, 60, *see also* Copepoda
Acropoma, 173, 174
Acropoma japonicum, 173
Acropora, 89, 91, 139, 140, *see also*
  Scleractinia
Actinia, 90, *see also* Zoanthid
Adioryx, see Holocentridae
Albacore, 61, 210, 211, 254
Albulidae, 185
  Albula, 185, 186
  Albula vulpes, 233, 234
  Pterothrissus belloci, 152

Alcyonacea, 90, *see also* Alcyonaria
Alcyonaria, 89, 91, 139, 243, *see also*
  Anthozoa
Alepisaurus, 252, 253
Alfonsin, 171, *see also Beryx splendens*
Algae, 11, 12, 14, 16, 19, 22, 23, 26, 61,
  72, 79, 88–93, 107, 116, 125, 129, 139,
  140–144, 177, 205, 232, 233, 246–248,
  250, 308, 312
Alosa, see Clupeidae
Alpheus, 234, *see also* Shrimp
Alutera, see Monacanthidae
Ambassis, see Centropomidae
Amphiodia, 74, *see also* Ophiuroidea
Amphioplus, 74, 76, 78, *see also*
  Ophiuroidea
Amphipholis, 78, *see also* Ophiuroidea
Amphipoda, 87, 95, 96, 104, 234, 252
Amphiprion, see Pomacentridae
Amphiura, 74, 76, *see also* Ophiuroidea
Anchoa, see Engraulidae
Anchovia, see Engraulidae
Anchoviella, see Engraulidae
Anchovy, 60, 61, 137, 138, 184, 188–190,
  192, 196, 197, 240, 275, 276, 292, 294,
  see also Engraulidae
Anemone, 72, 89, *see also* Actinia
Angelfish, 242, 243, *see also* Chaetodonti-
  dae, *Holacanthus*
Anisotremus, see Pomadasyidae
Anguilliform, 220

* Prepared by Victor Sambilay, Jr., College of Fisheries, University of the Philippines.